CHEMICAL ANALYSIS

CHEMICAL ANALYSIS

A SERIES OF MONOGRAPHS ON ANALYTICAL CHEMISTRY AND ITS APPLICATIONS

Editors

P. J. ELVING · I. M. KOLTHOFF

VOLUME 19

A WILEY-INTERSCIENCE PUBLICATION

JOHN WILEY & SONS

New York / London / Sydney / Toronto

THERMAL METHODS
OF ANALYSIS

SECOND EDITION

WESLEY WM. WENDLANDT

Department of Chemistry
University of Houston
Houston, Texas

A WILEY-INTERSCIENCE PUBLICATION

JOHN WILEY & SONS

New York / London / Sydney / Toronto

Library of Congress Cataloging in Publication Data:

Wendlandt, Wesley William.
    Thermal methods of analysis.

    (Chemical analysis, v. 19)
    "A Wiley-Interscience publication."
    Includes bibliographical references.
    1. Thermal analysis.   I. Title.   II. Series.

QD79.T38W46   543'.086   73-18444
ISBN 0-471-93366-X

Printed in the United States of America

10 9 8 7 6 5 4 3 2 1

This book is dedicated to the memories of Dr. Robert L. Stone and Professor Lazlo Erdey. Their many contributions helped lay the foundation of modern thermal analysis.

PREFACE TO
THE FIRST EDITION

The purpose of this monograph is to acquaint chemists and other investigators with the relatively new series of instrumental techniques which are
broadly classified as "thermal methods." In the past, many of these
techniques involved tedious, time-consuming, manual recording methods;
however, all of them are now completely automatic and employ either analog
(recorder) or digital readout devices. Thus, due to automation, the instruments become capable of self-operation, improving both the accuracy and
precision of the measurements as well as relinquishing both the investigator's
time and patience.

These thermal methods provide a new means of solving existing chemical
problems, as well as creating new ones. It is difficult for the author to think
of a modern chemical laboratory without a thermobalance or differential
thermal analysis apparatus. The former instrument can provide rapid
information concerning the thermal stability, composition of pyrolysis
intermediates, and composition of the final product, as a compound is heated
to elevated temperatures. The latter apparatus can provide information
concerning the enthalpy changes occurring during thermal decomposition
of the compound, as well as the detection of phase transitions of various
types. Both techniques yield a wealth of information in a very short period
of time.

This book is not intended to be a comprehensive survey of the literature
on each thermal technique. Rather, it is a critical review, as far as space
permits, on each method. It is felt that the investigator should be well
informed on both the *advantages* and *limitations* of each thermal technique
in order to use these techniques intelligently. It must be admitted that this
book is written primarily for the analytical chemist, although the techniques
are useful in other fields of investigation as well.

The author would like to acknowledge his gratitude to Professors P. J.
Elving and I. M. Kolthoff for their helpful advice and guidance during the
preparation of the manuscript; to helpful comments from his former
colleague, Dr. Edward Sturm; to Professor J. Jordan and S. T. Zenchelsky
for supplying him with their personal reprints; to Mr. Irwin Dosch and Dr.
Robert L. Stone for their assistance in supplying several of the badly needed
photographs; and to his present and former students who made this work
possible in the first place.

Also, the author would like to express his indebtedness to the Division of
Research, U.S. Atomic Energy Commission; the Air Force Office of Scientific
Research, U.S. Air Force; and to the Robert A. Welch Foundation, for their
continual support of the author's work in this field.

And finally, because of their efforts above and beyond the call of duty, the author would like to acknowledge with thanks his typists, Miss Sallie Hardin, Miss Sue Richmond, and Miss Kathryn White.

Wesley Wm. Wendlandt

Lubbock, Texas
January, 1964

PREFACE

The field of thermal analysis has undergone a period of tremendous growth since the first edition of this book was published. Indeed, it can probably be stated that the past nine years have been the prime years of maximum growth for the field. Numerous new techniques have been developed during this period and applications to many chemical and other types of problems have been carried out. Along with this growth, maturity has also developed, as is evidenced by the founding of three thermal analysis societies, the publication of two specialized journals, and the development of a systematic nomenclature for thermal analysis. All of these were very much needed.

The principal goal of this second edition is to serve as an introduction to the techniques and applications of thermal analysis. It should be possible for a novice to acquire a working knowledge of thermogravimetry, differential thermal analysis, and so on, just by studying the enclosed chapters. No attempt is made to develop a comprehensive treatment of any subject. As is obvious to any worker in thermal analysis, each chapter could be expanded to a full-length book (or books).

This revision consists of about 90% new material, most of which was developed between 1964 and 1972. As in the previous edition, dynamic thermogravimetry (TG), differential thermal analysis (DTA), and differential scanning calorimetry (DSC) are discussed in greater detail than the other techniques because of their fundamental importance. Three chapters are devoted to TG and three to DTA (and DSC). The chapter on thermal analysis (Chap. X) has undergone the least revision. Greater emphasis has been given evolved gas detection (or analysis) (Chap. VIII) and spectroscopic, photometric, and optical thermal techniques (Chap. IX). Two entirely new chapters are presented—one on application of digital and analog computers (Chap. XII) and one on nomenclature (Chap. XIII). The importance of these two subjects certainly justifies their inclusion in this edition. Because of page limitations, only a selected number of commercial instruments are described. However, as pointed out in their respective chapters, these instruments are adequately described elsewhere.

As in other creative endeavors of this magnitude, the execution, completion, and success of this book depend on the encouragement and assistance of many individuals. It is my pleasure to acknowledge the long suffering of my wife, Gay; the encouragement and indulgence of Professors P. J.

Elving and I. M. Kolthoff and Miss Dorothy A. White; the partial financial support given the author by the Robert A. Welch Foundation of Houston, Texas; and, last but not least, the typing ability of Miss Lessie Washington.

Wesley Wm. Wendlandt

Houston, Texas
July, 1973

CONTENTS

CHAPTER I

GENERAL INTRODUCTION

With the development of each new instrumental technique, the chemist has a new tool with which to attack and solve chemical problems. However, sometimes there is a fairly long interval between the date the technique is developed and the time it is applied *en masse* to chemical problems. Such has been the case with many of the thermal methods discussed in this book. For example, the first thermobalance was developed by Honda in 1915, yet it was not until 1947 that Duval called attention to its applications to the field of inorganic gravimetric analysis. A similar situation is noted with differential thermal analysis (DTA), which was originally conceived by Roberts-Austen in 1899. For many years DTA was an invaluable technique for the identification of minerals, clays, metallic alloys, and so on, but was virtually ignored by the chemists. In recent years, however, DTA has been successfully applied, either by itself or in conjunction with other thermal techniques, to the elucidation of problems of chemical interest.

The term *thermal analysis* will be defined in this book as those techniques in which some physical parameter of the system is determined as a function of temperature. The physical parameter is recorded as a *dynamic* function of temperature; measurements made at a fixed or isothermal temperature will, in general, not be included. Using this definition, the principal techniques of thermal analysis are dynamic thermogravimetry (TG) and differential thermal analysis (DTA). Other less widely employed but useful techniques include evolved gas detection or analysis (EGD or EGA), thermomechanical analysis (TMA), dynamic reflectance spectroscopy (DRS), electrical conductivity (EC), photothermal analysis (PTA), and so on. Each of these techniques will be discussed in detail. Two other important thermal techniques that do not fall under the above classification are cryoscopic analysis and thermometric titrimetry. One of these is included because it represents an important method of analysis in analytical chemistry and involves the measurement of the temperature of the system as a function of time.

The various thermal analysis techniques are summarized in Table I.1. Each technique is listed in terms of the parameter recorded, a typical recorded curve, the instrumentation involved, and the chapter in which it is described. The list in Table I.1 is certainly not comprehensive; obviously, almost any analytical instrumental technique can be considered a thermal

TABLE I.1
Some Thermal Analysis Techniques

Technique	Parameter measured	Instrument employed	Chapter	Typical curve
Thermogravimetry (TG)	Mass	Thermobalance	II–IV	
Derivative thermogravimetry (DTG)	dm/dt	Thermobalance	II	
Differential thermal analysis (DTA)	$T_s - T_r (\Delta T)$	DTA apparatus	V–VII	
Differential scanning calorimetry (DSC)	Heat flow, dH/dt	Calorimeter	V–VII	

TABLE 1.1 (*continued*)

Technique	Parameter measured	Instrument employed	Chapter	Typical curve
Evolved gas detection (EGD)	Thermal conductivity[a]	TC cell[a]	VIII	
Thermolumines-cence (TL)	Light emission	Photo detector[b]	XI	
Electrical conductivity (EC)	Current or resistance	Electrometer or bridge	XI	
Thermomechanical analysis (TMA) (dilatometry)	Volume or length	Dilatometer	XI	

TABLE I.1 (*continued*)

Technique	Parameter measured	Instrument employed	Chapter	Typical curve
Thermometric titrimetry	Temperature	Calorimeter		Titrant volume
Thermal analysis	Temperature[c]	Calorimeter	X	Time
Dynamic reflectance spectroscopy (DRS)	Reflectance	Spectrophotometer	IX	
Emanation thermal analysis (ETA)	Radioactivity	ETA apparatus	XI	

[a] Other detectors may be used.
[b] May be a photomultiplier tube, photodiode, photocell, or other instrument.
[c] DTA or DSC may also be used.

analysis method if the measured parameter is determined as a function of temperature. This would include proton nuclear magnetic resonance, electron spin resonance, electron diffraction, X-ray diffraction, mass spectrometry, ultraviolet, visible, and infrared spectrophotometry, and so on. It is beyond the scope of this book to discuss all of these techniques, although their high- (or low-) temperature applications are quite obvious.

It should be pointed out that, in many cases, the use of only a single thermal analysis technique may not provide sufficient information about a given system. As with many other analytical methods, complementary or supplementary information, as can be furnished by other thermal analysis techniques, may be required. For example, it is fairly common to complement all DTA or DSC data with thermogravimetry. If a gaseous product(s) is (are) evolved, evolved gas analysis may prove useful in solving the problem at hand. Simultaneous thermal techniques are useful in this respect in that several types of data are obtained from the same sample under identical pyrolysis conditions.

The methods of thermal analysis have been widely accepted in analytical chemistry during the past ten years. The field has nurtured the founding of three common interest societies—the North American Thermal Analysis Society (NATAS), the International Confederation of Thermal Analysis (ICTA), and the Society of Calorimetry and Thermal Analysis (Japan). Each society has sponsored meetings and symposia in which the main topic of discussion was thermal analysis techniques and their applications. Publications of thermal analysis investigations have been centralized into two widely accepted specialized journals—the *Journal of Thermal Analysis* (1), which was first published in 1969; and *Thermochimica Acta* (2), which first appeared in 1970. Also as an aid to workers in the field, *Thermal Analysis Abstracts* (3) became available in 1972.

During the past few years, much consideration has been given to the nomenclature of thermal analysis. The Nomenclature Committee of ICTA has been very active in this area; the results of their deliberation are summarized in Chap. XIV. However, much more work needs to be done along these lines to prevent further confusion with certain terms that are now being used.

References

1. E. Buzágh and J. Simon, eds., *J. Therm. Anal.*, Heyden-Akademiai Kiado.
2. W. W. Wendlandt, ed., *Thermochim. Acta*, Elsevier.
3. J. P. Redfern, ed., *Therm. Anal. Abst.*, Heyden.

CHAPTER II

THERMOGRAVIMETRY

A. Introduction

The thermal analysis technique of thermogravimetry (TG) is one in which the change in sample mass is recorded as a function of temperature. Three modes of thermogravimetry may be described: (*a*) *isothermal* or *static thermogravimetry*, in which the sample mass is recorded as a function of time at constant temperature; (*b*) *quasistatic thermogravimetry*, in which the sample is heated to constant mass at each of a series of increasing temperatures; and (*c*) *dynamic thermogravimetry*, in which the sample is heated in an environment whose temperature is changing in a predetermined manner, preferably at a linear rate. Most of the studies discussed here will refer to dynamic thermogravimetry, which will be designated as thermogravimetry.

The resulting mass-change versus temperature curve (which has various synonyms such as thermolysis curve, pyrolysis curve, thermogram, thermogravimetric curve, thermogravigram, thermogravimetric analysis curve, and so on) provides information concerning the thermal stability and composition of the initial sample, the thermal stability and composition of any intermediate compounds that may be formed, and the composition of the residue, if any. To yield useful information with this technique, the sample must evolve a volatile product, which can originate by various physical and chemical processes such as discussed in Chap. IV. Except for the mass-changes, much of the information obtained from the TG curve is of an empirical nature in that the transition temperatures are dependent on the instrumental and sample parameters. Thus, it is difficult to make meaningful comparisons between TG data obtained on different thermobalances in different laboratories. The use of commercially available thermobalances has done much to improve this situation, but it should still be noted that the curve transition temperatures are procedurally obtained temperatures and are not fundamental to the compound as are X-ray *d*-spacings and infrared absorption band minima.

The characteristics of a single-stage mass-loss curve are illustrated in Figure II.1. Two temperatures may be selected as characteristic of any single-stage nonisothermal reaction: T_i, the *initial temperature* or *procedural decomposition temperature* (pdt), which is the temperature at which the

6

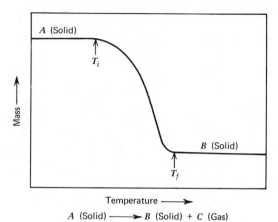

Figure II.1. Characteristics of a single-stage reaction TG curve (1).

cumulative mass-change reaches a magnitude that the thermobalance can detect; and T_f, the *final temperature*, which is the temperature at which the cumulative mass-change first reaches its maximum value, corresponding to complete reaction. Although the T_i may be the lowest temperature at which the onset of a mass-change may be observed in a given experiment, it is neither a transition temperature in the phase-rule sense nor a true decomposition temperature below which the reaction rate suddenly becomes zero. At a linear heating rate, T_f must be greater than T_i, and the difference, $T_f - T_i$, is called the *reaction interval*. For an endothermic decomposition reaction, T_i and T_f both increase with increasing heating rate, the effect being greater for T_f than for T_i.

The thermal stability is defined as a general term (2) indicating the ability of a substance to maintain its properties as nearly unchanged as possible on heating. From a practical point of view, thermal stability needs to be considered in terms of the environment to be imposed on the material and the functions it has to perform. The thermobalance is a useful technique for studying the ability of a substance to maintain its mass under a variety of conditions.

The historical aspects of thermogravimetry have been adequately described by Duval (3–5) and others (6–8). The technique apparently began about 1915 with the work of Honda (9) and was further developed during the 1920s by Guichard (10). Duval (3, 4) used the thermobalance to create an interest in gravimetric procedures in the late 1940s and early 1950s. The modern aspects of thermogravimetry began in the late 1950s with the introduction of high-quality, commercially available thermobalances.

B. Some Factors Affecting Thermogravimetric Curves

As with any instrumental technique, there are with thermogravimetry a large number of factors which affect the nature, precision, and accuracy of the experimental results. Thermogravimetry probably has a larger number of variables because of the dynamic nature of the temperature change of the sample. Duval (3, 4, 11) discussed in detail the precautions involved in using a thermobalance as well as the many other variables involved in thermogravimetry. No attempt will be made to include all of these in this discussion; only the most important parameters will be reiterated here. Basically, the factors that can influence the mass-change curve of a sample fall into the following two categories:

(*1*) Instrumental (thermobalance) factors
   (*a*) Furnace heating rate
   (*b*) Recording or chart speed
   (*c*) Furnace atmosphere
   (*d*) Geometry of sample holder and furnace
   (*e*) Sensitivity of recording mechanism
   (*f*) Composition of sample container
(*2*) Sample characteristics
   (*a*) Amount of sample
   (*b*) Solubility of evolved gases in sample
   (*c*) Particle size
   (*d*) Heat of reaction
   (*e*) Sample packing
   (*f*) Nature of the sample
   (*g*) Thermal conductivity

Unfortunately, definitive studies are lacking on some of the above factors; if some type of a study has been made, it has been limited to only one type of thermobalance or recording system and correlations cannot be easily made with other types of instruments. It is true, of course, that many of the above factors, such as sample-holder geometry, recording speed, balance sensitivity, and sample-container air buoyancy, are fixed with any given thermobalance. Factors which are variable and difficult to reproduce are the sample-particle size, packing, the solubility of evolved gases in the sample, furnace convection currents, and electrostatic effects. In view of the above variables, it is unfortunate that some type of standard sample is not available for comparing one given experimental apparatus with another. An insight into the use of standard compounds for temperature calibration, however, is given in Chap. III.

1. Instrumental (Thermobalance) Factors

a. Heating Rate

The effect of heating rate change on the procedural decomposition temperatures of a sample has been widely studied. Perhaps the only other parameter that has been studied more is that of the effect of atmosphere on the TG curve. For a single-stage endothermic reaction, Simons and Newkirk (1) have pointed out the following changes for T_i and T_f, as a function of fast (F) and slow (S) heating rates. For the initial procedural decomposition temperature, T_i,

$$(T_i)_F > (T_i)_S$$

For the final procedural temperature, T_f,

$$(T_f)_F > (T_f)_S$$

while the reaction interval, $T_f - T_i$, varies according to

$$(T_f - T_i)_F > (T_f - T_i)_S$$

For any given temperature interval, the extent of decomposition is greater at a slow rate of heating than for a similar sample heated at a faster rate. If the reaction involved is exothermic, the sample temperature will rise above that of the furnace, and it has been shown (8) that the difference between the furnace temperature and the sample temperature is greatest for the faster rate of heating when a reaction is occurring. When successive reactions are involved, the rate of heating may well determine whether or not these reactions will be separated. The appearance of a point of inflection in the TG curve at a faster heating rate may resolve itself into a horizontal plateau at a slow heating rate.

The effect of heating rate on the TG curve of a sample has been discussed by numerous authors. Mention should be made of the investigations or reviews by Duval (3, 4, 11), Newkirk (12), Redfern and co-workers (6, 8), Simmons and Wendlandt (13), DeVries and Gellings (14), Herbell (15), and others. The effect of three different heating rates on the TG curves of $CaC_2O_4 \cdot H_2O$ is shown in Figure II.2 (15). As can be seen, for each consecutive reaction, an increase in heating rate increases the procedural decomposition temperatures, T_i and T_f. This effect in general, is quite small for the three heating rates studied, 2.5, 5, and 10°C/min. It is probably more clearly shown by the qualitative moisture evolution curves, also illustrated in Figure II.2. It should be noted that although the procedural decomposition temperatures are changed with change in heating rate, the mass-losses remain unchanged.

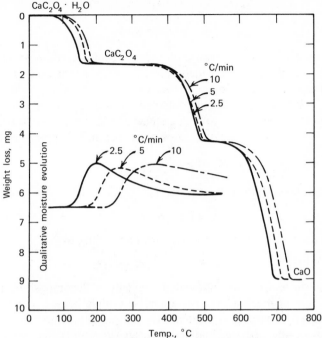

Figure II.2. Effect of heating rate on the TG curves of $CaC_2O_4 \cdot H_2O$ (15). Sample size is
14.8 mg; dynamic He atmosphere at 150 ml/min.

Simmons and Wendlandt (13) found that the procedural decomposition
temperatures for the polymerization reactions of

$$[Co(NH_3)_4(H_2O)(CN)(Co(NH_3)_4CN]\,(ClO_4)_4$$

varied not only with heating rate but also with water content in the surrounding atmosphere. Keeping the water partial pressure constant, the
decomposition temperatures (at the given heating rate) were as follows:
151°C (5°C/min), 159°C (10°C/min), and 161°C (15°C/min). Rather drastic
changes in the procedural decomposition temperatures and the formation
of horizontal mass plateaus were noted in the TG curves of $KHSO_4$ and
related compounds in changing the heating rate from 4 to $\frac{4}{3}$°C/min (15).
Parts of the curve were not reproducible because of various factors other
than the heating rate.

It should not be inferred that the use of high heating rates in thermogravimetry always has a deleterious effect on the TG curves obtained. If a
small sample is used, very fast heating rates may be employed and one will
still be able to detect the presence of intermediate compounds formed

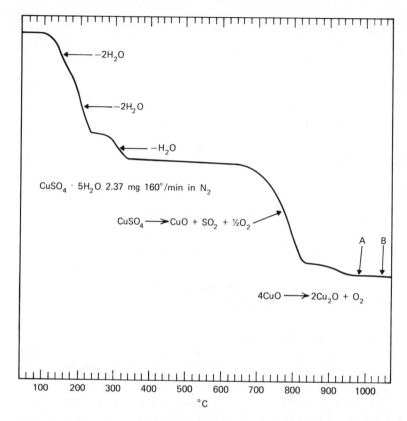

Figure II.3. TG curve of $CuSO_4 \cdot 5H_2O$ obtained at a very fast heating rate (16).

during the decomposition reaction.   Using an unprecedented heating rate of $160°C/min$, the TG curve of $CuSO_4$, shown in Figure II.3, was obtained (16). The entire curve was recorded in 6.5 min, with the intermediate compounds indicated either by curve inflection points or by horizontal mass plateaus.

The detection of intermediate compounds in the mass-loss curve is also dependent on the heating rate employed.   This is shown in Figure II.4, in which the curves do not reveal intermediate curve breaks at the fast heating rate; however, breaks are evident at the lower heating rate (5).   Fruchart and Michel (17) detected intermediate compounds, with the compositions, 6-, 4-, 2-, and 1-hydrate, when $NiSO_4 \cdot 7H_2O$ was heated at a rate of $0.6°C/$ min.   A previous study at $2.5°C/min$ revealed only the existence of the 1-hydrate (18).   A similar situation was observed with the monosalicyl-aldoxinezinc(II) chelate as studied by Rynasiewicz and Flagg (19) and

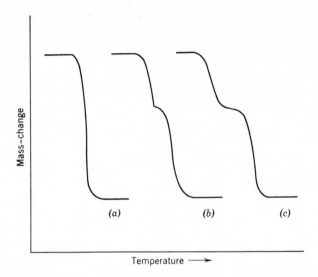

Figure II.4. The effect of heating rate on the formation of intermediate compounds. (a) fast heating rate; (b) slow heating rate; (c) slow heating rate and recording rate (4).

DeClerq and Duval (20). On heating at 300°C/h a wet precipitate containing a 250% excess of water, a horizontal mass region was obtained in the curve from 215 to 290°C (19). DeClerq and Duval (20), using a higher heating rate of 380°C/h, did not detect a horizontal mass level and hence rejected the method for the determination of zinc. The latter results indicated that when samples which contain a large amount of water are studied, a slow heating rate should be employed. It should also be noted that a sudden inflection in the mass-loss curve may be caused by a sudden variation in the rate of heating and hence be false (6). One method used to detect this phenomenon is to always record the furnace temperature as a function of time on a strip-chart recorder. Temperature perturbations are discussed further in Chap. III.

The recording of more pronounced horizontal mass plateaus in the TG curve is possible by use of a quasistatic heating-rate mode, as previously mentioned. This method was first used by Honda (9) and also by Lukaszewski and Redfern (6) and Paulik and Paulik (21). With this technique, provision is made for the interruption of the linear temperature rise cycle and continuation of the heating at a constant fixed temperature. This method gives mass-loss curves that are, in general, steeper than those obtained under dynamic conditions and provides more accurate data on the final decomposition temperatures.

b. Recording or Chart Speed

The recording of the mass-loss curves for either rapid or slow reactions can have a pronounced effect on the shape of the curves. The effect of chart speed on the recording of the curves of various reactions is illustrated in Figure II.5. In curve (*a*), there is a definite flattening of the curve as the chart speed is increased for a slow thermal decomposition reaction.

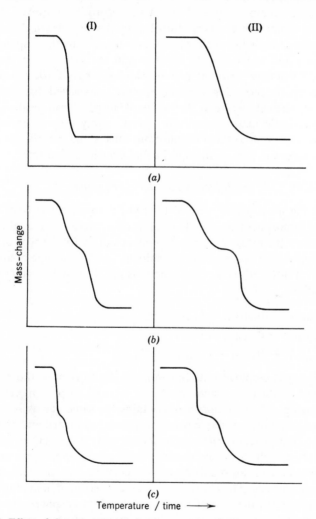

Figure II.5. Effect of chart speed upon the shapes of mass-loss curves: (I) low chart speed; (II) high chart speed (6).

In the case of a slow reaction followed by a rapid one, curve (*b*), the lower-chart-speed curve shows less separation of the two steps than the higher-chart-speed curve.   For a fast reaction followed by a slower one, curve (*c*), an effect similar to that of curve (*b*) was observed, namely, shorter curve plateaus.

An excessive chart speed will tend to minimize differing rates of mass-loss. It is recommended that a chart speed of 6–12 in./h for a heating rate of 1–6°C/min be employed (6).   With *X–Y* recorders, however, the chart speed on the temperature axis is controlled by the response of the instrument and the heating rate, while the mass axis is controlled by the responses of the recorder pen and the recording balance and the rate of the thermal decomposition reaction.   Simons and Newkirk (1) have criticized the use of *X–Y* recorders for recording TG data because unexpected disturbances that can occur in the heating rate or thermocouple response can produce spurious perturbations in the mass-temperature record.   Only a separate temperature-time record can disclose these adventitious effects and permit effects caused solely by changes in sample mass to be distinguished from them.

c. Effect of Furnace Atmosphere

Perhaps the most widely studied instrument variable has been the effect of furnace atmosphere on the TG curve of a sample.   The effect of the atmosphere on the mass-change curve depends upon (*a*) the type of reaction, (*b*) the nature of the decomposition products, and (*c*) the type of atmosphere employed.   For (*a*), three types of reactions may be studied, either reversible or irreversible:

(*i*) $A_{solid(1)} \rightleftarrows B_{solid(2)} + C_{gas}$

(*ii*) $A_{solid(1)} \rightarrow B_{solid(2)} + C_{gas}$

(*iii*) $A_{solid(1)} + B_{gas(1)} \rightarrow C_{solid(2)} + D_{gas(2)}$

If an inert gas is employed, its function will be to remove the gaseous decomposition products in reactions (*i*) and (*ii*) and to prevent reaction (*iii*) from occurring   If the atmosphere contains the same gas as is evolved in the reaction, only the reversible reaction (*i*) will be affected and no effect will be observed on reaction (*ii*).   In reaction (*iii*), if gas (1) is changed in composition, the effect on the reaction will depend upon the nature of the gas introduced (e.g., an oxidizing or reducing gas will probably affect the mass-change curve).   The above discussion concerns a dynamic (or flowing) gas atmosphere; in the case of a static (or fixed) atmosphere, the following behavior probably takes place.   If the sample evolves a gaseous product reversibly, as the temperature of the furnaces (and sample) increases, it will begin to dissociate as soon as its dissociation pressure exceeds the partial pressure of the gas or vapor in its immediate vicinity. Since a dynamic temperature system is employed, the specific rate of the decomposition reaction will increase, as well as the concentration of the ambient gas surrounding the sample, due to the decomposition of the sample. If the ambient gas concentration increases, the rate of the reaction will decrease. However, due to convection currents in the furnace, the gas concentration around the sample is continuously changing, which is one of the reasons why static atmospheres are not recommended; for reproducible results, dynamic atmospheres under rigidly controlled conditions are used.

A good illustration of the effect of atmosphere on reversible and irreversible reactions, such as are illustrated in reactions (*i*) and (*iii*), is shown in Figure II.6 (1). The curves show the effect of heating $CaC_2O_4 \cdot H_2O$ in both dry N_2 and O_2 atmospheres. The dehydration step, which is reversible,

$$CaC_2O_4 \cdot H_2O(s) \rightleftarrows CaC_2O_4(s) + H_2O(g)$$

is unaffected because both gases are equally effective in sweeping evolved water vapor away from the sample surface. For the second reaction,

$$CaC_2O_4(s) \rightarrow CaCO_3 + CO \qquad (if\ N_2)$$

the curves diverge because the oxygen reacts with the evolved CO, giving a secondary oxidation reaction which raises the temperature of the unreacted solid. This higher temperature produces a marked acceleration in the

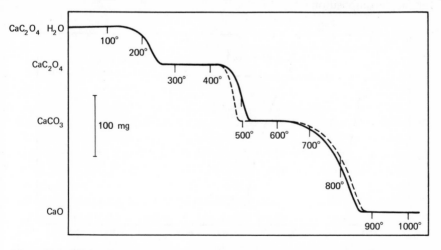

Figure II.6. Effect of atmosphere on the mass-loss curve of $CaC_2O_4 \cdot H_2O$ (500-mg sample heated at 300°C/h) (1). $- - - -$, dry O_2; ———, dry N_2.

decomposition rate. Thus, the decomposition of the compound occurs more rapidly and is completed at a lower furnace temperature in dry O_2 than in an inert atmosphere of dry N_2.

The third step in the decomposition reaction,

$$CaCO_3(s) \rightarrow CaO(s) + CO_2(g)$$

is also a reversible reaction and hence should not be influenced by either the oxygen or the nitrogen. As can be seen, however, there is a slight difference for the two gases studied. This small difference was attributed to the difference in the composition of the $CaCO_3$ formed during the second step of the decomposition reaction. The $CaCO_3$ formed in an oxygen atmosphere is slightly different from that formed in nitrogen. This difference was not described (1), but it was stated that the mass-change curves cannot disclose differences in particle size, surface area, lattice defects, or other characteristics of the sample.

In the case of reversible reactions such as previously described, increasing the partial pressure of carbon dioxide in the furnace atmosphere will increase the T_i of the curve, as is shown in Figure II.7. The initial procedural decomposition temperature can range from about 400°C at reduced pressure to 900°C in a carbon dioxide atmosphere at a pressure of 760 Torr. These are rather pronounced changes, due to the extremes in furnace atmosphere used, but they illustrate dramatically the effect of the furnace atmosphere on the TG curve. Paulik et al. (23) described a similar series of TG curves for the decomposition.

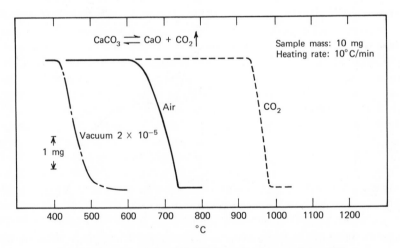

Figure II.7. TG curves of $CaCO_3$ in various atmospheres (22).

Many of the problems involved in obtaining reproducible TG curves of a sample can be solved by using the technique of self-generated atmospheres. This technique is discussed in Section D of the present chapter.

The effect of water vapor in the furnace atmosphere has been rather widely studied (1), especially in reactions involving dehydration and hydration. Feldman and Ramachandran (24) constructed a balance and sample preparation enclosure which enabled them to study reactions under controlled amounts of water vapor. Numerous reactions could be studied in the furnace containing a saturated water vapor atmosphere, as described in Chap. III. Wendlandt and Simmons (25) studied the thermal decomposition of $BaCl_2 \cdot 2H_2O$ and $BaBr_2 \cdot 2H_2O$ in water-saturated atmospheres and in dry nitrogen atmospheres, while Herbell (15) studied the reduction of NiO with dry and water-saturated hydrogen.

The effects of reduced pressure on the TG curve of various compounds have been reported. Guenot and Manoli (26) reported the effect on the dehydration of $CuSO_4 \cdot 5H_2O$, while Nicholson (27) studied the effects of low pressure on the thermal decomposition of $FeC_2O_4 \cdot 2H_2O$. Numerous other studies have been reported.

The effect of an increase in pressure on the TG curve has been described by Brown et al. (28). High-pressure effects would be the opposite of those encountered in low-pressure atmospheres; the T_i for the reaction would be shifted to higher temperatures as well as the increase in the reaction interval, $T_f - T_i$. Using a high-pressure thermobalance, the TG curves of

$$CuSO_4 \cdot 5H_2O,$$

as shown in Figure II.8, were obtained. The curve obtained at 20 atm of pressure was similar to that recorded at atmospheric pressure, after a correction was made for buoyancy effects. Procedural decomposition temperatures were almost identical to those obtained from the curve at 1 atm pressure, but this may be due to the limited pressure range studied. One of the adverse effects of increased pressure is the large increase in buoyancy observed; the curve is usually corrected for this change.

Apparent mass gains are occasionally observed in the thermal decomposition of a sample under high vacuum conditions if the sample layer is of a critical thickness and if a certain type of sample holder is employed. Such an effect is shown for the dehydration of $CaC_2O_4 \cdot H_2O$ in Figure II.9, as recorded by Wiedemann (29). The broken-line curve shows the dehydration reaction under normal atmospheric pressure, while the solid-line curve is that recorded under high-vacuum conditions. The "apparent" gain in sample mass at the beginning of the reaction is due to the collision of the water molecules with the sample container during pumping.

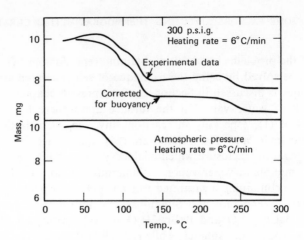

Figure II.8. TG curves for the dehydration of $CuSO_4 \cdot 5H_2O$ (28).

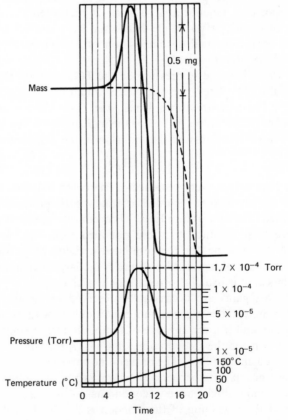

Figure II.9. TG curves of $CaC_2O_4 \cdot H_2O$ dehydration at − − − −, normal pressure and ─────, high vacuum condition (29).

Friedman (30) has discussed this effect in the thermal dissociation of Teflon *in vacuo*. He showed that the magnitude of this momentum-transfer effect is expressed by the equation

$$w = m - \frac{1}{g} \alpha v \left(\frac{dm}{dt}\right) \tag{II.1}$$

where w is the mass of the sample as determined by the thermobalance, m is the actual sample mass, g is the acceleration due to gravity, α is a geometric factor, v is the velocity of the ejected gas, and dm/dt is the rate of change of actual mass.

Numerous other studies on the effect of the furnace atmosphere on TG curves have been reported (12, 32–37) while more recent studies have also been described (21, 38–41).

d. Sample Holder

The large number of sample holders used in thermogravimetry are described in Chap. III. Sample holders range from flat plates to deep crucibles of various capacities. Materials used in their construction may vary from glass, alumina, and ceramic compositions to various metals and metallic alloys.

Simons and Newkirk (1) have shown that for $CaC_2O_4 \cdot H_2O$ the geometry of the sample holder is immaterial if no interaction is possible between the sample and the gaseous atmosphere or products. As seen in the TG curves in Figure II.10, the curves for $CaC_2O_4 \cdot H_2O$ heated in a carbon dioxide atmosphere are identical above 275°C. As expected, the loss of water occurred more readily from the shallow quartz dish than from the crucible. The shape of the crucible had no effect upon the decomposition of anhydrous CaC_2O_4 because this reaction is not reversible, and in a CO_2 atmosphere no important diffusion-controlled reaction can occur. The geometry of the sample holder also had no effect on the dissociation of $CaCO_3$ because this reaction is reversible, and the atmosphere used was solely the gas involved in the reaction.

When the thermal-decomposition reaction was carried out in a dynamic atmosphere of nitrogen, both the loss of water and the loss of carbon dioxide were affected by the geometry of the container. Likewise, the decomposition of CaC_2O_4 was unaffected. The marked effect of the geometry of the sample holder provides evidence that a significant pressure of water vapor and carbon dioxide must have existed in the interior of the crucible during dissociation, even when the atmosphere that flowed over the crucible entered the thermobalance free from either water or carbon dioxide or both (1).

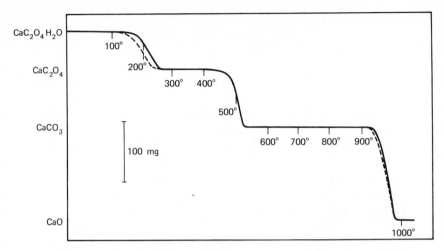

Figure II.10. Effect of sample-holder geometry on the TG curve of $CaC_2O_4 \cdot H_2O$ in a dynamic CO_2 atmosphere; – – – –, quartz dish; ———, porcelain crucible (1).

The difference in TG curves for $CuSO_4 \cdot 5H_2O$ obtained using a crucible and the multiplate sample holder is illustrated in Figure II.11 (42). Curve (1) was obtained by use of a crucible, while in obtaining curve (2) the sample was placed as a thin layer on the surfaces of a multiplate holder. The latter type of holder yielded better separation of overlapping reactions and also resulted in lower procedural decomposition temperatures. A similar effect was reported by Paulik and Paulik (21).

The effect of the size and the heat-sink properties of the sample holder has been illustrated by Garn (34). The effect is shown on the thermal decomposition of lead carbonate in Figure II.12. The sample holder, which was placed on an aluminum block 1 in. in height with cylindrical surface of 3.1 sq. in., was employed. The shallow aluminum pan was of the same diameter but had an overall height of $\frac{1}{16}$ in. A factor of 16 in the area directly exposed to the heated furnace wall permitted more rapid heat transfer to the middle of the sample and hence more uniform temperature throughout. It should be noted that the sample on the massive sample holder decomposed over a smaller temperature range. The conventional sample holders employed in thermogravimetry are far from being infinite heat sinks (34). Generally, the heat of reaction of the sample is the principal consumer of heat energy, yet the sample holders are not designed to supply that heat rapidly. Garn (34) suggests that sample holders should be designed to supply this heat to the sample as rapidly as possible.

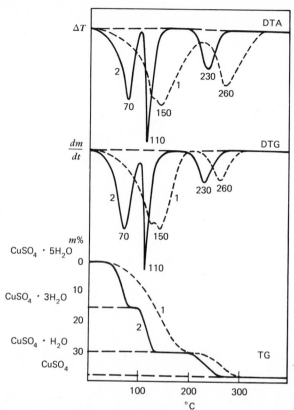

Figure II.11. Effect of crucible and multiplate sample holder on TG curve of $CuSO_4 \cdot 5H_2O$ (24); – – – –, crucible, 500 mg; ————, multiplate, 200 mg; heating rate of 10°C/min.

Sestak (43) has calculated the maximum temperature gradient between the wall of the sample holder and the center of the sample. For a disk

$$Y_m = \left(\frac{\Delta H G \phi}{\lambda}\right)^{1/2} \cdot \frac{S}{2} \qquad (II.2)$$

For a cylinder

$$Y_m = \left(\frac{\Delta H G \phi}{2\lambda}\right)^{1/2} \cdot \tau \qquad (II.3)$$

where S is the sample thickness, τ is the diameter, ΔH is the enthalpy of the reaction, G is the heat capacity of the sample, ϕ is the heating rate, and λ is the thermal conductivity. For a cylindrical sample 1 mm in diameter contained in a silver block at a heating rate of 5°C/min, the maximum

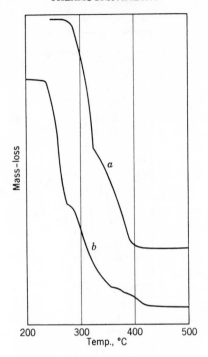

Figure II.12. Effect of heat sink on the TG curves of lead carbonate. (*a*) massive aluminum block; (*b*) thin aluminum pan. Heating rate is about 450°C/h (34).

temperature gradients found were 4.8°C for the dehydration of kaolin, 13.2°C for the decomposition of $MgCO_3$, and 3.1°C for the dehydration of α-$CaSO_4$·$0.5H_2O$. Numerous other factors concerning sample holders and the effects of mass transfer were discussed in detail.

Although Paulik et al. (42) present TG curves of "loose" and "packed" samples of CaC_2O_4·H_2O, a more vivid illustration of the effect of sample packing on the evolution of water is shown in Figure II.13 (44). The evolution of water was monitored by a mass spectrometer ($m/e = 18$) rather than a thermobalance, but the effect should be similar in the latter. Packings A and B lead to an almost symmetrical peak with rapid dehydration, while packing C shows that dehydration is slower and spread out over a larger temperature range.

The material of construction of the sample holder should have little effect on the TG curves if it does not react with the sample. Newkirk and Aliferis (45) have shown that sodium carbonate can react with porcelain or alumina crucibles at high temperatures and hence reveal a mass-loss in the TG curve (see Chap. IV). The catalytic properties of platinum may affect

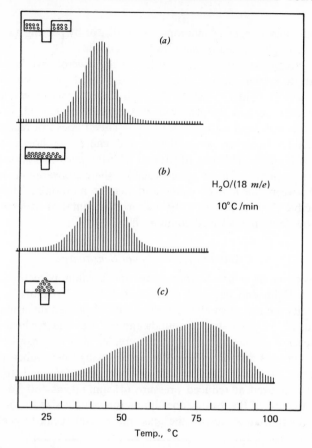

Figure II.13. Effect of sample packing and sample-holder geometry on the evolution of water from $CaC_2O_4 \cdot H_2O$ (43).

the TG curves of certain metal sulfides, as was shown by Ramakrishna Udupa and Aravamudan (46). The platinum crucible catalyzed the oxidation of zinc sulfide to zinc sulfate, a process that did not occur in alumina crucibles.

In the study of reversible reactions or of reactions in which a component of the atmosphere can react either with the original sample or with a solid or gaseous decomposition product, the possible existence of partial pressure gradients throughout the mass of the powdered sample should be recognized (11). These gradients can effect both the shape of the TG curves and the magnitude of the thermal effects that accompany the reactions. They can be reduced by packing the powder loosely in a shallow container, by using crucibles with microporous or macroporous walls, or by passing a controlled atmosphere through the bed of the powdered sample.

Duval (3) suggested that since the walls of the crucible are heated more strongly than the center, the use of a plate and a thin layer of sample would be the best sample holder, whereas the high-walled crucible would be the worst. However, certain samples swell or spatter when heated, so that the use of crucibles with high walls is necessary. Duval does not recommend a covered crucible, however, since this would cause the horizontal mass plateaus to be longer. This was illustrated with the pyrolysis of magnesium ammonium phosphate. In an open crucible, there appeared to be a discontinuity between the loss of water and that of ammonia, while in the covered crucible there appeared to be a short horizontal or at least a break as soon as the ammonia stopped coming off.

e. Conditions for Optimum Sensitivity

Mass sensitivity as a critical parameter in thermogravimetry has been considered by Cahn and Peterson (47). Greater sensitivity of the thermobalance permits the use of smaller samples, with improved determination of mass plateaus of intermediate compounds and the use of faster heating rates. However, thermobalances with sensitivities greater than 1 μg can be attained only under two conditions (48): (a) sample hangdown tubes of 9 mm inside diameter or less if used at atmospheric pressure, and (b) larger-diameter tubes if used at reduced pressure (41 mm i.d. at 150 Torr). Unfortunately, the use of the 9-mm tubes limits the sample size to 15–20 mg.

The effect of tube diameter on the mass-noise level (peak-to-peak in μg)

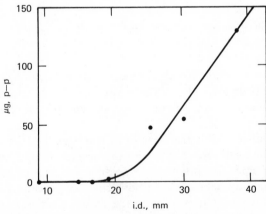

Figure II.14. Mass noise, μg peak-to-peak versus inside diameter of sample tube, in air at atmospheric pressure (47).

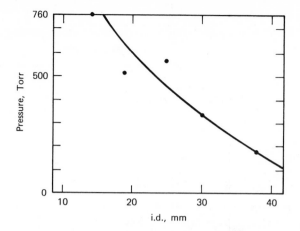

Figure II.15. Pressure in Torr for 1 μg peak-to-peak noise versus tube diameter (47).

is shown in Figure II.14 (47). At larger diameters substantial noise is observed, while at 19 mm there are only 3 μg of noise. At 16 mm, the noise is about 0.5 μg peak-to-peak, which is readable to 0.1 μg.

A similar study was made of the effect of pressure and tube diameter on the noise level. For 1 μg peak-to-peak noise, the tube diameter varied, as shown in Figure II.15, with the pressure. As can be seen, the lower the pressure employed, the greater the tube diameter that can be used for an equivalent amount of noise. With the larger tube diameter, it was found that for a given pressure maximum noise usually occurred between 150 and 650°C. Higher temperatures were usually less noisy than the lower values studied. Noise as a function of tube diameter was about the same with flowing gases as with static atmospheres. With the former, the noise was nearly independent of gas velocity, at least from 5 to 500 ml/min; extremes of noise varied from 1 to 2 μg peak-to-peak. The 16-mm-diameter tube appeared to be about ideal for both static- and dynamic-gas applications. It also appears to be a good selection for low-pressure operation, except where exceptionally large samples must be accommodated.

2. SAMPLE CHARACTERISTICS

a. Sample Mass

According to Coats and Redfern (8), the sample mass can affect the TG curve in three ways:

(1) the extent to which endothermic or exothermic reactions of the sample will cause sample temperature to deviate from a linear temperature change (the larger the sample mass, the greater the deviation);

(2) the degree of diffusion of the product gas through the void space

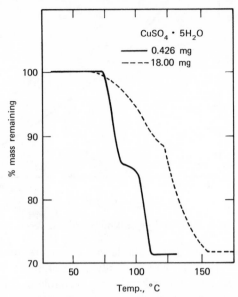

Figure II.16. Comparison of sample masses in the TG of $CuSO_4 \cdot 5H_2O$ (48).   Heating rate of 13°C/min in a static air atmosphere.

around the solid particles (under static conditions, the atmosphere immediately surrounding the reacting particles will be somewhat governed by the bulk of the sample);

(3) the existence of large thermal gradients throughout the sample, particularly if it has a low thermal conductivity.

In order to detect the presence of intermediate compounds, a small sample is preferred to a larger one, as is seen by the curves in Figure II.16 (48). The presence of a $CuSO_4 \cdot 3H_2O$ mass plateau is clearly indicated when a sample mass of 0.426 mg is used, in contrast with the 18.00 mg sample. This is probably an extreme case because samples as small as 0.426 mg are not commonly investigated.

Concerning the effect of sample mass on the T_i and T_f values, Richer and Vallett (32) found that T_i was virtually constant for calcium carbonate in the mass range 0.25 to 1 g, both in nitrogen and carbon dioxide. On the other hand, once the decomposition reaction has begun (1), it generally does not occur uniformly in every particle throughout the entire mass of the sample. Under such nonhomogeneous conditions, it would be expected that the time required for complete decomposition of a powdered solid would increase with increased sample mass. Because the furnace heating rate is linear, there would be a resultant increase in the observed value of T_f.

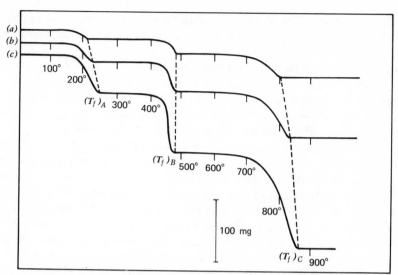

Figure II.17. Effect of sample mass on the TG curves of $CaC_2O_4 \cdot H_2O$ in a static atmosphere at a heating rate of 300°C/h (1). (*a*) 126 mg; (*b*) 250 mg; (*c*) 500 mg.

Such a change in T_f is observed in the thermal decomposition of $CaCO_3$ formed during the dissociation of $CaC_2O_4 \cdot H_2O$, as seen in Figure II.17. There was a regular increase in the T_f values for the dehydration reaction and for the decomposition of $CaCO_3$ with increased sample mass. However, if the decomposition reaction is exothermic, as is the case with the decomposition of CaC_2O_4 in air, the T_f values do not change with change in sample mass. In this case, the sample temperature increases more rapidly than does the measured furnace temperature, and the resultant acceleration in specific reaction rate may compensate, at least in part, for the increase in sample mass. Simons and Newkirk (1) found that if the decomposition reaction was carried out in an inert nitrogen atmosphere, all three T_f values for $CaC_2O_4 \cdot H_2O$ were shifted to higher temperatures with an increase in sample mass.

b. Sample Particle Size

The effect of sample particle size on the TG curve has been comparatively little studied. Various particle sizes will cause a change in the diffusion of product gases, which will alter the reaction rate and hence the curve shape. Most of the studies in this area that have been reported have been concerned with the effect of particle size on the kinetics parameters (8, 43). Large crystals of the sample may decrepitate, causing sudden mass-losses in the TG curve. The smaller the particle size, the greater the extent to which equilibrium is reached, and at any given temperature, the greater the extent of decomposition will be (8).

In comparing the thermal decomposition curves of calcium carbonate and calcite, Richer and Vallet (32) found that the empirical decomposition temperatures obtained at the heating rate of 150°C/h in a stream of nitrogen gas were the following: powdered calcium carbonate, 783°C; powdered calcite, 802°C; cube of calcite weighing about 350 mg, 891°C.

Likewise, for a chrysotile sample, Martinez (49) found that the decomposition temperature decreased with a decrease in sample particle size. For the ground material, there was a continuous loss in mass from about 50 to 850°C, with the most rapid decomposition between 600 and 700°C. For the massive material, there was little mass-loss until a temperature of about 600°C was attained. Similar results were obtained for serpentine and a brucine–carbonate mixture. In general, a decrease in particle size of the sample lowers the temperature at which thermal decomposition begins, as well as the temperature at which the decomposition reactions are completed.

c. Miscellaneous Sample Effects

The effect of the heat of reaction of the sample on the mass-loss curve has been studied by Newkirk (12). The heat of reaction will affect the difference between the sample temperature and the furnace temperature, causing the sample temperature to lead or lag behind the furnace temperature, depending on whether the heat effect is exothermic or endothermic. Since these temperature changes may be 10°C or more, depending on the heating rate employed, the calculation of kinetic constants from mass-loss curves may be unavoidably and significantly in error. This effect is more thoroughly discussed in Section C.3 of this chapter.

The solubility of gases in solids imposes a serious limitation on the thermogravimetric method, as discussed by Guiochon (50). It is difficult to eliminate or even measure, and is generally unknown. This was shown by the heating of solid ammonium nitrate initially containing 1% nitric acid at 200°C for 3 h. At the end of this period, the sample contained 0.6% nitric acid. This acid has no catalytic effect on the decomposition of the sample, which gives no nitric acid under these conditions, so that only the slowness of its evaporation can explain these results. The concentration of this dissolved substance may be decreased to a small value by the use of wide crucibles without covers, a thin layer of sample, and a flow of inert gas through the furnace. According to Guiochon (50), this gas flow through the furnace is almost always necessary to facilitate the diffusion of gases to and from the sample.

The effect of sample packing, though difficult to reproduce, has been discussed in Section B.1.d of this chapter.

C. Sources of Error in Thermogravimetry

The sources of error in thermogravimetry can lead to considerable inaccuracies in the temperature and mass-change data obtained. Accurate thermogravimetry requires that a correction be applied for these errors or that at least some recognition be made of their magnitude. Many of these errors are interrelated and hence cannot be considered separately. Full consideration must be given to all of these factors in thermogravimetry.

The possible sources of error in thermogravimetry are many; among them can be listed the following:

(*1*) Sample-container air buoyancy
(*2*) Furnace convection currents and turbulence
(*3*) Random fluctuations in the recording mechanism and balance
(*4*) Furnace induction effects
(*5*) Electrostatic effects on balance mechanism
(*6*) Environment of the thermobalance
(*7*) Condensation on sample support
(*8*) Temperature measurement and calibration
(*9*) Weight calibration of recording balance
(*10*) Chart-paper rulings
(*11*) Reaction of the sample with sample container
(*12*) Temperature fluctuations
(*13*) Momentum-transfer effects in vacuum TG

1. SAMPLE-CONTAINER BUOYANCY

The effect of air buoyancy changes on the sample holder and certain balance components has been studied (3, 4, 12, 15, 29, 37, 48, 51, 52). Most of the studies have used the Chevenard thermobalance (3, 4, 37, 51, 52), but other balances have been studied as well (12, 15, 29, 37, 48).

Wiedemann (29) discussed the effect of buoyancy on the sample and certain parts of the thermobalance as a function of temperature. It should be kept in mind that the density of the gas phase decreases with temperature also. At about 300°C the density, and therefore the buoyancy exerted on

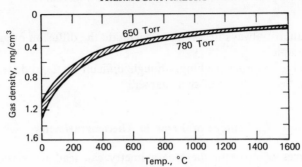

Figure II.18. Changes in gas density or buoyancy versus temperature at various pressures (29).   Curves are given for air.

the sample, is about one-half as great as at 25°C.   In air, this results in an apparent mass variation of about 0.6 mg/cm³.   This variation in gas density and buoyancy (mg/cm³) versus temperature is shown in Figure II.18.   The area lying between the curves roughly corresponds to the normal pressure fluctuations expected while working at atmospheric pressure.

The apparent mass-gain curve for the Chevenard thermobalance, as a function of temperature, is given in Figure II.19.   In general, the mass-gain is also a function of the heating rate and load, as well as temperature.   On the basis of a factorial design, Simons et al. (51) worked out a table giving the apparent mass-gain versus temperature for different volumes of load on the balance.   The apparent mass-gain of a single porcelain crucible with

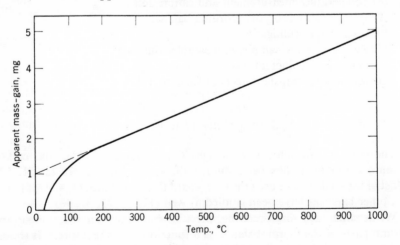

Figure II.19. Apparent mass-gain as a function of temperature for the Chevenard thermo-balance (51).

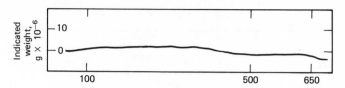

Figure II.20. Change in mass of platinum sample holder using the Cahn balance (48).

mass ~4 g and volume ~1.5 ml ranged from 1.8 mg at 200°C to 4.2 mg at 1000°C at a heating rate of 300°C/h. For a pair of nested crucibles weighing ~8 g and having a volume of ~3 ml, the range was 2.5–5.4 mg. The corresponding numbers at the heating rate of 150°C/h were 2.0–5.3 mg and 2.7–6.2 mg. These values are consistent with those reported by Mielenz et al. (52).

For a platinum sample holder, 0.7 cm square, 0.5 cm deep, and weighing 1.6 g, Lukaszewski (37) found that the increase in mass from ambient to 350°C was 0.3 ± 0.05 mg, and that from 350 to 1400°C was 0.2 ± 0.05 mg, at a heating rate of 1–3°C/min. The effects of different heating rates and load sizes were also studied.

Using the Cahn Model RG Electrobalance converted to a thermobalance, Cahn and Schultz (48) recorded the buoyancy curve for a platinum sample holder, as shown in Figure II.20. The change in mass is very small ($< 2 \times 10^{-6}$ g) in the temperature range from 25 to 650°C, using an 8-mm-diameter hangdown tube. This correction is much smaller than that for the Chevenard balance, by a factor of a thousand or so.

The sample buoyancy changes using high-pressure thermogravimetric techniques have already been mentioned (28).

2. FURNACE CONVECTION CURRENTS AND TURBULENCE

The apparent mass-gain or mass-loss due to convection currents in the furnace has been studied by Newkirk (12) and by Lukaszewski (37). The apparent mass-loss caused by the upflowing stream of air on the sample container and the apparent mass-gain due to air turbulence are determined largely by the sample crucible size and shape (12). The apparent mass-changes as a function of furnace top openings are given in Figure II.21. It was found that except for a large opening, there was always an initial mass-gain even when on further heating there was an overall mass-loss. It was not possible to choose an opening that would give no apparent mass-gain on heating over the entire temperature range. This effect is also dependent on the furnace heating rate. When using a flowing gas atmosphere in the furnace, Newkirk (12) found an additional apparent mass-gain,

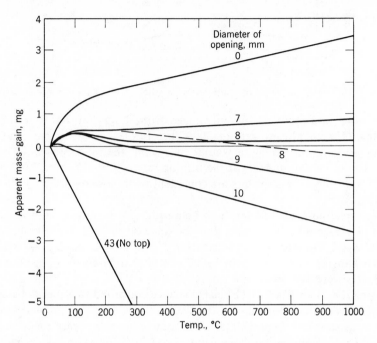

Figure II.21. Effect of furnace top opening on apparent mass-change of Chevenard thermobalance at a heating rate of 300°C/h (12); ——, one porcelain crucible, Coors 230-000 (about 4 g); – – –, two crucibles.

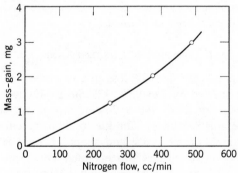

Figure II.22. Effect of gas velocity on apparent mass-gain of a porcelain crucible at room temperature on the Chevenard thermobalance (12).

as shown in Figure II.22.   Its magnitude was governed by the molecular mass of the gas employed.

3. Temperature Measurement

If the temperature of the sample is taken as the temperature measured by a thermocouple located just above or below the sample container, then the true sample temperature will either lead or lag behind the furnace temperature.   The magnitude of this difference depends upon the nature of the reaction (whether it is endothermic or exothermic), the heating rate, the sample thermal conductivity, the geometry of the sample holder, and so on. This effect is illustrated by the curves for the sample and thermowell temperatures of $CaC_2O_4 \cdot H_2O$, as shown in Figure II.23.   There are definite inflections at three places on the sample temperature curve caused by the decomposition reactions of the compound.

The temperature difference between the sample and the furnace for $CaC_2O_4 \cdot H_2O$ has also been studied by Newkirk (12).   The difference in temperature for this compound at a fairly high heating rate, 600°C/h, is illustrated in Figure II.24.   Curve (a) showed a 10–14° lag in the range of 100–1000°C.   The endothermic loss due to water evolution resulted in a 25° lag at 200°C.   With the larger sample, these effects are accentuated.

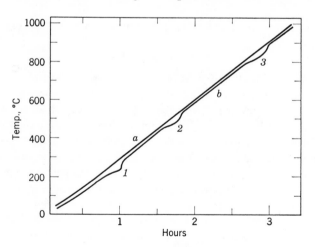

Figure II.23. Thermowell and sample temperature in the decomposition of $CaC_2O_4 \cdot H_2O$ in nitrogen on a Chevenard thermobalance.   One-gram sample heated at 300°C/h at a nitrogen flow rate of 400 ml/min.   (a) Thermowell temperature (+10° cal); (b) sample temperature (36).   (1) $CaC_2O_4 \cdot H_2O \rightarrow CaC_2O_4 + H_2O$; (2) $CaC_2O_4 \rightarrow CaCO_3 + CO$; (3) $CaCO_3 \rightarrow CaO + CO_2$.

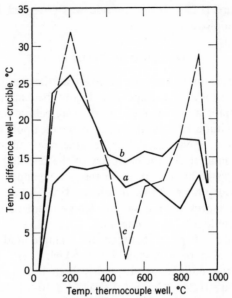

Figure II.24. Temperature differences between sample and furnace for $CaC_2O_4 \cdot H_2O$ (12).
(a) Crucible only; (b) crucible + 0.2 g of sample; (c) crucible + 0.6 g of sample.

It should be noted that Newkirk (12) observed an exothermic heat effect for the reaction

$$CaC_2O_4 \rightarrow CaCO_3 + CO$$

while Soulen and Mockrin (36) stated that it was an endothermic reaction. The discrepancy is that in an air atmosphere, which Newkirk presumably employed, carbon monoxide was oxidized to carbon dioxide by air, the oxidation reaction being highly exothermic. The latter investigators used a nitrogen furnace atmosphere in which the oxidation reaction did not take place. This again emphasizes the importance of furnace atmosphere and its effect on the pyrolysis reactions. By lowering the heating rate from 600 to 150°C/h, the temperature difference between the sample crucible and the furnace thermocouple decreased, as was shown by Newkirk (12). The effect of heating rate on this temperature difference is illustrated in Figure II.25. The lag varied from 3 to 14° and was roughly proportional to the heating rate.

The uncertainty in the actual temperature of the sample can be greatly reduced by use of a sample holder which contains a thermocouple as an integral part of it. Small temperature differences will still be observed, however, due to the size of the sample, the geometry of the sample holder,

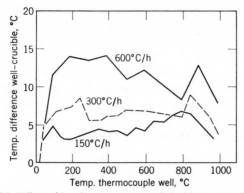

Figure II.25. Effect of heating rate on sample-holder temperature (12)

and so on, but they will be far smaller than in the previous type of temperature detection. One approach which has been used with some success is to position the thermocouple very close to the sample but not in contact with it, as is done with the DuPont thermobalance (53). The thermocouple is placed within the sample holder itself but does not touch it. A rapid temperature rise in the sample temperature can be detected by this method, as shown by the curves for $CaC_2O_4 \cdot H_2O$ in Figure II.26. As has been previously shown, the decomposition of CaC_2O_4 in air is an exothermic reaction. This exothermic reaction causes the sample temperature to rise very rapidly at about 500°C (Curve B) to a maximum value of 630°, where

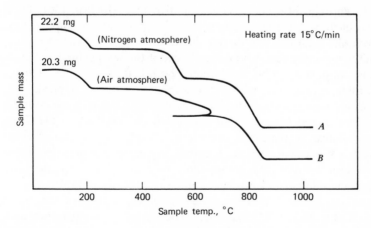

Figure II.26. Detection of temperature change of sample with a close-proximity thermocouple detector (53).

reaction is complete. The sample temperature then drops back to the temperature of the furnace, which is still slightly below 500°C.

Gayle and Egger (54) showed that the mode of heat-rise is important in determining the amount of mass that is lost after a given time, but it is unimportant in its influence on the mass-loss rate and kinetic analysis. The symmetrical temperature fluctuations do not result in a cancellation of errors when the rate behavior is an exponential rather than a linear function of temperature.

The calibration of the temperature axis in thermogravimetry is discussed in Chap. III.

4. OTHER ERRORS

A well-designed thermobalance should reduce several of the other errors to negligible values. The errors that are caused by random fluctuations of the recording mechanism, furnace induction effects, electrostatic effects, changes in thermobalance environment, and so on, can be eliminated by proper thermobalance design, construction, and location in the laboratory.

Newkirk (12) found that if the balance mechanism of the Chevenard thermobalance was not properly thermally shielded, the oil in the dash pots became warm, causing an apparent mass-gain due to the decreased buoyancy of the oil. In the latest model of this balance, the oil dash pots have been replaced by a magnetic damping device.

Condensation on the cool part of the sample-holder support rod is another source of error. The condensate may reevaporate as the temperature is increased and may again condense still lower on the support. This can lead to entirely false conclusions. Soulen and Mockrin (36) stated that this problem is intensified when a rapid inert gas flow is employed because the volatile materials are driven downward onto the support rod. The magnitude of this effect can be ascertained if the sample holder, the sample, and the support assembly are weighed both before and after each run. If they differ appreciably in mass, a correction must be applied. This, of course, gives no information about the correction during the course of the run. Soulen and Mockrin (36) eliminated this problem in the Chevenard thermobalance by placing a ceramic or nickel sleeve around the crucible support. Without the sleeve, a completely erroneous mass-loss curve was obtained for this particular compound. Of course, for compounds involving noncondensable gaseous products, this will present no problem.

Cahn and Schultz (55) have discussed the elimination of weighing errors in the Cahn Model RG balance. The effects of temperature and of a corrosive atmosphere, as well as electrical and magnetic effects, were considered.

Periodic calibration of the thermobalance will prevent errors on the mass axis of the recorder. Many investigators calibrate the instrument before each run by adding a known weight to the sample container.

D. Self-Generated Atmosphere Thermogravimetry

A "self-generated" atmosphere is one which is composed of the gaseous decomposition products of the reaction and which is in intimate contact with the sample by virtue of the type of sample holder employed. The thermogravimetry of various compounds in such an atmosphere is of importance because of the reproducibility of the composition of the atmosphere and, hence, the mass-loss curve. This technique was suggested simultaneously in 1960 by Garn and Kessler and (33) and Forkel (56) has been the subject of an extensive review by Newkirk (57). The technique is not very widely used, especially since the introduction of high-quality commercial thermobalances in which a reproducible furnace atmosphere can be maintained.

The most important aspect of the technique is in the design of the sample holders, two of which are described in Chap. III. Other sample-holder designs have been described by Newkirk (57, 58). The evolution of sample holders for use in self-generated atmospheres is illustrated in Figure II.27. A listing of the different types of compounds that have been studied by the self-generated atmosphere technique is shown in Table II.1 (57).

The primary influence of the use of the self-generated atmosphere technique is to increase the pressure of the gas evolution to 1 atm, which gives rise to

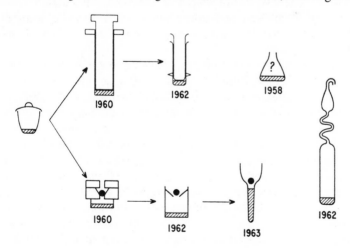

Figure II.27. Evolution of sample holders for self-generated atmospheres (57).

favorable thermodynamic, physical, and kinetic effects. The control is not exact, and some investigators consider the use of the technique to be a makeshift or last resort (57). In many instances, however, precise atmosphere control is not available, or is difficult or impossible because the reaction products are complex or unknown. The technique has a sound theoretical basis and, in such instances, would seem to be a good choice for the initial TG study of a complex solid–gas system.

The sample holder should have as small a vapor volume as possible. A large vapor volume will allow gas pressure gradients in the sample, may result in different reactions, and, it is claimed, may cause nonstoichiometric mass-losses. A large vapor volume will also make it more difficult to locate T_i and hence more difficult to compare results by different investigators. Also in this technique, the atmosphere produced by the first reaction may have a beneficial or detrimental effect on the following reaction; for example, water can accelerate the decomposition of anhydrous CaC_2O_4.

To illustrate the use of the self-generated atmosphere, the thermal decomposition of manganese(II) acetate 4-hydrate is discussed (57). A two-step decomposition sequence has been proposed, similar to the following:

$$Mn(CH_3CO_2)_2 \cdot 4H_2O(s) \rightarrow Mn(CH_3CO_2)_2(s) + 4H_2O(g)$$
$$Mn(CH_3CO_2)_2(s) \rightarrow MnO(s) + (CH_3)_2CO(g) + CO_2(g)$$

The mass-loss curves of $Mn(CH_3CO_2)_2 \cdot 4H_2O$ are illustrated in Figure II.28 (57). In curve A, the sample loses mass immediately at room temperature which is water containing some acetic acid, the latter being detected by its odor. The two major stages of mass-loss on heating correspond approximately to the loss of hydrate-bound water and the decomposition of the anhydrous salt to manganese(II) oxide. Both stages of mass-loss show curve inflections.

The effect of the self-generated atmosphere, curve B, is to increase the initial mass-loss temperature, T_i, and to decrease the reaction interval, $T_f - T_i$. The increase in T_i has the beneficial effect of eliminating the initial mass-loss at room temperature with its resulting uncertainty about the starting point of the curve. The inflection point during loss of water is located at about 135°C and a mass fraction of 0.9. The second stage of mass-loss occurs in two approximately equal parts, a very rapid and uniform initial mass-loss followed by a less rapid but still fairly uniform second loss. The nearly horizontal mass plateaus had to be corrected for a buoyancy effect which is observed for all self-generated atmosphere sample holders.

The two curves are compared in a quantitative manner in Table II.2. An advantage of a close-fitting piston crucible is that a determination can be stopped at any point, the crucible cooled and removed, and the gas

TABLE II.1
Compounds Studied by the Self-Generated-Atmosphere Technique (57)

Crucible[a] type	Compound
P	Ammonium carbonate monohydrate
MR	Anthracite
P	Brucite [$Mg(OH)_2$]
C	n-Butylammonium tetrachloroborate
C	t-Butylammonium tetrachloroborate
C	i-Butylammonium tetraphenylborate
C	s-Butylammonium tetraphenylborate
P	Cadmium carbonate
BV	Cadmium(II) sulphate (8/3) hydrate
P	Calcite ($CaCO_3$)
BV	Cerussite ($PbCO_3$)
MR	Charcoal (wood)
P	Chrysotile
MR	Coal
BV	Cobalt(II) acetate tetrahydrate
CC	Cobalt(II) oxalate dihydrate
P	Cobalt oxalate hydrate
BV	$CuSO_4 \cdot 3Cu(OH)_2 \cdot H_2O$
BV	$CuSO_4 \cdot 2Cu(OH)_2$
P, BV	Copper sulfate pentahydrate
C	Ethylammonium tetrachloroborate
BV	Gypsum ($CaSO_4 \cdot 2H_2O$)
DC	Iron(II) carbonate
P	Lead(II) carbonate
C (powder), P, CC	Lead(II) carbonate
CC, C	Lead(II) oxide
MR	Lignite
P	Magnesite ($MgCO_3$)
CC	Magnesium sulfate heptahydrate
CC, C, P	Manganese(II) acetate tetrahydrate
DC, P	Manganese(II) carbonate
C	n-Octylammonium tetrachloroborate
C	n-Propylammonium tetrachloroborate
C	n-Propylammonium tetraphenylborate
DC	Rhodochrosite ($MnCO_3$)
DC	Siderite ($FeCO_3$)
P	Silver carbonate
P	Sodium oxalate
P	Talc
P	Thorium(IV) nitrate pentahydrate
P	Thorium(IV) oxalate hexahydrate
P	Uranyl sulfate hydrate
DC, BV	Zinc sulfate heptahydrate

[a] Key to crucibles: BV, ball valve; C, covered; CC, capillary crucible; DC, deep crucible; MR, micro-retort; P, piston.

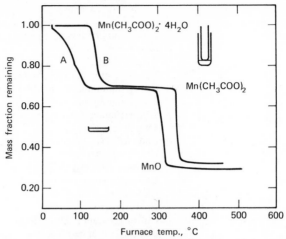

Figure II.28. TG curves of $Mn(CH_3CO_2)_2 \cdot 4H_2O$ in (A) shallow dish containing 153.8 mg in dynamic N_2 and (B) piston crucible containing 102.6 mg in air. Heating rate was 150°C/h (57).

analyzed. The disadvantage of this type of crucible is that if the piston is in contact with the molten sample it will stick, then subsequently be expelled from the cylinder.

A rather dramatic effect of the use of the self-generated atmosphere technique plus quasistatic heating-rate mode is shown in Figure II.29. Paulik and Paulik (21) found that the thermal decomposition of $Ni(NH_3)_6Cl_2$ dissociated in three separate steps, at 180, 320, and 360°C, respectively.

TABLE II.2
Comparison of Self-Generated and Dynamic N_2 Mass-Loss Curves for $Mn(CH_3CO_2)_2$-$4H_2O$ (57)

	Dynamic N_2		Self-generated atm.	
Reaction	Obs.	Calc.	Obs.	
T_i, first stage, °C	25		112	
T_f, first stage, °C	130		205	
Loss to $Mn(CH_3CO_2)_2$, %	30.6	29.40	29.0[a]	(30.3)[b]
T_i, second stage, °C	260		340	
T_f, second stage, °C	350		367	
Loss to MnO, %	71.8	71.06	70.6[a]	(68.8)[b]

[a] Corrected for buoyancy effect.
[b] Uncorrected for buoyancy effect.

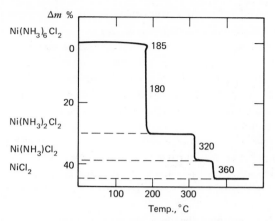

Figure II.29. TG curve of $Ni(NH_3)_6Cl_2$ in self-generated sample holder and quasistatic heating (21).

The slight overshoot of the curve during the first dissociation reaction was said to be due to an induction period caused by delayed nucleus formation. The sample holder employed is described in Chap. III.

The advantages and the limitations of the self-generated atmosphere technique have been described by Newkirk (57); they are the following.

Limitations

(*1*) Buoyancy corrections vary depending on the molecular weight of the gas filling the crucible.

(*2*) Large, heavy crucibles will cause a greater uncertainty in sample temperature.

(*3*) In dehydration of hydrates, the chances of melting and the appearance of pseudo-plateaus may be enhanced.

(*4*) Poorer resolution may result if the first reaction is delayed to a temperature at which a subsequent reaction begins.

(*5*) Secondary reactions with the evolved gas may make interpretation difficult.

Advantages

(*1*) The reaction interval will be narrower, overlapping reactions will be more clearly resolved, and intermediates more accurately identified.

(*2*) New phases will be revealed.

(*3*) Reactions will proceed, for the most part, at a fixed pressure of the gaseous products equal to atmospheric pressure. The course of reactions, except at the start, will not be affected by varying partial pressure.

(*4*) The observed initial decomposition temperature will be more closely related to an equilibrium decomposition temperature.

(*5*) Experiments can by performed on materials subject to oxidation at elevated temperatures with little interference from oxidation.

(*6*) Very fast reactions can be studied without loss of solid product.

(*7*) Better results will be obtained on materials with an appreciable vapor pressure at room temperature. The sample can be weighed more accurately and will yield a horizontal baseline on the TG curve.

(*8*) The effects of particle size difference will be reduced and the effects of crucible geometry standardized. This is particularly important with inhomogeneous materials such as rocks and minerals.

(*9*) The recrystallizations of new phases from hydrates or hydroxides will be facilitated.

(*10*) It has been claimed that irreversible decompositions will show better resolution and a smaller reaction interval in some instances, though it is not known why.

Recommended Uses

Thermogravimetry in self-generated atmospheres may be useful for studies of the following:

(*1*) Consecutive reactions, and particularly for hydroxides, hydrates, ammoniates, carbonates, acetates, oxalates, and sulfates.

(*2*) Inhomogeneous materials.

(*3*) Compounds that decrepitate or explode.

(*4*) Air sensitive materials.

(*5*) Volatile materials.

(*6*) Materials that decompose to yield several gaseous products.

(*7*) Destructive distillation.

E. Derivative Thermogravimetry (DTG)

In conventional thermogravimetry, the mass of a sample, m, is continuously recorded as a function of temperature, T, or time, t,

$$m = f(T \text{ or } t) \tag{II.4}$$

Quantitative measurements of the mass-changes are possible by determination of the distance, on the curve mass axis, between the two points of interest or between the two horizontal mass levels. In derivative thermogravimetry, the derivative of the mass-change with respect to time, dm/dt, is recorded as a function of temperature or time, or

$$\frac{dm}{dt} = f(T \text{ or } t) \qquad (\text{II.5})$$

The curve obtained is the first derivative of the mass-change curve. A series of peaks are now obtained, instead of the stepwise curve, in which the areas under the peaks are proportional to the total mass-change of the sample.

De Keyser (59, 60) first suggested this technique in 1953, followed by Erdey et al. (61) and Waters (62). Further work in this area of thermogravimetry has been by Erdey (63, 64), Paulik et al. (65), Waddams and Gray (66), Waters (67), Campbell et al. (68), and Erdey et al. (69).

A comparison between a conventional (a) and a derivative (b) mass-loss curve is given in Figure II.30. The derivative curve may be obtained either from the TG curve by manual differentiation methods or by electronic differentiation of the TG signal. Accessory equipment is available for most thermobalances so that the DTG curve can be easily recorded along with the TG curve. The DTG curve, whether derived mathematically or recorded directly, contains no more information than does an integral TG curve

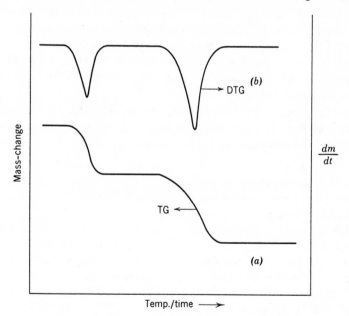

Figure II.30. Comparison between (a) integral (TG) and (b) derivative (DTG) mass-loss curves.

obtained under the same experimental conditions; it simply displays it differently (70).

The advantages of derivative thermogravimetry have been summarized by Erdey et al. (61):

(*1*) The curves may be obtained in conjunction with TG and DTA measurements.

(*2*) The curves for DTA and derivative thermogravimetry (DTG) are comparable, but the results of the former method indicate even those changes of state that are not accompanied by loss in mass. The curves by the latter method are more reproducible.

(*3*) While the curves for DTA extend over a wider temperature interval, due to subsequent warming of the material after reaction, the DTG measurements indicate exactly the temperatures of the beginning, the maximum rate, and the end of the change.

(*4*) On the TG curves, changes following each other very closely cannot be distinguished, as the corresponding stages coincide. The DTG curves of the same change indicate by sharp maxima that the thermogravimetric stages can be divided into two parts.

(*5*) The DTG curves are exactly proportional to the derivatives of the TG curves; therefore, the area under the curves gives the change in mass precisely. Accordingly, DTG can give exact quantitative analyses.

(*6*) The DTG method can be used for the investigation of materials which for some reason or another cannot be analyzed by DTA. For example, some organic compounds melt during heating, but even so the DTG method yields fairly good results.

The DTG technique has been criticized by Newkirk (58, 70), who claimed that the technique has not been subjected to extensive critical analysis. Some claims and comments of DTG compared to TG are given below (58):

Claim	*Comment*
(*1*) DTG is more precise in showing onset of reaction accompanied by a small loss in mass.	(*1*) Not unless a more sensitive balance is used.
(*2*) DTG shows overlapping changes better.	(*2*) This seems to be a real advantage to most people.
(*3*) DTG permits reaction temperatures to be defined more exactly.	(*3*) Only the temperature of the maximum rate.
(*4*) Comparison of DTG and DTA enables distinction between DTA peaks due to mass-loss, and those due to other thermal changes.	(*4*) No advantage over TG.

The practice of designating the peak temperature of a DTG curve as the "decomposition temperature" should not be used. The peak temperature represents the temperature at which the rate of mass-change is at a maximum, and it clearly is not the temperature at which the sample begins to lose mass or T_i (70).

F. Reaction Kinetics

1. NONISOTHERMAL METHODS

Dynamic thermogravimetry has been widely used during the past ten or so years to study the kinetics of thermal decomposition reactions. As pointed out by Doyle (71), one mass-loss curve is equivalent to a large number of isothermal mass-loss curves; also, this large amount of information is gained without sample-to-sample error, since the same sample is used throughout the determination. It should be pointed out, however, that thermogravimetric data are only narrowly definitive, and consequently, merely going through the motions of kinetic analysis can lead only to trivial kinetic parameters for mass-loss under a particular set of experimental conditions. Empirical data treatments, no matter how augustly clothed in the trappings of kinetics, are still empirical, although admittedly on a higher level of sophistication (71). As a general rule, definitive kinetic parameters can be derived from mass-loss data only in the light of a large amount of additional evidence. In the case of polymers, there is usually not enough pertinent information available to warrant undertaking meaningful kinetic analysis. Thus, mass-loss data for these materials are commonly subjected to empirical kinetic treatments.

The advantages of determining kinetic parameters by nonisothermal methods rather than by conventional isothermal studies are (8) (*a*) that considerably fewer data are required; (*b*) that the kinetics can be calculated over an entire temperature range in a continuous manner; (*c*) that when a sample undergoes considerable reaction in being raised to the required temperature, the results obtained by an isothermal method are often questionable; and (*d*) that only a single sample is required. A decided disadvantage of nonisothermal compared with isothermal methods is that the reaction mechanism cannot *usually* be determined, and hence, the meanings of the activation energy, order of reaction, and frequency factor are uncertain. The use of nonisothermal kinetic methods has been seriously questioned (72) and much criticized (43, 73–79). As a result of this criticism, and also in order to keep within the scope of this book, the discussion of kinetic methods given here will only be superficial in nature. For more comprehensive treatments, the reader should consult other sources (71, 80–82).

The foundation for the calculations of kinetic data from a TG curve is based on the formal kinetic equation (43)

$$-\frac{dX}{dt} = kX^n \tag{II.6}$$

where X is the amount of sample undergoing reaction, n is the order of reaction, and k is the specific rate constant. This equation describes very well the kinetics of the thermal decomposition of solids, such as the endothermic reactions of metal oxalates, permanganates, perchlorates, and azides. The temperature dependence of the specific rate constant, k, is expressed by the Arrhenius equation,

$$k = Ae^{-E/RT} \tag{II.7}$$

where A is the pre-exponential factor, E is the activation energy, and R is the gas-law constant, which generally applies to only a narrow temperature range.

The mathematical treatment of kinetic equations makes use of one of the following three methods of evaluation: (a) differential, (b) integral, or (c) approximate.

As discussed by Sestak (43), the relationship of X to mass-loss, w, is given by the equation

$$-dX = \frac{m_0}{w_\infty} dw \tag{II.8}$$

where m_0 is the initial mass of the sample and w_∞ is the maximum mass-loss. By integration of the left-hand side of equation (II.8) from m_0 to X and by integration of the right-hand side from zero to w, the following is obtained:

$$X = \frac{m_0}{w_\infty}(w_\infty - w) \tag{II.9}$$

By substitution of equations (II.9) and (II.7) into equation (II.6) and by differentiating the logarithmic form, an expression is obtained which is one of the differential methods: the Freeman and Carroll (83) method.

Integral methods use the integrated form of equation (II.6) after the transposition of the mass-loss, w, in equations (II.8) and (II.9),

$$\left(\frac{m_0}{w_\infty}\right)^{1-n} \int_0^w (w_\infty - w)^{-n} \, dw = \frac{A}{\phi} \int_{T_1}^{T_2} e^{-E/Ru} \, dT \tag{II.10}$$

The right-hand side of this equation can be solved by various methods, and the final solution to the equation is an infinite series of which the first two terms are of interest generally. These methods are used by Doyle (84) and Coats and Redfern (85) as well as others (86, 87).

In the approximation methods, the right-hand side of equation (II.10) is solved by an approximation using the temperature, T_i, corresponding to the maximum rate of decomposition. This method was used by Horowitz and Metzger (88) and others (87, 89–91).

a. Newkirk Method

From a single TG curve, Newkirk (12) obtained rate constants for the decomposition reaction, as illustrated in Figure II.31. For a series of

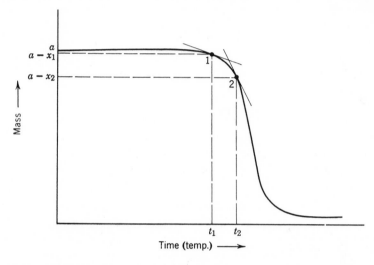

Figure II.31. Determination of reaction rate and extent of reaction from a TG curve (12).

temperatures, T_1 and T_2, the sample mass remaining, $a - x_1$ and $a - x_2$, and the reaction rates, $(dx/dt)_1$ and $(dx/dt)_2$, were obtained by tangents to the curve at points 1 and 2. If the reaction is of first order, then the logarithm of the reaction rate constant k, $dx/dt = k(a - x)$, when plotted against $1/T$, should yield a straight line. The results of measurements under isothermal and nonisothermal conditions yielded straight lines that were approximately parallel.

b. Freeman and Carroll Method

Perhaps the most widely used kinetics method is that developed by Freeman and Carroll (83) in 1958. The advantage of this method is that considerably fewer experimental data are required than in the isothermal method, and that the kinetics can be obtained over an entire temperature range in a continuous manner without any missing regions. In addition, where a sample undergoes considerable reaction in being raised to the temperature of interest, the results obtained by an isothermal method are often questionable. The order of reaction, n, and the activation energy of the reaction are calculated from the equation

$$\frac{-(E/2.3R)\Delta(1/T)}{\Delta \log w_r} = -n + \frac{\Delta \log (dw/dt)}{\Delta \log w_r} \qquad (II.11)$$

where $w_r = w_c - w$, in which w_c is the maximum mass-loss, and w is the total loss in mass up to time t.

c. Horowitz and Metzger Method

For the example in which the rate constant involves concentrations expressible as mole fractions and the total number of moles is constant, the expression (88)

$$\ln \ln \left(\frac{W_0 - W_t}{W - W_t}\right) = \frac{E\theta}{RT_s^2} \qquad (II.12)$$

may be used, in which W is the mass remaining at a given temperature, W_0 and W_t are the initial and final masses, respectively, and θ is a reference temperature.

d. Coats and Redfern Method

For a reaction in which the order is unknown, Coats and Redfern (85) derived the following expression:

$$\log \left\{\frac{1 - (1 - \alpha)^{1-n}}{T^2(1 - n)}\right\} = \log \frac{AR}{aE}\left[1 - \frac{2RT}{E}\right] - \frac{E}{2.3RT} \qquad (II.13)$$

where α is the fraction of the sample decomposed at time t, and a is the heating rate. A plot of either $\log [1 - (1 - \alpha)]^{1-n}/T^2(1 - n)$ against $1/T$, or, where $n = 1$, $\log [-\ln (1 - \alpha)]/T^2$ against $1/T$, should result in a straight line of slope $-E/2.3R$ for the correct value of n. The quantity $\log (AR/aE)1 - (2RT/E)$ appears to be reasonably constant for most values of E and in the temperature range over which most reactions occur.

e. Doyle Method

The kinetics of volatilization of polymers have been discussed in detail by Doyle (71, 84). If w is the apparent residual mass fraction calculated on the initial mass, the apparent mass fraction volatilized, v, is

$$v = (1 - w) \qquad \text{(II.14)}$$

The apparent volatilization rate is found by multiplying the TG curve slope, $-dw/dT$, by the constant heating rate, B:

$$\frac{dv}{dt} = -B\frac{dw}{dT} \qquad \text{(II.15)}$$

For a particular volatilization step, however, the appropriate residual mass fraction is the true one, h, calculated on the total fraction volatilized during the step, rather than on the total initial mass

$$h = \frac{w - G}{H} \qquad \text{(II.16)}$$

where H is the total apparent mass fraction volatilized during the step and G is the apparent weight fraction remaining after the step has been completed. From equations (II.14) and (II.15),

$$\frac{dv}{dt} = -H\frac{dh}{dt} \qquad \text{(II.17)}$$

It should be noted that H and G are seldom clearly defined; their estimation, in many cases, constitutes one of the major difficulties of kinetics calculations.

Another difficulty arises from the need to take into account the nature of the kinetic process. In general,

$$-\frac{dh}{dt} = rf(h) \qquad \text{(II.18)}$$

where r is the empirical rate constant for volatilization and where the specific form of $f(h)$ depends on the reaction order. The constant r must be treated as empirical because its value for a particular substance is not always uniquely determined by temperature, but may depend on the nature and geometry of the sample holder, the nature of the environmental atmosphere, and other factors. The potential triviality of r was constantly emphasized by Doyle (84) by use of the symbol r instead of the specific rate constant, k. Using this terminology, the Arrhenius equation for the volatilization process is

$$r = ae^{-b/RT} \qquad \text{(II.19)}$$

The constant b, at least over part of the experimental temperature range, had the same value as the activation energy.

When evaluating the constants in equation (II.15), usually only a small portion of the TG curve is used, the region where the slope is neither too shallow nor too steep to be measured with sufficient precision.   In fact, in the range of volatilization rates that are small compared to the heating rate, the slopes found from the TG curve are not only imprecise but also inherently inaccurate, being consistently greater than those found by an isothermal method.   This effect is due to the fact that, in thermogravimetry, the dwell time at each temperature is so brief that no evidence of volatilization can accumulate in the range of small volatilization rates.

An approximate integral method, similar to that proposed by Doyle (84), was derived by Ozawa (92, 93) and was said to have wider application than the former.

Zsako (94) has also attempted to simplify Doyle's trial-and-error method and to find new applications to TG curves.   Starting with the basic equation previously described by Doyle (84),

$$g(\alpha) = \frac{AE}{Rq} p(x) \qquad (\text{II.20})$$

and taking the logarithm

$$\log \frac{AE}{Rq} = \log g(\alpha) - \log p(x) = B \qquad (\text{II.21})$$

where B depends only upon the nature of the compound studied and upon the heating rate, but not the temperature.   The value of $g(\alpha)$ for a given temperature can be calculated from the TG curve if $f(\alpha)$ is known.

The method consists of the following: Presuming the validity of the function $f(\alpha)$ and using TG data, $g(\alpha)$ values are calculated for different temperatures.   By means of a trial-and-error method, the apparent activation energy can be estimated by finding the E value which ensures the maximum constancy of

$$B = \log g(\alpha) - \log p(X) \qquad (\text{II.22})$$

f. Ingraham and Marier Method

For a reaction such as the thermal decomposition of calcium carbonate, which obeys linear kinetics, the rate constant may be expressed as

$$k = \frac{dw}{dt} \qquad (\text{II.23})$$

in which dw represents the loss in mass from unit area in period of time, dt. When the temperature of the sample is increased at a linear rate, the temperature at any time, t, is

$$T = b + at \qquad (\text{II.24})$$

where a is the heating rate and b the initial temperature. From this relationship, Ingraham and Marier (95) showed that if dt is replaced by dT/a, the following equation can be written:

$$\log \frac{dw}{dT} = \log T - \log a + \log C - \frac{E}{2.303R} \qquad (\text{II}.25)$$

The activation energy is calculable from the slope of a plot of $[\log (dw/dT) - \log T + \log a]$ versus $1/T$. The logarithm of the heating rate, although constant for any particular experiment, was retained in the equation to permit correlation of TG curves carried out at different heating rates.

g. Vachuska and Voboril Method

In this method (96), the kinetic constants are calculated from the TG curve by a differential method. It takes into account also the thermal effects of reactions which result in a deviation of the sample temperature from the programmed values of the linear heating. Starting with the differential equation for the thermal decomposition of a solid,

$$\frac{d\alpha}{dt} = k(1 - \alpha)^n \exp\left(-\frac{E}{RT}\right) \qquad (\text{II}.26)$$

where α is the degree of transformation, the logarithm of this equation is

$$\ln \left(\frac{d}{dt}\right) = \ln k + n \ln (1 - \alpha) - \frac{E}{RT} \qquad (\text{II}.27)$$

Considering that α and T are functions of time and differentiating with respect to time,

$$\frac{d^2\alpha/dt^2}{d\alpha/dt} = -\frac{n}{1 - \alpha}\left(\frac{d\alpha}{dt}\right) + \frac{E}{RT^2}\left(\frac{dT}{dt}\right) \qquad (\text{II}.28)$$

By rearrangement, a linear equation is obtained,

$$\frac{(d^2\alpha/dt^2)T^2}{(d\alpha/dt)(dT/dt)} = -n\frac{d\alpha/dt}{1 - \alpha}\frac{T^2}{dT/dt} + \frac{E}{R} \qquad (\text{II}.29)$$

by means of which the reaction order, n, can be calculated from the slope and the intercept gives the activation energy.

h. Miscellaneous Methods

Simplified methods for the calculation of specific rate constants from TG curves have been described by various investigators. Dave and Chopra (97) calculated the specific rate constant by use of the equation

$$k = \frac{(A/m_0)^{n-1}(dx/dt)}{(A - a)^n} \tag{II.30}$$

where A is the total area under the derivative TG curve (DTG); a is the area for the reaction up to time t; dx/dt is the height of the curve at time t; m_0 is the initial mole fraction of the reactant; and n is the order of reaction.

A similar approach was used by Papazian et al. (98) and Adoniji (100) in which the specific rate constant was calculated from a DTG curve by use of the expression

$$k = \frac{dx/dt}{(a - x)^n} \tag{II.31}$$

where $(a - x)$ is the mass of reactant not decomposed. This simple method was said to yield kinetic results which were similar to those obtained from isothermal measurements.

Farmer (99) used the expression

$$\log(-\dot{w}w^{-n}) = \frac{-E}{R \ln 10}\left(\frac{1}{T}\right) + \log A \tag{II.32}$$

where w is the reactant mass fraction, calculating E and A from the slope and intercept of a parametric plot of $\log(-\dot{w}w^{-n})$ versus $1/T$.

Magnuson (101) discussed in detail the individual steps needed to calculate the kinetics of the thermal decomposition of a high-molecular-weight dimethylsiloxane polymer. Use was made of the simple first-order rate equation

$$-\frac{dW}{dt} = k_p W \tag{II.33}$$

where $-dW/dt$ is the rate at which sample mass is decreasing, k_p is the procedural rate constant, and W is the sample mass.

The method derived by Achar et al. (102) is a differential one and applies to all reaction mechanisms, provided that the correct mechanism is known (76). It is based on the use of the expression

$$\ln\left(\frac{1}{f(\alpha)} \cdot \frac{d\alpha}{dT}\right) = \ln\frac{A}{B} - \frac{E}{RT} \tag{II.34}$$

where α is the fraction of sample reacted in time t, and B is the heating rate. When the left-hand side of the equation is plotted against $1/T$, a straight line is obtained from which E and A can be determined. The form of $f(\alpha)$ depends on the nature of the reaction; for example, for the parabolic law, $\alpha^2 = kt$, which applies to many diffusion controlled reactions, $f(\alpha) = \frac{1}{2}\alpha$.

The method developed by Broido (103), which was applied to the pyrolysis of cellulose, is based on the equation

$$\ln\ln\frac{1}{y} = -\left(\frac{E}{R}\right)\left(\frac{1}{T}\right) + \text{constant} \tag{II.35}$$

where y is the number of initial molecules not yet decomposed. Thus, a plot of $\ln\ln(1/y)$ versus $(1/T)$ should yield a straight line from which E and Z can be calculated from the slope.

2. COMPARISON OF DIFFERENT METHODS

Sestak (43) compared the kinetic results calculated by five different methods for a system corresponding to the dehydration of α-$CaSO_4 \cdot 0.5H_2O$. The five methods evaluated mathematically were (a) Freeman and Carroll (83); (b) Doyle (84); (c) Coats and Redfern (85); (d) Horowitz and Metzger (88); and (e) Van Krevelen et al. (87). From these calculations it was found that the deviations of computed values of E did not differ by more than 10%. Thus, all of the methods appear to be satisfactory for the calculation of E within the limits of accuracy required. The errors of each method due to the inaccuracy of visual deduction of values from the TG curves were calculated also. These errors, ϵ_E and ϵ_n (errors in calculation of E or n, respectively), were as follows: (a) Freeman and Carroll method, $\epsilon_E = 4\%$ and $\epsilon_n = 12\%$; (b) Horowitz and Metzger method, $\epsilon_E = 2\%$ (when the correct value of n is assumed); (c) Doyle method, $\epsilon_E = 4\%$. However, the magnitude of this error depends primarily on the position of the point on the TG curve on which the calculations are being performed. In the case of differential methods, the most accurate data are calculated from the medium-steep parts of the curve. For the approximation method, the accuracy depends on the determination of the curve inflection point temperature.

From the viewpoint of ease of computation, Doyle's method seems to be very simple because the kinetic data are obtained from a single point on the curve. The necessity of knowing the reaction order ahead of time appears to be a disadvantage which finds a partial remedy in the Coats and Redfern method.

Three kinetic methods were evaluated by Sharp and Wentworth (76) using the thermal decomposition of calcium carbonate under various conditions. The physical state of the sample was as a pellet, a powder, or as 1:1 molar ratios with α-aluminum oxide or α-iron(III) oxide. The three

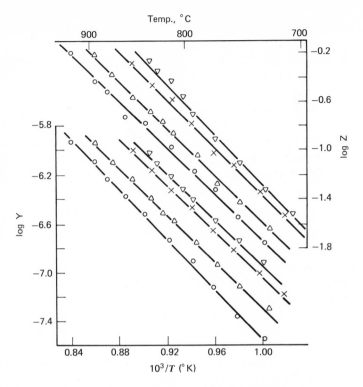

Figure II.32. Data calculated from Methods II and III (76). ○ $CaCO_3$ (pellet); △ $CaCO_3$ (powder); × $CaCO_3$: Al_2O_3; ▽ $CaCO_3$: Fe_2O_3.

methods used were: Method I, Freeman and Carroll; Method II, Coats and Redfern; and Method III, Achar et al. (102). The kinetic data calculated by Methods II and III are presented graphically in Figure II.32 and in Table II.4. In every case a linear plot was obtained over a wide range of α with $n = \frac{1}{2}$. When these methods were applied with $n = \frac{2}{3}$, the range of α was less, especially in the case of Method III, which led to noticeable curvature at high values of α. With $n = 1$, this curvature became more marked for both methods. Activation energies with $n = \frac{1}{2}$ averaged about 44 kcal/mole. The addition of iron or aluminum oxides did not affect the activation energy.

The results obtained by means of Method I, from a combination of data for the pellet and powder, are given in Figure II.33 and Table II.5.

The points are more scattered than those given in Figure II.32, leading to much uncertainty in the value of E. The line shown is a theoretical line drawn to pass through the ordinate axis at $+0.5$ with a slope which leads

TABLE II.4
Kinetic Parameters Using Methods II and III (76)

Material	Method II $n = \frac{1}{2}$ Range	E kcal/mole	Method III $n = \frac{1}{2}$ Range	E kcal/mole	Method II $n = \frac{2}{3}$ Range	E kcal/mole	Method III $n = \frac{2}{3}$ Range	E kcal/mole
$CaCO_3$ pellet	0.03–0.98	46	0.01–0.98	43	0.03–0.81	49	0.01–0.81	45
$CaCO_3$ powder	0.05–0.96	43	0.01–0.96	44	0.05–0.79	44	0.01–0.79	47
$CaCO_3:Al_2O_3$	0.06–0.87	43	0.03–0.87	44	0.06–0.87	44	0.03–0.67	46
$CaCO_3:Fe_2O_3$	0.11–0.83	42	0.07–0.83	48	0.11–0.83	45	0.07–0.72	48

to the value of 44 kcal/mole for the activation energy. Although this line seems satisfactory, other lines leading to different values for n and E could be easily drawn.

Methods II and III lead to substantially the same results; the latter has the advantages of leading directly to values for A from the intercept and being equally applicable to both diffusion controlled and order-of-reaction kinetics. The disadvantage of this method is the necessity of determining $d\alpha/dT$, the tangents to the TG curve, at a series of values for T. Method II is slightly less tedious to calculate than Method III and avoids the determination of tangents.

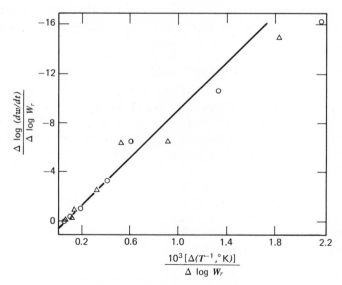

Figure II.33. Data calculated from Method I (93). ○ $CaCO_3$ (pellet); △ $CaCO_3$ (powder).

TABLE II.5
Kinetic Parameters from Method I (76)

Number of points used	Range	n	E kcal/mole
8	0.25–0.84	+0.55	43
14	0.05–0.84	−0.07	36

The greater scatter in the data obtained by Method I is due largely to the determination of three difference functions over short temperature intervals. One of these involves the tangents dw/dt; therefore, any inaccuracy in determining a single value affects the position of two points in the plot used to determine the activation energy. As a result of these analyses, Method I was considered less satisfactory than the other two methods and was not recommended for use.

Carroll and Manche (104) criticized the use of trial-and-error methods of calculations, such as those previously described in Methods II and III. The question arises whether the criterion of constancy of kinetic parameters is appropriate even for the case of a simple chemical reaction, since a solid-state reaction is a very complex process. They argued that in our present state of knowledge, constancy of kinetic parameters in general, for reactions in the solid state, is an unwarranted assumption. The Freeman and Carroll method can be used if the random errors in the slope measurements are smoothed out, particularly for the initial and final stages of the process. Precision of the method can be enhanced considerably by this simple procedure.

3. MECHANISM OF REACTION FROM NONISOTHERMAL KINETICS

Deduction of the mechanism of the reaction by use of nonisothermal kinetic methods has been discussed by Sestak and Berggren (73) and Satava (105). The procedure used by Satava is based on the assumption that the nonisothermal reaction proceeds in an infinitesimal time interval isothermally, where the rate may be expressed as

$$\frac{d\alpha}{dt} = Ae^{-E/RT}f(\alpha) \tag{II.36}$$

where α is the fraction decomposed in time t, and, $f(\alpha)$ depends on the mechanism of the process. With a constant temperature increase, $dT/dt = q$, integration of equation (II.36) leads to

$$\int_0^\alpha \frac{d\alpha}{f(\alpha)} = g(\alpha) = \frac{E}{Rq} p(x) \tag{II.37}$$

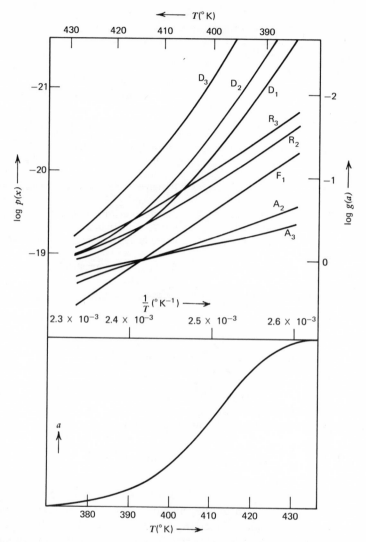

Figure II.34. Procedure for obtaining the mechanism from a TG curve (105). Lower curve complies with kinetic equation F_1 (see Table II.6); $E = 30$ kcal/mole; $Z = 10^{13/S^{-1}}$. mole^{-1}; and $q = 1°C \cdot$ min^{-1}. Upper curves are plots of $\log g(\alpha)$ versus $1/T_\alpha$) calculated from TG curve for various kinetic equations. The straight line for kinetic equation F_1 coincides with the plot of $-\log p(x)$ versus $1/Tf$ or $E = 30$ kcal/mole.

TABLE II.6
Commonly Used Reaction Mechanisms (105)

Function	Equation	Rate-controlling process
D_1	$x^2 = kt$	One-dimensional diffusion
D_2	$(1 - \alpha) \ln (1 - \alpha) + \alpha = kt$	Two-dimensional diffusion, cylindrical symmetry
D_3	$[1 - (1 - \alpha)^{1/3}]^2 = kt$	Three-dimensional diffusion, spherical symmetry; Jander equation
D_4	$(1 - \tfrac{2}{3}\alpha) - (1 - \alpha)^{2/3} = kt$	Three-dimensional diffusion, spherical symmetry; Ginstling-Brounshtein equation
F_1	$-\ln (1 - \alpha) = kt$	Random nucleation, one nucleus on each particle
A_2	$[-\ln (1 - \alpha)]^{1/2} = kt$	Random nucleation; Avrami equation I
A_3	$[-\ln (1 - \alpha)]^{1/3} = kt$	Random nucleation; Avrami equation II
R_2	$1 - (1 - \alpha)^{1/2} = kt$	Phase boundary reaction, cylindrical symmetry
R_3	$1 - (1 - \alpha)^{1/3} = kt$	Phase boundary reaction, spherical symmetry

where $p(x)$ is defined as

$$p(x) = \frac{e^{-x}}{x} - \int_{+x}^{\infty} \frac{e^{-u}}{u}\, du = \frac{e^{-x}}{x} + Ei(-x) \qquad (II.38)$$

where $u = E/RT$, $x = E/RT$, and T is the temperature at which the fraction α of the sample has reacted.

From the logarithmic form of equation (II.36),

$$\log g(\alpha) - \log p(x) = \log \frac{AE}{Rq} \qquad (II.39)$$

it can be seen that the right-hand side is independent of temperature, while the left-hand side is temperature-dependent. To a first approximation, the function, $\log p(x)$ is a linear function of $1/T_\alpha$ if x is sufficiently large, and thus $\log g(x)$ must also be a linear function of $1/T_\alpha$. For the correct mechanism, $\log p(x)$ versus $1/T_\alpha$ should be a straight line. For other incorrect mechanisms, this will not be true. The sensitivity of this procedure for mechanism determination is not high, but still it yields useful information.

The types of mechanisms most frequently encountered are shown in Table II.6, while the procedure used to evaluate the reaction mechanism from a TG curve is given in Figure II.34. As can be seen, only the curve corresponding to the F_1 mechanism gives a straight-line curve.

References

1. Simons, E. L., and A. E. Newkirk, *Talanta*, **11**, 549 (1964).
2. Newkirk, A. E., and E. L. Simons, private communication.
3. Duval, C., *Inorganic Thermogravimetric Analysis*, 2nd ed., Elsevier, Amsterdam, 1963.
4. Duval, C., *Inorganic Thermogravimetric Analysis*, 1st ed., Elsevier, Amsterdam, 1953.
5. Duval, C., *Anal. Chem.*, **23**, 1271 (1951).
6. Lukaszewski, G. M., and J. P. Redfern, *Lab. Pract.*, **10**, 469 (1961).
7. Jacque, L., G. Guiochon, and P. Gendrel, *Bull Soc. Chim. France*, **1961**, 1061.
8. Coats, A. W., and J. P. Redfern, *Analyst*, **88**, 906 (1963).
9. Honda, K., *Sci. Rept. Tohoku Univ.*, **4**, 97 (1915).
10. Guichard, M., *Bull. Soc. Chim. France*, **2**, 539 (1935).
11. Duval, C., *Anal. Chim. Acta*, **31**, 301 (1964).
12. Newkirk, A. E., *Anal. Chem.*, **32**, 1558 (1960).
13. Simmons, E. L., and W. W. Wendlandt, *Thermochim. Acta*, **2**, 465 (1971).
14. DeVries, K. J., and P. J. Gellings, *J. Inorg. Nucl. Chem.*, **31**, 1307 (1969).
15. Herbell, T. P., *Thermochim. Act*, **4**, 295 (1972).
16. Perkin-Elmer Model TGS-1 Thermobalance Brochure, Perkin-Elmer Co., Norwalk, Conn.
17. Fruchart, R., and A. Michel, *Compt. Rend.*, **246**, 1222 (1958).
18. Demassieux, N., and C. Malard, *Compt. Rend.*, **245**, 1514 (1957).
19. Rynasiewicz, J., and J. F. Flagg, *Anal. Chem.*, **26**, 1506 (1954).
20. DeClerq, M., and C. Duval, *Anal. Chim. Acta*, **5**, 282 (1951).
21. Paulik, F., and J. Paulik, *Thermochim. Acta*, **4**, 189 (1972).
22. Mettler Thermal Techniques Series, *Tech. Bull.* **T-106**.
23. Paulik, J., F. Paulik and L. Erdey, *Anal. Chim. Acta*, **34**, 419 (1966).
24. Feldman, R. F., and V. S. Ramachandran, *Thermochim. Acta*, **2**, 393 (1971).
25. Wendlandt, W. W., and E. L. Simmons, *Thermochim. Acta*, **3**, 171 (1972).
26. Guenot, J., and J. M. Manoli, *Bull. Soc. Chim. France*, **1969**, 2663.
27. Nicholson, G. C., *J. Inorg. Nucl. Chem.*, **29**, 1599 (1967).
28. Brown, H. A., E. C. Penski, and J. J. Callahan, *Thermochim. Acta*, **3**, 271 (1971).
29. Wiedemann, H. G., Achema Congress paper, Frankfurt, June 26, 1964.
30. Friedman, H. L., *Anal. Chem.*, **37**, 768 (1965).
31. Saito, H., *Sci. Rept. Tohoku Univ.*, **16**, 1 (1927).
32. Richer, A., and P. Vallet, *Bull. Soc. Chim. France*, **1953**, 148.
33. Garn, P. D., and J. E. Kessler, *Anal. Chem.*, **32**, 1563, 1900 (1960).
34. Garn, P. D., *Anal. Chem.*, **33**, 1247 (1961).
35. Rabatin, J. G., and C. S. Card, *Anal. Chem.*, **31**, 1689 (1959).
36. Soulen, J. R., and I. Mockrin, *Anal. Chem.*, **33**, 1909 (1961).
37. Lukaszewski, G. M., *Nature*, **194**, 959 (1962).
38. Newkirk, A. E., and D. W. McKee, *J. Catalysis*, **11**, 370 (1968).
39. Macklen, E. D., *J. Inorg. Nucl. Chem.*, **29**, 1229 (1967).
40. Steger, H. F., *J. Inorg. Nucl. Chem.* **34**, 175 (1972).
41. Newkirk, A. E. *Thermochim. Acta*, **2**, 1 (1971).
42. Paulik, F., J. Paulik and L. Erdey, *Talanta*, **13**, 1405 (1966).
43. Sestak, J., *Talanta*, **13**, 567 (1966).
44. Wiedemann, H. G., in *Thermal Analysis*, R. F. Schwenker and P. D. Garn, eds., Academic, New York, 1969, p. 229.
45. Newkirk, A. E., and I. Aliferis, *Anal. Chem.*, **30**, 982 (1958).

46. Ramakrishna Udupa, M., and G. Aravamudan, *Current Sci.*, **39**, 206 (1970).
47. Cahn, L., and N. C. Peterson, *Anal. Chem.*, **39**, 403 (1967).
48. Cahn, L., and H. Schultz, *Anal. Chem.* **35**, 1729 (1963).
49. Martinez, E., *Am. Minerologist*, **46**, 901 (1961).
50. Guiochon, G., *Anal. Chem.*, **33**, 1124 (1961).
51. Simons, E. L., A. E. Newkirk, and I. Aliferis, *Anal. Chem.*, **29**, 48 (1957).
52. Mielenz, R. C., N. C. Schieltz and M. E. King, *Clays and Clay Minerals*, NAS-NRC, Washington Pub. 327, 1954, pp. 289–296.
53. Sarasohn, I. M., and R. W. Tabeling, Pittsburgh Conf. on Analytical and Applied Spectroscopy, Pittsburgh, Pa., March 5, 1964.
54. Gayle, J. B., and C. T. Egger, *Anal. Chem.*, **44**, 421 (1972).
55. Cahn, L., and H. R. Schultz, in *Vacuum Microbalance Techniques*, K. H. Behrndt, ed., Plenum, New York, 1963, Vol. 3, p. 29.
56. Forkel, W., *Naturwissenschaften*, **47**, 10 (1960).
57. Newkirk, A. E., *Thermochim. Acta*, **2**, 1 (1971).
58. Newkirk, A. E., in *Proceedings of the First Toronto Symposium on Thermal Analysis*, H. G. McAdie, ed., Chemical Institute of Canada, Toronto, 1965, p. 29.
59. De Keyser, W. L., *Nature*, **172**, 364 (1953).
60. De Keyser, W. L., *Bull. Soc. France Ceram.*, **20**, 1 (1953).
61. Erdey, L., F. Paulik, and J. Paulik, *Nature*, **174**, 885 (1954).
62. Waters, P. L., *Nature*, **178**, 324 (1956).
63. Erdey, L., *Periodica Polytech.*, **1**, 35, 91 (1957).
64. Erdey, L., *Chem. Zvesti*, **12**, 352 (1958).
65. Paulik, F., J. Paulik, and L. Erdey, *Z. Anal. Chem.*, **160**, 241, 321 (1958).
66. Waddams, J. A., and P. S. Gray, *Nature*, **183**, 1729 (1958).
67. Waters, P. L., *J. Sci. Instr.*, **35**, 41 (1958).
68. Campbell, C., S. Gordon, and C. L. Smith, *Anal. Chem.*, **31**, 1188 (1959).
69. Erdey, L., B. Liptay, G. Svehla, and F. Paulik, *Talanta*, **9**, 489 (1962).
70. Newkirk, A. E., and E. L. Simons, *Talanta*, **13**, 1401 (1966).
71. Doyle, C. D., in *Techniques and Methods of 'Polymer Evaluation'*, P. E. Slade and L. T. Jenkins, eds., Marcel-Dekker, New York, 1966, Chap. 4.
72. Clarke, T. A., E. L. Evans, K. G. Robins, and J. M. Thomas, *Chem. Comm.*, **1969**, 266.
73. Sestak, J., and G. Berggren, *Thermochim. Acta*, **3**, 1 (1971).
74. Hill, R. A. W., *Nature*, **227**, 703 (1970).
75. Simmons, E. L., and W. W. Wendlandt, *Thermochim Acta*, **3**, 498 (1972).
76. Sharp, J. H., and S. A. Wentworth, *Anal. Chem.*, **41**, 2060 (1969).
77. Sestak, J., *Silikaty*, **11**, 153 (1967).
78. MacCallum, J. R., and J. Tanner, *European Polymer J.*, **6**, 1033 (1970).
79. Draper, A. L., and L. Sveum, *Thermochim. Acta*, **1**, 345 (1970).
80. Flynn, J. H., and L. A. Wall, *J. Res. Natl. Bur. Stand.*, **70A**, 487 (1966).
81. Carroll, B., and E. P. Manche, *Thermochim. Acta*, **3**, 449 (1972).
82. Reich, L., and S. Stivala, *Elements of Polymer Degradation*, McGraw-Hill, New York, 1971.
83. Freeman, E. S., and B. Carroll, *J. Phys. Chem.*, **62**, 394 (1958).
84. Doyle, C. D., *J. Appl. Polymer Sci.*, **5**, 285 (1961).
85. Coats, A. W., and J. P. Redfern, *Nature*, **201**, 68 (1964).
86. Turner, R. C., J. Hofman, and D. Chen, *Can. J. Chem.*, **41**, 243 (1963).
87. van Krevelen, W., C. van Heerden, and F. Hutjens, *Fuel*, **30**, 253 (1951).

88. Horowitz, H. H., and G. Metzger, *Anal. Chem.*, **35**, 1464 (1963).
89. Fuoss, R. M., I. O. Sayler, and H. S. Wilson, *J. Polymer Sci.*, **2**, 3147 (1964).
90. Berlin, A., and R. T. Robinson, *Anal. Chim. Acta*, **27**, 50 (1962).
91. Reich, L., *J. Polymer Sci.*, **B2**, 621 (1964).
92. Ozawa, T., *Bull. Chem. Soc. Japan*, **38**, 1881 (1965).
93. Ozawa, T., *J. Thermal Anal.*, **2**, 301 (1970).
94. Zsako, J., *J. Phys. Chem.*, **72**, 2406 (1968).
95. Ingraham, T. R., and P. Marier, *Can. J. Chem. Eng.*, **1964**, 161.
96. Vachuska, J., and M. Voboril, *Thermochim Acta*, **2**, 379 (1971).
97. Dave, N. G., and S. K. Chopra, *Z. Physik. Chem.*, **48**, 257 (1966).
98. Papazian, H. A., P. J. Pizzolato, and R. R. Orrell, *Thermochim. Acta*, **4**, 97 (1972).
99. Farmer, R. W., Air Force Materials Lab. Tech. Report AFML-TR-65-246, 1967, Part II.
100. Adonyi, Z., *Periodica Polytech.*, **11**, 325 (1967).
101. Magnuson, J. A., *Anal. Chem.*, **36**, 1807 (1964).
102. Achar, B. N. N., G. W. Brindley and J. H. Sharp, *Proc. Int. Clay Conf., Jerusalem*, **1**, 67 (1966).
103. Broido, A., *J. Polymer Sci.*, Part A-2, **7**, 1761 (1969).
104. Carroll, B., and E. P. Manche, *Anal. Chem.*, **42**, 1296 (1970).
105. Satava, V., *Thermochim. Acta*, **2**, 423 (1971).

CHAPTER III

THERMOBALANCES AND ACCESSORY EQUIPMENT

A. Introduction

The thermobalance is an instrument that permits the continuous weighing of a sample as a function of temperature. The sample may be heated or cooled at some selected rate or it may be isothermally maintained at a fixed temperature. Perhaps the most common mode of operation is heating the sample at furnace heating rates from 5 to 10°C/min. Almost all of the modern thermobalances are automatically recording instruments, although manual recording is still used occasionally for long-term isothermal measurements (for example, with helix-type thermobalances).

The modern thermobalance is illustrated schematically in Figure III.1. It consists, generally, of the following component parts: (a) recording balance; (b) furnace; (c) furnace temperature programmer or controller; and (d) recorder, either of the strip-chart or X-Y function type. The specific details of each component depend on the particular application that is required of the instrument. For example, furnaces can be obtained that operate up to 2400°C or more, and employ air, inert gases, hydrogen, nitrogen, vacuum, and other atmospheres. Likewise, for the recording balance, sensitivities from as low as 0.02 mg full scale deflection to 100 g or more are available.

Some factors that must be considered in the construction or purchase of an automatic thermobalance have been given by Lukaszewski and Redfern (1):

(1) The instrument should be capable of recording the mass-loss or -gain of the sample as a function of temperature and time.

(2) The thermobalance furnace should have a wide range of operation, such as from ambient temperature to 1000, 1600, or 2400°C.

(3) The mass-loss of the sample should be recorded to an accuracy of better than ±0.01%, while the temperature should be recorded to an accuracy of ±1%.

(4) The physical effects due to the normal functioning of the instrument, such as radiation and convection currents and the magnetic effects due to the furnace heaters, should not affect the accuracy of the balance. The

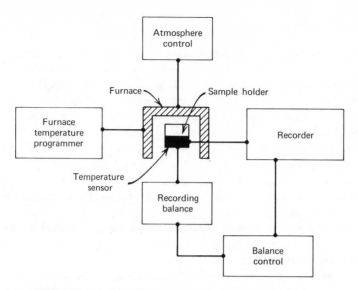

Figure III.1. Schematic diagram of a modern thermobalance.

latter effect does not interact with any conducting or magnetic materials that may be studied.

(*5*) The position of the crucible within the furnace of the thermobalance should always be the same, so that the temperature recorded corresponds to the sample temperature.

(*6*) The furnace should be equipped to allow for the heating of samples in various atmospheres.

(*7*) The instrument should be as versatile as possible, providing for easy changes in heating rates together with automatic control of temperature programming.

(*8*) The balance should be adequately protected from the furnace, and care should also be taken to keep the wear of the knife edges and other moving parts to a minimum to ensure accuracy of weighing.

(*9*) The balance should be capable of simple, periodic calibration to ensure accuracy of operation.

(*10*) The chart used to record mass-loss and temperature-rise should be capable of various speeds, and there should be provision for accurate recording of a suitable time interval.

Obviously, all these requirements cannot be met in every thermobalance. However, a number of commercially available instruments do incorporate these features.

1. Recording Balances

Perhaps the most important component of the thermobalance is the recording balance. The requirements for a suitable recording balance are essentially those for a good analytical balance, namely, accuracy, reproducibility, sensitivity, capacity, rugged construction, and insensitivity to ambient temperature changes. In addition (2), the balance should have an adjustable range of mass-change and a high degree of electronic and mechanical stability, be able to respond rapidly to changes in mass, be relatively unaffected by vibration, and be of sufficiently simple construction to minimize the initial cost and need for maintenance. From a practical viewpoint, the balance should be simple to operate and versatile in that it can be used for varied applications.

Recording balances can be divided basically into two general classifications, based on their mode of operation: (a) deflection-type instruments, and (b) null-type instruments (2).

The automatic null-type instrument is based upon the principle given in Figure III.2. The balance incorporates a sensing element which detects a deviation of the balance beam from its null position; horizontal for beam balances and vertical for electromagnetic-suspension types. A restoring force, of either electrical or mechanical mass loading, is then applied to the beam through the appropriate electronic or mechanical linkages, restoring it to the null position. This restoring force, proportional to the change in mass, is recorded directly or through an electromechanical transducer of some type.

The various deflection-type balances are shown in Figure III.3. These instruments involve the conversion of the balance-beam deflections about the fulcrum into the appropriate mass-change curves by (a) photographic recording, (b) recording electrical signals generated by an appropriate displacement measurement transducer, and (c) using an electromechanical device. The types of deflection balances are the following:

(a) The helical spring, in which changes of mass are detected by contraction or elongation of the spring and which may be recorded by suitable transducers.

(b) The cantilevered beam, constructed so that one end is fixed and the other end, from which the sample is suspended, is free to undergo deflection.

(c) The suspension of a sample by an appropriately mounted strain gauge that stretches and contracts in proportion to mass-changes.

(d) The attachment of a beam to a taut wire which serves as the fulcrum and is rigidly fixed at one or both ends so that deflections are proportional to the changes in mass and the torsional characteristics of the wire.

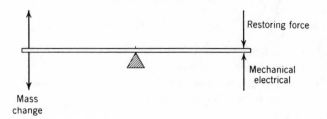

Figure III.2. Null-type balance (2).

Gordon and Campbell (2) have summarized the various methods that have been employed to detect the deviation of a balance beam from its horizontal or vertical position in the null-point balances. They are the following:

Optical

(a) Light source—mirror—photographic paper
(b) Light source—shutter—photocell

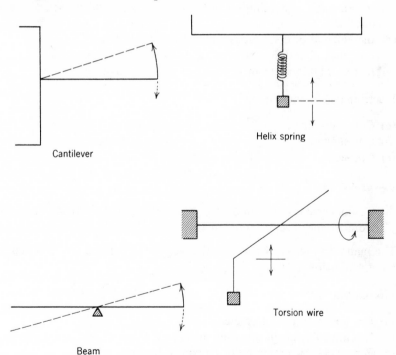

Figure III.3. Deflection-type balance (2).

Electronic

(*a*) Capacitance bridge
(*b*) Mutual inductance: coil-plate, coil-coil
(*c*) Differential transformer or variable permeance transducer
(*d*) Radiation detector (Geiger tube)
(*e*) Strain gauge

Mechanical

(*a*) Pen electromechanically linked to balance beam or coulometer

After the departure of the beam from its rest position has been detected, some method of restoring the beam back to the rest position must be employed. These methods are the following (2):

Mechanical

(*a*) Addition or removal of discrete weights; or beam rider positioning
(*b*) Incremental or continuous application of torsional or helical spring force
(*c*) Incremental or continuous chainomatic operation
(*d*) Incremental addition or withdrawal of a liquid (buoyancy)
(*e*) Incremental increase or decrease of pressure (hydraulic)

Electromagnetic interaction

(*a*) Coil-armature
(*b*) Coil-magnet
(*c*) Coil-coil

Electrochemical

(*a*) Coulometric dissolution or deposition of metal at electrode suspended from balance beam or coulometer

The manner in which the weight-changes of the balance are recorded are summarized below (2):

Mechanical

(*a*) Pen linked to potentiometer slider
(*b*) Pen linked to chain-restoring drum
(*c*) Pen or electric arcing-point on end of beam
(*d*) Pen(s) linked to servo-driven photoelectric beam-deflection follower

Photographic

(a) Light source—mirror—photographic paper using either a drum, time base, or flat bed; temperature base—mirror galvanometer

Electronic

(a) Current generated in a transducing circuit such as photocell, differential transformer, strain gauge, bridge, radiation detector, capacitor, or inductor
(b) Current passing through the coil of an electromagnet

In general, electronic recording is more versatile and convenient than a mechanically linked system because of the many transducers that can be used to obtain the electrical signal proportional to the change in mass as determined by either the deflection-type or the null-type balance. The continuously recorded analog data from the primary curve can be simultaneously translated into other useful forms such as derivatives, integrals, logarithms, or any desired function, many of which lend themselves to the digital operations associated with automatic computation and automatic processes (2).

a. Cahn Balances

About half of the commercially available thermobalances employ one of the Cahn-type electromagnetic balances. This is a sensitive, accurate, reliable, and easily operated balance which is based on a D'Arsonval galvanometer, with the sample suspended from the indicating needle (beam). Changes in sample mass cause the beam to deflect, changing the photocell current, which is then amplified and applied to the coil attached to the beam. This coil is in a magnetic field so that current flowing through it exerts a moment on the beam restoring it to a null position. The coil current is thus an exact measure of sample mass. A schematic diagram of the balance is shown in Figure III.4. The entire balance can be placed in a glass enclosure which can be used to pressures down to $\sim 10^{-6}$ Torr or lower. Various models are available, each differing in sensitivity, mass capacity, and other features.

2. SAMPLE HOLDERS

One of the most important components of a thermobalance is the sample holder. The geometry, size, and material of construction have a rather important effect on the mass-loss curve obtained. A large variety of sample holders have been described, a representative number of which are shown in Figure III.5. As in the case of DTA sample holders (Chap. VI), the type of holder used depends upon the size and nature of the sample and the

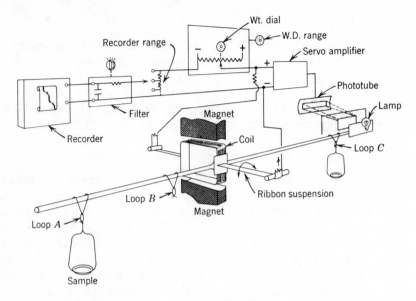

Figure III.4. Schematic diagram of the Cahn Electrobalance.

maximum temperature range to be employed. Materials of construction include glass, quartz, alumina, graphite, aluminum, stainless steel, platinum, and many others. Sample size may vary from 1 mg to hundreds of grams, but the most commonly used sample mass is from 5 to 100 mg.

A sample holder which is widely used with Cahn Electrobalances is shown in (a). It is 5–9 mm in diameter and is constructed of platinum or quartz with the geometrical shape of a hemisphere. Sample sizes from 1 to 20 mg may be investigated. The sample holder in (b) is used in horizontal thermobalances such as are available from Linseis or Netzsch. Various sizes and materials of construction are available. A large number of sample holders are available for use in the Mettler thermobalance system (3), as shown by the examples in (c)–(f). In (c), the open-type crucible, various sizes are available from 0.1 to 5 ml in volume; they are constructed from a platinum-rhodium alloy, alumina, quartz, and graphite. The flat-plate-type holder, (d), is used for high-vacuum studies. Thin layers of sample are used to prevent decripitation during dissociation. Use of the polyplate sample holder, (e), provides a large surface area for the sample. Similar holders have been described by Erdey et al. (4, 5) for use in the Derivatograph; one holder consisted of 10–20 plates spaced at 2-mm intervals, permitting the use of samples from 0.2 to 1 g in mass. Effervescent samples can be studied conveniently in the holder shown in (g), while strips of sample (for example,

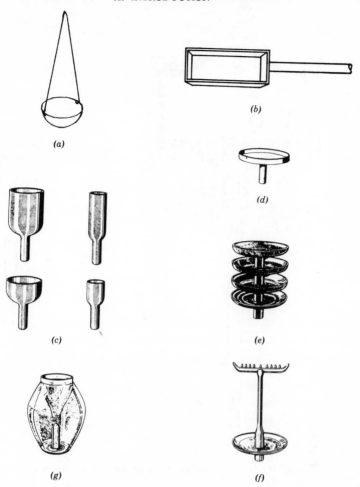

Figure III.5. Representative sample holders for thermogravimetry.

metal foil strips) can be suspended on the sample holder illustrated in (*f*). For simultaneous TG-DTA measurements, the sample holders discussed in Chap. VI, Figure VI.2, may be employed.

For vapor pressure measurements, the sample holders illustrated in Figure III.6 may be used. The holder in (*a*) is a schematic cross section of a Knudsen cell described by Wiedemann and Vaughan (6). The main body of the cell, A, is constructed of aluminum and is suspended on the thermobalance by a ceramic tube, G. A copper-constantan thermocouple, E, brought into the cell through a vacuum-tight connection, detects the sample

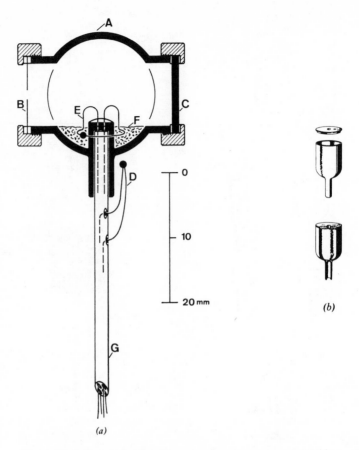

Figure III.6. Sample holders for vapor-pressure measurements (6).

temperature, F, or the temperature of the vapor phase.   A second thermo-couple, D, serves to control the furnace temperature.   The orifice, B, is made from 0.01-mm-thick Nichrome foil and contains a small hole.   A screw fitting on the cell permits rapid removal or exchange of orifices of different hole diameters (1–3 mm in diameter).   The other end plate, C, is used to load the sample into the cell.   Both end plates are sealed to the body by Teflon O-rings.   Sample volume is ⩽1.5 ml and the cell can be used in the temperature range from −100 to 200°C.   A simpler version of a vapor pressure cell, which contains a removable lid, is shown in (b).

For studies involving dissociation reactions in "self-generated" atmospheres (7), the sample holders in Figure III.7 may be used.   In the piston and cylinder arrangement, shown in (a), the clearance between the two provides

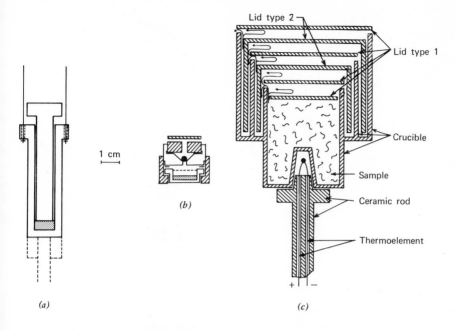

Figure III.7. Sample holders for "self-generated" atmospheres (7–10).

a long diffusion path and effectively prevents contamination from the atmosphere. The Forkel crucible (8), as shown in (b), has a lid containing a ball valve to ensure separation of the sample atmosphere and the atmosphere in the furnace. Various other glass and quartz sample holders have been discussed by Newkirk (9). Perhaps the most elaborate sample holder is the one shown in (c), which was described by Paulik and Paulik (10). The sample crucible is covered with six close-fitting lids or covers in which the gaseous decomposition products are forced to escape through a long and narrow labyrinth.

Aids to ensure uniform sample packing are shown in Figure III.8 (3). A vibrator with spatula attached, as shown in (a), is used to load small samples into the sample holder. In order to pack the samples in a uniform manner, the vibrator and sample holder device in (b) may be employed. The packing of voluminous, low-density samples into the appropriate sample holder is accomplished by the small press illustrated in (c).

The loading of moisture- or oxygen-sensitive samples into the Mettler thermobalance sample holder is conveniently carried out by use of the controlled-atmosphere enclosure shown in Figure III.9 (3). The sample is introduced, via an enclosed sample holder, into the enclosure and loaded

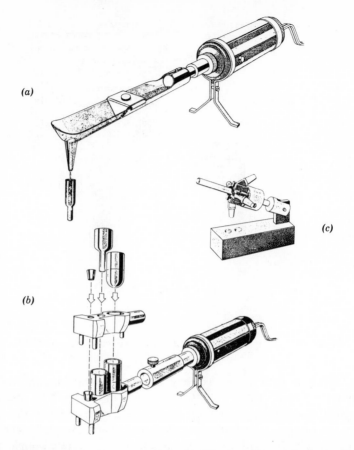

Figure III.8. Sample-loading accessories. (a) Vibrator with spatula for loading small samples; (b) crucible vibrator for uniform packing of samples; (c) sample press for compressing low-density samples into the sample holder (3).

into the furnace chamber after the controlled atmosphere has been introduced. After loading, the furnace chamber is closed and the enclosure removed.

Direct visual examination of the sample during heating is shown schematically in Figure III.10 (11). The viewing hole is covered by a quartz plate to prevent furnace heat loss and thermal gradients.

3. Furnaces and Furnace Temperature Programmers

The types of furnaces employed in TG measurements are similar to those discussed in Chap. VI.A.4. The choice of furnace heating element and

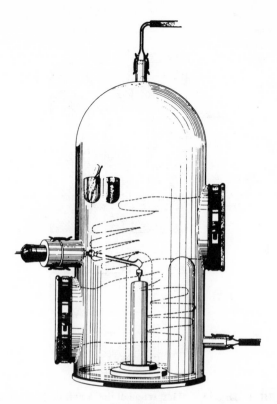

Figure III.9. Enclosure for sample loading in controlled atmosphere (3)

type of furnace depends upon the temperature range desired. Furnaces have been described which operate in the temperature range from ambient room temperature to 1000, 1600, or even 2400°C. The furnace may be mounted vertically or horizontally, and positioned either above or below the balance mechanism. Mounting above the balance is generally recommended because of a number of factors, most of them due to thermal gradients in the furnace chamber and temperature changes of the balance. Resistance heater element furnaces are the most widely used, the temperature limits of which are given in Chap. VI (Table VI.2). Very few TG studies require-low temperature furnaces, although low-temperature TG-magnetic susceptibility measurements may employ a low-temperature cryostat of some type.

The furnace thermal symmetry requirements are perhaps less stringent than those required with DTA measurements. It is desirable to know the temperature distribution curves of the furnace, such as illustrated in Figure

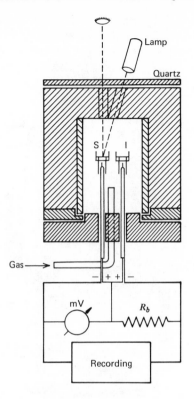

Figure III.10. Direct visual inspection of the sample during heating (11).

VI.6 in Chap. VI.   There should be a zone of homogeneous temperature flux in the sample holder area.

The temperature rise of the furnace is controlled by a temperature programmer, the requirements of which are similar to those used in DTA (see Chap. VI).   The heating rates should be linear and reproducible, as a nonlinear heating rate will influence the resulting TG curve, especially if a time-based recording system is used.   The linearity requirements are not as important as in DTA, however.   The heating rates used are generally from 5 to 10°C/min, although some measurements have been made at 160°C/min. An average value which is widely used is 5°C/min, but this can vary depending upon the size of sample and the TG information desired.

TEMPERATURE DETECTION AND RECORDING SYSTEMS

As previously mentioned, in thermogravimetry the mass-change of the sample is continuously recorded as a function of temperature.   The temperature, in this definition, may be that of the furnace chamber, the temperature near the sample (that is, in close contact with the sample container), or the temperature of the sample. These three sources of temperature detection are shown in Figure III.11. In (a), the thermocouple is near the sample container but not in contact with it. There is a correlation between the temperature of the container and that detected by the thermocouple, but the thermocouple will either lead or lag behind the sample temperature, depending upon the thermochemistry of the reaction. Most thermobalances use this type of thermocouple arrangement even though it is a poor one. It is even worse when low-pressure atmospheres are employed, as in high-vacuum thermogravimetry; in this case, heat transfer is by radiation alone rather than by convection and conduction.

The thermocouple is in close proximity to the sample in (b), and is positioned inside, but not in contact with, the sample holder. This arrangement is better than (a) because the thermocouple will respond to small changes of sample temperature. The best method of sample temperature detection is to have the thermocouple either in contact with the sample or with the sample container, as shown in (c). In the latter, the temperature detected will be the integrated temperature. However, the main problem is that with sensitive recording balances the thermocouple leads can cause weighing errors, or at least interference with the balance mechanism. One way to detect the actual sample temperature and yet not interfere with the balance mechanism is to suspend an electronic device near the sample holder which will transmit the sample temperature to a fixed receiver located near the sample container. Manche and Carroll (13) described a unijunction transistor relaxation oscillator which used a thermistor as the temperature detector. The frequency of oscillation, which is a function of sample

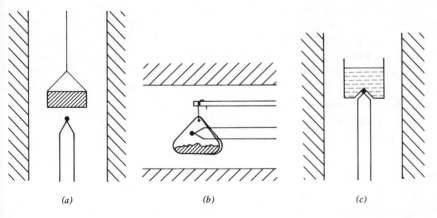

(a)   (b)   (c)

Figure III.11. Location of temperature-detection thermocouple in a thermobalance (12).

temperature, was transmitted via a mutual inductance between two suspended coils to a receiver and counter. The device was limited, however, to a maximum temperature of about 150°C.

The calibration of the temperature of the furnace and/or sample chamber has been discussed by Stewart (12) and Norem et al. (14, 15). Stewart (12) used a conventional thermobalance which contained a thermocouple mounted external to the sample, while Norem et al. (14, 15) calibrated a furnace which used a resistance element for temperature detection.

Stewart (12) discussed three approaches to temperature calibration:

(1) The use of standard materials with reproducible mass-loss points that could be referred to the temperature.

(2) The use of materials having known reproducible (and reversible) temperature transitions and direct measurement of temperature.

(3) The use of materials with magnetic transitions which could be displayed on a mass-loss curve and be referred to the temperature.

The first approach is the most appealing, but the evolution of a volatile product is dependent not only on the temperature and rate of temperature change but also upon the type and nature of the furnace atmosphere. The second method was used by Stewart (12) in his temperature calibration scheme but required that the thermocouple be in contact with the sample or the sample container during the calibration procedure. Compounds chosen for standards were those containing $solid_1 \rightleftarrows solid_2$ or $solid \rightleftarrows liquid$ type transitions, which were not atmosphere-dependent. The standards used and their transition temperatures were potassium nitrate (129.5 and 333°C), potassium chromate (665°C), and tin (231.9°C).

Norem et al. (14, 15) used the third method, as previously discussed, to calibrate the temperature of their type of furnace and/or sample container. A ferromagnetic material was placed in the sample container and suspended within a magnetic field. At the material's Curie-point temperature, its equivalent magnetic mass diminishes to zero and the thermobalance indicates an apparent mass-loss. For calibration over the temperature range from ambient temperature to 1000°C, it is obvious that a number of ferromagnetic materials must be used. The criteria which were considered characteristic of an ideal standard were the following (15):

(1) The transition must be sharp; that is, its natural or true width should extend over a small temperature range.

(2) The energy required to effect the transition should be small (under the dynamic scanning conditions of TG, the "sharpness" of a transition is inversely proportional to transition energy).

(*3*) The transition temperature should be unaffected by the chemical nature of the atmosphere and independent of pressure.

(*4*) The transition should be reversible so that the sample can be run repetitively to optimize or check the calibration.

(*5*) The transition should be unaffected by the presence of other standards so that several samples can be run simultaneously to obtain a multipoint calibration in a single experiment.

(*6*) The transition should be readily observable using standard samples in the milligram range—comparable to normal sample sizes investigated with the apparatus.

A typical calibration curve using four standard ferromagnetic samples is shown in Figure III.12. The actual transition temperatures are indicated in parentheses. Standards must be used in the heating mode because they exhibit hysteresis behavior on cooling. Transition temperatures are also slightly dependent on the heating rate of the furnace, but this dependency is quite small in the 5–20°C/min range. The effect of heating rate on the

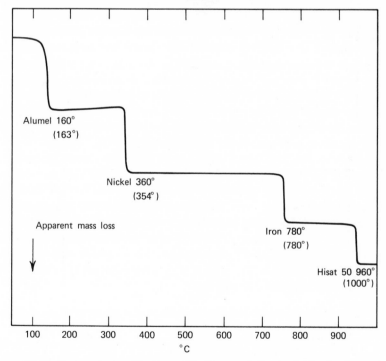

Figure III.12. Temperature calibration with ferromagnetic standards; heating rate of 10°C/min (15).

observed initial transition temperature is given by

$$T_{\text{indicated}} = T_{\text{isothermal}} + RC_s\dot{T}_n \qquad (\text{III.1})$$

where R is the effective thermal resistance between heat source and sample container, C_s is the effective heat capacity of the sample and its container, and $\dot{T}_n$ is the heating rate. The parameter RC_s is the instrumental time constant, a most important parameter characterizing the performance of the instrument (or any thermal analysis instrument).

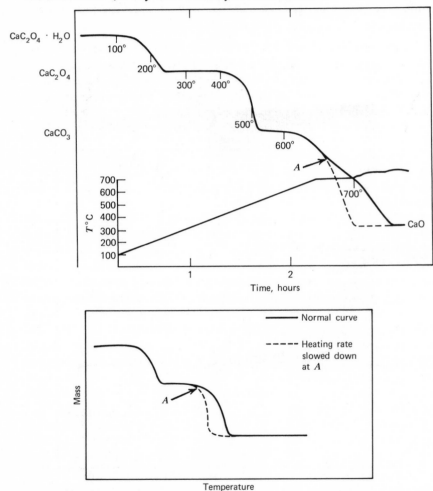

Figure III.13. Effect of temperature perturbation on mass-loss curve. (*a*) Time base recorder; heating rate slowed down at *A*; (*b*) *X–Y* recorder.

The two types of recording systems that are normally employed with thermobalances are time-base potentiometric strip-chart recorders and $X-Y$ recorders. In several commercial instruments, light-beam-galvanometer–photographic-paper type recorders are still used. In the case of the strip-chart recorder, the temperature of the system must be recorded as well as the mass-change curve. Either two single-pen recorders or one recorder with two or more pens (or multipoint type) may be employed. One advantage of this type of recording system is that the heating rate of the furnace can also be observed and checked for linearity.

The effects of a temperature perturbation on the mass-loss curves using both types of recording systems are shown in Figure III.13. In (a), using a time-base recorder, the temperature perturbation occurred at A, resulting in the recording of the solid-line curve. The normal curve, if no change in the heating rate took place, is shown by the dashed-line curve. Similarly, for the $X-Y$ recorder, (b), the temperature perturbation occurred at A, resulting in the change to the dashed-line curve. The normal curve is indicated by the solid curve. However, in the time-base recorder, a curve of the system temperature was also recorded so that the change in heating rate could be immediately detected. With the $X-Y$ recorder, this change in heating rate would probably not be detectable unless the curve was duplicated several times.

Most commercial thermobalances, with several exceptions, record the change in sample mass rather than the percent mass-change; the latter is a more convenient form of presentation. From direct mass-change data, the sample mass must be known in order to calculate the percent mass-change and/or other stoichiometry calculations. In percent mass-change data recording, the data are readily available for stoichiometry calculations of various types. Wendlandt (72) found that by the proper choice of strip-chart recorder, the percent mass-change can be recorded directly. The recorder must contain a variable span adjustment (such as on a Sargent Model SRG recorder) which permits the pen to be positioned to the extreme left of the chart and corresponds to 0% change. Any mass-loss of the sample is then recorded as percent mass-loss; it is not even necessary to know the exact sample mass.

The other characteristics and requirements of recording systems are similar to those used in DTA systems (Chap. VI).

B. Thermobalances

1. INTRODUCTION

A large number of reviews and books have been written describing various commercial and noncommercial thermobalances. Mention should be made of the reviews by Gordon and Campbell (2), Duval (16), Lewin (17), Jacque et al. (18), Saito (19), Vaughan (20), Wendlandt (21, 22), and others. Books containing descriptions of thermobalances include those by Duval (23, 24), Garn (25), Keattch (26), Anderson (27), Wendlandt (28), Saito (29), and others. Due to the limited space available, only those thermobalances containing novel or important features will be discussed here. Likewise, only a few of the many commercial instruments will be mentioned as these are adequately described elsewhere in recent reviews or books (21, 22, 28).

The first thermobalance was probably the instrument described by Honda (30) in 1915. This instrument, as shown in Figure III.14, consisted of a balance with a quartz beam. The sample was placed in a procelain or magnesia crucible, G, which was suspended in an electrically heated furnace, J. Attached to the opposite end of the balance beam was a thin steel wire helix, E, which was immersed in oil contained in a Dewar flask, H. The Dewar-flask–helix assembly was adjusted by a screw mechanism to maintain the balance beam in a null position. A rather low heating rate was employed, as it took 10–14 hours to attain a temperature of 1000°C. However, Honda used a quasi-isothermal heating cycle in that during a mass-loss transition the furnace temperature was maintained at a constant temperature until the transition was completed. This procedure alone sometimes required 1–4 hours. Convection currents were evident above 300°C, as

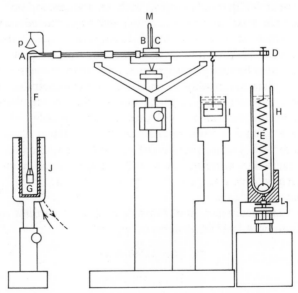

Figure III.14. The thermobalance as described by Honda (30).

might be expected from the furnace-sample arrangement.   A sample mass of about 0.6 g was normally employed.

The work of Honda did indeed lay the foundation for practically all of the future work in thermogravimetry.   His thermobalance enabled the investigator to continuously weigh the sample as it was heated and also employed the feature of quasiisothermal heating.   The latter technique cannot be carried out automatically by present-day thermobalances.   He modestly concluded his paper by saying, "All of the results above given are not altogether original; the present investigation with the thermobalance has however revealed the exact positions of the change in structure and also the velocity of the change in respective temperatures.   The investigation also shows the great convenience of using such a balance in similar investigations in chemistry."

Numerous other Japanese workers modified the Honda thermobalance and also developed new instruments.   The results of their studies have been summarized in a monograph by Saito (19).

The French school of thermogravimetry was started by Guichard (31) in 1923.   He apparently was unaware of Honda's work, but then he never claimed to be the discoverer of thermogravimetry.   He improved the technique, brought it to a high state of development for that time. and critically examined each phase of it.   His original thermobalance (23) contained a gas-fired furnace in which gas was metered to the burner via a constant-level device consisting of a valve with a float attached to it.   The float rested on the surface of a tank filled with water.   This water drained into another container via an adjustable valve, thus causing a slow but gradual opening of the gas metering valve and hence increasing the furnace temperature.   Equally interesting is the manner in which the mass-change of the sample was detected by the balance.   This was achieved with a hydrostatic device in which small amounts of oil were added to a "U" tube to exactly compensate for the mass-change.   A loss of mass of 100 mg corresponded to the addition of 9 ml of oil.   Mass-loss curves recorded on this balance agreed well with the results obtained by Duval (23) some 40 years later.

Guichard's work was followed by the investigations of Vallet (32), Dubois (33), and others (23).   Perhaps the greatest impetus to the French school of thermogravimetry was the development of the Chevenard (34) recording thermobalance.   This balance had been under development since 1936 and became commercially available in 1945; it was the first automatic (photographically) recording instrument.   In the hands of Duval and co-workers (23, 24), it became the standard instrument for work in this field.   This thermobalance will be described in the next section.

Two other important milestones in the development of the modern thermobalance occurred in 1958 and 1964. A multifunctional instrument, called the Derivatograph, was described by Paulik (35) et al. in 1958. Not only could the TG curve be recorded, but also its first derivative (DTG) and the differential thermal analysis (DTA) curve. In 1964, Weidemann (3) described the Mettler system, which is perhaps the most sophisticated thermobalance available today.

2. Chevenard Thermobalance

Perhaps more studies have been conducted on the Chevenard automatic thermobalance than on all the other instruments combined. The instrument was first described by Chevenard et al. (34) in 1944, and was used by them for studying the corrosion of metals at elevated temperatures with great success. The balance has been modified in recent years; the model now available is the TH-59, which can be used to study samples in a high vacuum to temperatures of 1500°C. Provision has been made for automatic pen recording of the mass-versus-time or mass-versus-temperature curves, using a spot-follower photocell system.

The basic principles of the balance have not been changed very much. The balance is a deflection-type instrument containing a wire-supported beam. Deflections may be recorded photographically, as was done originally or by some type of phototransducer. The basic design is illustrated in Figure III.15. The sample is placed at the end of a rod, which is suspended from one arm of a sensitive suspension balance; a movable counterpoise on the other arm is used for adjustment of the balance at the beginning of the run. Parasitic oscillations of the beam are magnetically damped. The rod and sample holder are inserted into a bifilarly wound furnace, which is mounted above the balance. The recording unit consists of a resistance-type photocell mounted on a motor-driven carriage and a revolving drum. A beam of light, reflected from a mirror on the balance beam, is allowed to impinge on the photocell. Any movement of the light beam, caused by changes of sample mass, is automatically followed by the photocell-servo-motor arrangement and recorded on the drum by means of a pen. The pen is coupled to the photocell through a "thin-line device," the adjustable backlash of which enables the pen to draw a thin line corresponding to the vertical movement of the light spot without following the purely oscillatory movements of the photocell. The rotation of the drum is either a function of time or a function of temperature. The maximum range of the balance is about 50 g, and in the lower ranges it is possible to have a sensitivity of 1 mg per 5 mm of chart distance.

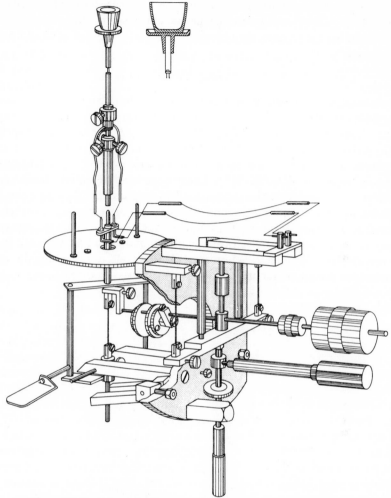

Figure III.15. Schematic diagram of the Chevenard thermobalance (34).

A number of modifications have been described for the Chevenard thermobalance.   Gordon and Campbell (36) converted a photographic recording
instrument into an electronic recording instrument by use of a linear variable-
differential transformer to convert balance-beam displacements into linear
electrical signals.   After demodulation, the signal was recorded on a strip-
chart potentiometric recorder.   The range of mass-change that could be
linearly recorded was about 400 mg, with an accuracy over a range of 200 mg
of about ±0.25%.   The response time for a 200-mg mass-change was about

2 sec. A recorder temperature-marking system has been described by Griffith (37). Other changes to the balance have been mainly to the sample holders and to the furnace enclosure to allow better control of the furnace atmosphere.

3. HELICAL SPRING THERMOBALANCES

Various types of helical-spring recording thermobalances have been described (38–53), all of which are modifications of the original design proposed by McBain and Bakr (54) in 1926. A typical helical-spring thermobalance is shown in Figure III.16. This instrument, called the Thermo-Grav, consists of a deflection-type balance in which the extension of a helical spring is detected by a linear variable-differential transformer (LVDT). The helical spring is enclosed in a glass chamber and can be operated under a vacuum or a controlled gaseous atmosphere, at or below atmospheric pressure. Spring deflections, proportional to changes in sample mass, are converted into electrical signals by movement of the core in a linear variable-differential transformer. These signals, after passing through the amplifier and demodulator, are presented as mass changes on the Y axis

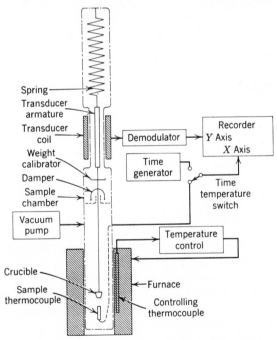

Figure III.16. Schematic diagram of a helical-spring-type thermobalance (Thermo-Grav).

of an $X-Y$ recorder. The X axis records an input signal, which corresponds either to furnace temperature or elapsed time. The suspension system carries a weight calibrator pan, an oil-dashpot damper, and, at the bottom, fused quartz rings to hold the sample and tare crucibles. Two furnaces are contained in the cabinet to allow consecutive runs to be made. Changes from 0 to 200 mg in mass can be measured with an accuracy of $\pm 1\%$ of full scale with sample masses up to 10 g. Furnace heating rates of 5–100°C/h are provided, while X-axis spans are adjustable for temperature increments of 200, 500, and 1000°C.

An elaborate modification of this balance has been described by Olson (50, 51) and Farmer (45). A $X-Y^1Y^2$ recorder was used in which the Y^2 axis recorded the time generated by a remote synchronous time generator. A small, spherically shaped furnace capable of operation up to 1400°C replaced the above furnace.

Franck and Harmelin (53) modified a helix-type balance so that concurrent TG–DTA studies could be performed. The modification consisted of the addition of a micro-DTA cell to the furnace chamber so that the sample holder was directly below the balance sample holder. Two samples were required, one each for the TG and the DTA data.

4. DuPont Thermobalance

The DuPont thermobalance has been described by Sarasohn and Tabeling (55). The balance, which is illustrated schematically in Figure III.17, contains a null-balancing (similar to the Cahn Model RG Electrobalance), taut-band electric meter movement with an optically actuated servoloop. The sample is placed in a container which is suspended directly on the balance beam. In normal operation, the temperature-sensing thermocouple is positioned within 1 mm of the sample, and hence indicates very close to the sample temperature. Mass-changes of the sample are plotted as a function of temperature on an internal $X-Y$ recorder. Balance sensitivity is reported to be 2 μg. The horizontal beam-in-furnace design and the position of the sample container permit axial flushing of the furnace tube with various inert gases. The sample presents minimum cross section to the gas flow, which results in a negligible torque perpendicular to the beam. Hence, there is a minimum of tubulence and noise in the balance. A maximum furnace temperature of 1200°C is attainable.

An atmosphere control system for the thermobalance, as described by Williams (56), permits the rapid changing of the dynamic gas atmosphere between oxygen, hydrogen, and argon.

Enclosure of the DuPont thermobalance in a high-pressure chamber has been described by Brown et al. (57) and Williams et al. (58). Pressures of

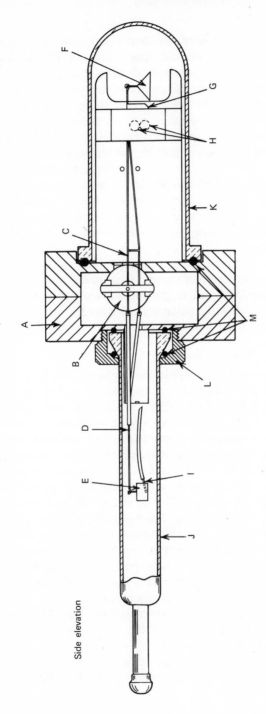

Side elevation

Figure III.17. DuPont thermobalance (55). A, balance housing; B, taut-band meter movement; C, rear beam; D, quartz beam—hot member (removable); E, sample boat; F, counterweight pan; G, signal flag; H, photovoltaic cells; I, "floating" thermocouple; J, quartz furnace tube; k, pyrex envelope (cold); L, threaded collar; M, O-rings.

up to 20 atm and a maximum temperature of 350°C were possible with the former, while the latter instrument could be operated up to 68 atm and a maximum temperature of 450°C. The modification by Williams et al. (58) also permitted simultaneous magnetic-susceptibility measurements on the sample (see Chap. XI). Chiu (59) described a modification of the DuPont thermobalance in which concurrent TG, DTA, and ETA (electrothermal analysis) curves could be recorded. This instrument is described in Chap. XI. Other combinations, such as the coupling of the balance to a gas chromatograph or mass spectrometer, are described in Chap. VIII.

5. DERIVATOGRAPH

The Derivatograph, which was first described by Paulik et al. (35) in 1958, is a multifunction thermal-analysis system which can record the TG, DTG, DTA, and T curves of a sample on a single chart. By means of accessory attachments, the TD (thermal dilation) and DTD (derivative of TD) (5) curves may be recorded as well as the evolved-gas curves (see Chap. VIII).

The instrument, as shown in Figure III.18, consists of an analytical balance, two furnaces, a furnace-temperature programmer, sample and

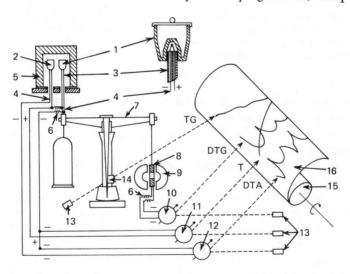

Figure III.18. Schematic diagram of the Derivatograph. 1, Crucible for the sample; 2, crucible for the inert material; 3, porcelain tube; 4, thermocouples; 5, electric furnace; 6, torsionless lead; 7, balance; 8, coil; 9, magnet; 10, DTG-galvanometer; 11, T-galvanometer; 12, DTA-galvanometer; 13, lamps; 14, optical slit; 15, photorecording cylinder; 16, photographic chart.

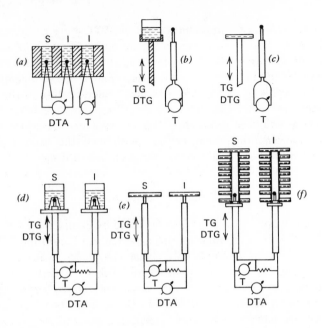

Figure III.19. Sample holders for Derivatography (60). (*a*) Sample holder of DTA apparatus; (*b*) crucible of derivative thermobalance; (*c*) monoplate sample holder of the derivative thermobalance; (*d*) crucible of the D-graph: (*e*) monoplate holder of the D-graph; (*f*) polyplate holder of the D-graph.

reference crucibles, a voltage regulator, and a galvanometric light-beam–photographic-paper recorder. The balance is of the air-damped analytical type, with a basic sensitivity of 20 ± 0.2 mg full-scale deflection and a working mass range of 10 mg to 10 g. The derivative of the TG curve is obtained by means of a simple device consisting of a magnet and an induction coil. The former is suspended on one arm of the balance beam, with both of its poles surrounded by two induction coils. When a change of mass occurs, movement of the magnet induces a voltage in the coils which is proportional to the rate of mass-change. This induced coil voltage is measured by one of the light-beam galvanometers and recorded on the chart as one of the curves. Maximum temperature of the furnace is 1100°C; N_2, CO_2, Ar, O_2, etc. may be used as the furnace atmosphere at atmospheric pressure only.

Various sample holders, several of which are illustrated in Figure III.19, have been described (60) and their applications to specific problems discussed. The labyrinth, self-generated-atmosphere type of sample holder has previously been discussed (Chap. II.D).

6. METTLER THERMOANALYZER

The Mettler Thermoanalyzer, as described by Wiedemann (3), is perhaps the most elaborate, versatile, and expensive of all of the thermobalances commercially available today.   Two models are available, each of which is a universal research instrument that will simultaneously plot, on a single recorder chart, the TG, DTG, and DTA curves of a single sample, as well as the furnace-chamber pressure and gas-flow velocity, if so desired.

The instrument is shown schematically in Figure III.20.   The balance is of the aluminum beam-substitution type with sapphire knife edges and planes.   Change in the sample mass causes a beam deflection which moves a light shutter interrupting a light-beam display on two photodiodes.   The unbalance in photo-diode current is amplified and fed back to a coil attached to the balance beam as a restoring force.   The electrical mass indication has a dual weighing range with three different sensitivities as standard; a

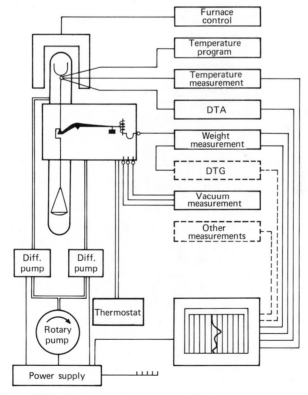

Figure III.20. Schematic diagram of the Mettler thermoanalyzer (3).

fourth is optional.   Two consecutive sensitivities, in the ratio of 1:10, are always recorded.   One range is 0–1000 mg, recorded as 100 mg/in;   the second is also 0–1000 mg but is recorded as 100 mg full-scale deflection, or 10 mg/in.   A more sensitive mass range of 0–10 mg is recorded in an identical manner.   A unique design feature is the gas-flow control system.   Corrosive and noncorrosive gases may be employed, with provisions in the case of the former to keep it from coming in contact with the balance mechanism. Two vacuum and diffusion pumps are employed to evacuate the furnace chamber to pressures of the order 5×10^{-6} Torr.

Three furnaces are available: a low-temperature unit with a maximum temperature of 1000°C; a high-temperature unit for use up to 1600°C, and a super high-temperature model with a maximum temperature of 2400°C. These furnaces are illustrated in Chap. VI (Figure VI.5).   The sample holders are also discussed in Chap. VI because of their simultaneous use for DTA and thermogravimetry.

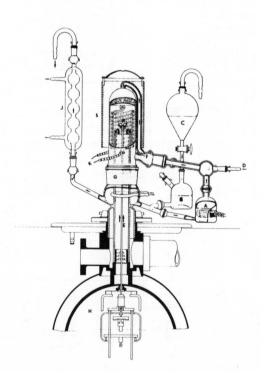

Figure III. 21. Vapor atmosphere furnace for the Mettler system.

Although a large number of different types of inert and oxidizing furnace atmospheres can be used with this system, a controlled-vapor furnace system may also be employed, as shown in Figure III.21. It replaces the standard Mettler 25–1000°C quartz tube furnace. In operation, the sample is placed in the sample holder, P, after which the furnace unit is put in place and sealed by ground glass joint, G. The sample chamber is heated by heater element, K, with the aid of reflector shield, S. A liquid (water or an organic compound) is heated to boiling at A by means of a small cartridge heater. Carrier gas (usually nitrogen) is introduced at D to transfer the vapor through inlet F. Another gas stream, which is brought through the balance mechanism, H, and the baffle, R, is used to prevent vapors from condensing on the sample-holder support rod or the balance chamber. Containers B and C are reservoirs for adding liquid to A. A water-cooled condenser at I condenses the vapor to a liquid which is then returned to boiler A.

The coupling of the Mettler thermobalance to a gas chromatograph or mass spectrometer is discussed in Chap. VIII.

7. MISCELLANEOUS THERMOBALANCES

The conversion of Cahn electrobalances to thermobalances has been described by a number of investigators. For the Model RG balance, mention should be made of the instruments described by Gulbransen et al. (61), Feldman and Ramachandran (62), Pedersen (63), Etter and Smith (67), and numerous others (72); the Model RM is described by Scott and Harrison (64); and the Model RTL is described by Bradley and Wendlandt (65) and others (66).

The hangdown tube and furnace assembly used by Etter and Smith (67) is illustrated in Figure III.22. A small platinum resistance heater located inside a relatively short quartz sample tube is used as the furnace. The temperature-sensing thermocouple is located inside the furnace chamber with entry into the chamber from the bottom. This design is similar to that of the Perkin-Elmer thermobalance except that the latter does not use a thermo-couple for temperature detection.

8. QUARTZ BALANCE

The use of resonating quartz crystals for mass-change determinations is very attractive because a sensitivity of about 10^{-12} g/cm² is feasible. Mass determinations with resonating quartz crystals have been discussed by Plant (68) and Van Empel et al. (69). The principle (69) of the determination

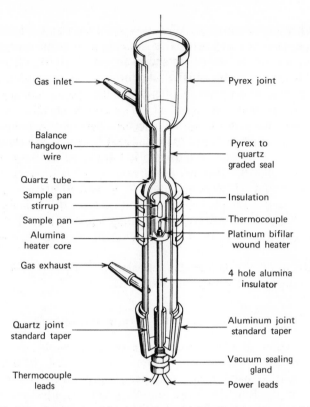

Gas inlet

Pyrex joint

Balance hangdown wire

Pyrex to quartz graded seal

Quartz tube

Sample pan stirrup

Insulation

Sample pan

Thermocouple

Alumina heater core

Platinum bifilar wound heater

Gas exhaust

4 hole alumina insulator

Quartz joint standard taper

Aluminum joint standard taper

Thermocouple leads

Vacuum sealing gland

Power leads

Figure III. 22. Microfurnace used in thermobalance described by Etter and Smith (67).

is based upon the relationship between the variation, Δf, of the resonant frequency, f, and the mass added, Δm, or

$$\frac{\Delta f}{f} = -\frac{\Delta m}{m} \qquad (III.2)$$

where m is the effective mass of the quartz between two electrodes. Measurement must be made at a fixed temperature due to the variation of the frequency with temperature. At relatively high temperatures, a variation of $1°C$ corresponds to a relative frequency change of 2×10^{-6}, which results in an apparent mass-change of 8×10^{-9} g. However, by use of a dual electrode system in which one electrode is used for temperature sensing and the other for mass-changes, a resonating quartz balance can be constructed (69).

Another approach is to use a thin quartz fiber, 5–20 μ in diameter and about 1 cm long, covered with a thin coating of evaporated gold (68). The fiber is held rigidly at one end and rests between two parallel metal plates to which a d.c. potential is applied. When an audio frequency signal is applied to the fiber, it displays vibrating reed resonances whose frequency depends on the physical parameters of the fiber. If mass is added to the free end of the fiber, the resonant frequencies shift downward, being dependent on the amount of mass.

Other types of balances using quartz crystals have been described also (68).

9. AUTOMATED THERMOBALANCES

The modern thermobalance is an automatic instrument in that the mass-change of a sample can be recorded over a wide temperature range. Little attention has been given to the introduction of a new sample automatically into the furnace chamber or of studying multiple samples in a sequential manner. The automated instrument (65, 70) is capable of automatic sample changing and temperature programming. Eight samples, contained in the rotatable sample holder disk, can be studied in an individual manner. A schematic diagram of the balance, furnace, and sample changer mechanism is shown in Figure III.23a, while a diagram of the furnace and sample holder configuration is given in Figure III.23b.

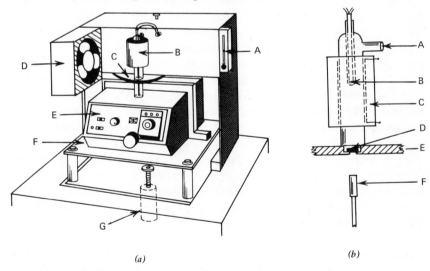

(a)

(b)

Figure III.23. An automated thermobalance. (a) Balance, furnace, and sample changer mechanism; (b) furnace and sample holder (70). (a) A, Gas flow-meter; B, Furnace; C, Sample holder disk; D, Cooling fan; E, Cahn Model RTL recording balance; F, balance platform. (b) A, Gas inlet tube; B, thermocouples; C, furnace heater windings and insulation; D, sample container; E, sample holder disk; F, ceramic sample probe.

The thermobalance is conventional in design in that it consists of a top-loading recording balance (Cahn Model RTL balance), a Leeds and Northrup four-channel multipoint potentiometric recorder (0 to 5 mV full scale), a small tube furnace, a sample-changer mechanism, and an automatic furnace-temperature programmer. Perhaps the most novel feature of the instrument is the automatic sample-changing mechanism, which operates in the following manner: The samples to be investigated are placed into small cylindrical platinum containers, Figure III.23b (D), 5.0 mm in diameter by 2.00 mm in height. Eight such containers are placed in the circular indentations cut in the periphery of the 0.25-in.-thick by 8.0-in.-diameter aluminum sample-holder disk, Figure III.23b (E). The sample containers are positioned directly below the opening of the tube furnace, Figure III.23b (C), by the rotation of a small electric motor connected to a microswitch which is tripped by an indentation in the circumference of the disk. The positioned sample is picked up by the ceramic sample probe, Figure III.23b (F), which is attached to the beam of the balance. Movement of the entire balance and the balance platform, Figure III.23a (E and F), is controlled by a motor-driven screw in the base of the platform. The motor is reversible so that the platform can be raised or lowered, with limits of movement in both directions controlled by microswitches. After the sample is positioned in the central part of the furnace, the furnace is flooded with nitrogen or some other gas and the furnace temperature programmer activated. On attaining a preselected furnace maximum temperature limit, the balance is lowered and the sample container retained by the sample-holder disk. The disk then rotates to position a new sample at the base of the furnace. A cooling fan, Figure III.23a (D), is activated, which cools the furnace to a pre-selected lower temperature limit, at which point the entire cycle is repeated, using a new sample. The heating and cooling cycles are performed on eight successive samples. Each sample is preweighed into the sample containers using a Mettler semimicro printing balance. The individual sample containers are tared to within ± 1 mg (empty weight is about 130 mg); each sample is kept under 10.0 mg so that the recorder pen deflection remains on the recorder scale. The recorder mass range is 0 to 10 mg at 1.00 mg per in. on a 10-in.-wide chart; a chart speed of $\frac{1}{15}$ or $\frac{1}{6}$ in. per min was normally used.

The obvious advantage of the automated thermobalance system over existing instruments is the ability to determine the mass-loss curves of eight successive samples. Operation of the instrument is completely automatic, and once the cycle is begun the instrument does not require the attention of the operator until the eighth sample curve is completed. The instrument should find use for the routine TG examination of a large number of samples,

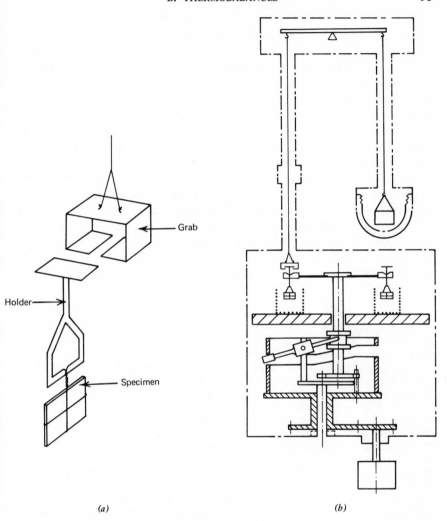

Figure III.24. Multisample weighing balance of Ferguson et al. (71). (*a*) Sample carrier; (*b*) Schematic diagram of balance.

each to be studied under identical thermal conditions. Because the system is completely automated, data reduction or control by a small digital computer could easily be accomplished (see Chap. XIV).

A somewhat different approach was taken by Ferguson et al. (71) in which they described a multispecimen weighing device which was capable of studying samples over long periods of time. The samples were heated in flowing CO_2-CO-O_2 atmospheres in the temperature range from 620 to

770°K.   The sample carrier, which is attached to the "carousel" beam of the balance, and the carousel mechanism, are shown in Figure III.24. The carriers were constructed of punched nickel sheet and were fitted with flat horizontal heads by spot welding.   The sample weighing is carried out by lifting a sample off the carousel beam by engaging the carrier with the grab.   The carousel mechanism is mounted so that the sample carriers, which rest in V-blocks near the ends of the radial arms, pass through the grab at the appropriate stage of the weighing cycle.   When six samples have passed through the grab the rotation of the carousel stops with the seventh sample in position for weighing.   The radial arms are lowered so that the head of the sample carrier is completely supported by the grab.   This position is held for 90 sec while the balance settles and the sample mass is recorded.   The radial arms are then raised to reengage with the sample holder and lift it clear of the grab before the carousel rotates once again to bring another sample to the weighing position.   Twenty samples may be heated at one time in which each sample is weighed once in a complete cycle lasting 1.5 h.

References

1. Lukaszewski, G. M., and J. P. Redfern, *Lab. Pract.*, **10**, 469 (1961).
2. Gordon, S., and C. Campbell, *Anal. Chem.*, **32**, 271R (1960).
3. Wiedemann, H. G., *Achema Congress*, Frankfurt, Germany, June 26, 1964.
4. Erdey, L., F. Paulik, and J. Paulik, *Mikrochim. Acta*, **1966**, 699.
5. Paulik, F., J. Paulik, and L. Erdey, *Talanta*, **13**, 1405 (1966).
6. Wiedemann, H. G., and H. P. Vaughan, *Thermochim Acta*, **3**, 355 (1972).
7. Garn, P. D., and J. E. Kessler, *Anal. Chem.*, **32**, 1563 (1960).
8. Forkel, W., *Naturwissenschaften*, **47**, 10 (1960).
9. Newkirk, A. E., *Thermochim. Acta*, **2**, 1 (1971).
10. Paulik, F., and J. Paulik, *Thermochim. Acta*, **4**, 189 (1972).
11. Erdey, L., J. Simon, and S. Gal, *Talanta*, **15**, 653 (1968).
12. Stewart, L. N., *Proceedings of the Third Toronto Symposium on Thermal Analysis*, H. G. McAdie, ed., Chemical Institute of Canada, Toronto, Feb. 25–26, 1969, p. 205.
13. Manche, E. P., and B. Carroll, *Rev. Sci. Instr.*, **35**, 1486 (1964).
14. Norem, S. D., M. J. O'Neill and A. P. Gray, *Proceedings of the Third Toronto Symposium on Thermal Analysis*, H. G. McAdie, ed., Chemical Institute of Canada, Toronto, Feb. 25–26, 1969, p. 221.
15. Norem, S. D., M. J. O'Neill, and A. P. Gray, *Thermochim. Acta*, **1**, 29 (1970).
16. Duval, C., *Anal. Chem.*, **23**, 1271 (1951).
17. Lewin, S. Z., *J. Chem. Educ.*, **39**, A575 (1962).
18. Jacque, L., G. Guiochon, and P. Gendrel, *Bull. Soc. Chim. France*, **1961**, 1061.
19. Saito, H., in *Thermal Analysis*, R. F. Schwenker and P. D. Garn, eds., Academic, New York, 1969, p. 11.
20. Vaughan, H. P., *Am. Lab.*, Jan., 10 (1970).
21. Wendlandt, W. W., *Lab. Management*, Oct., 26 (1965).
22. Wendlandt, W. W., *J. Chem. Educ.*, **49**, A571, A623 (1972).
23. Duval, C., *Inorganic Thermogravimetric Analysis*, Elsevier, Amsterdam, 1953.

24. Duval, C., *Inorganic Thermogravimetric Analysis*, second Ed., Elsevier, Amsterdam, 1963.
25. Garn, P. D., *Thermoanalytical Methods of Investigation*, Academic, New York, 1965, Chap. 10.
26. Keattch, C., *An Introduction to Thermogravimetry*, Heyden, London, 1969.
27. Anderson, H. C., in *Technique and Methods of Polymer Evaluation*, P. E. Slade and L. T. Jenkins, eds., Marcel-Dekker, New York, 1966, Chap. 3.
28. Wendlandt, W. W., in *Handbook of Commercial Scientific Instruments*, C. Veillon and W. W. Wendlandt, eds., Marcel-Dekker, New York, Vol. 2, in press.
29. Saito, H., *Thermobalance Analysis*, Gijitsu Shoin, Tokyo, 1962.
30. Honda, K., *Sci. Repts. Tohoku Univ.*, **4**, 97 (1915).
31. Guichard, M., *Bull. Soc. Chim. France*, **33**, 258 (1923).
32. Vallet, P., *Bull. Soc. Chim. France*, **3**, 103 (1936).
33. Dubois, P., *Bull. Soc. Chim. France*, **3**, 1178 (1936).
34. Chevenard, P., X. Wache, and R. de la Tullaye, *Bull. Soc. Chim. France*, **10**, 41 (1944).
35. Paulik, F., J. Paulik, and L. Erdey, *Z. Anal. Chem.*, **160**, 241 (1958).
36. Gordon, S., and C. Campbell, *Anal. Chem.*, **28**, 124 (1956).
37. Griffith, E. J., *Anal. Chem.*, **29**, 198 (1957).
38. Hooley, J. G., *Can. J. Chem.*, **35**, 374 (1957).
39. Izvekov, I. V., *Trudy Krymsk. Filiala, Akad. Nauk SSSR*, **4**, 81 (1953).
40. Van Norstrand, R. A., U.S. Patent No. 2,692,497, Oct. 26, 1954.
41. Rabatin, J. G., and R. H. Gale, *Anal. Chem.*, **28**, 1314 (1956).
42. Stephenson, J. L., G. W. Smith, and H. V. Trantham, *Rev. Sci. Instr.*, **28**, 380 (1957).
43. Fujii, C. T., C. D. Carpenter, and R. A. Meussner, *Rev. Sci. Instr.*, **33**, 362 (1962).
44. Satava, V., *Collection Czech. Chem. Commun.*, **24**, 2172 (1959).
45. Farmer, R. W., *Thermochim. Acta*, **4**, 203 (1972).
46. Guenot, J., J. M. Manoli, and J. M. Bregeault, *Bull. Soc. Chim. France*, **1969**, 2666.
47. Landsberg, A., *J. Sci. Instr.*, **41**, 337 (1964).
48. Pannetier, J. Guenot, and J. M. Manoli, *Bull. Soc. Chim. France*, **1964**, 2829.
49. Harrison, R. W., and E. J. Delgrosso, *J. Sci. Instr.*, **41**, 222 (1964).
50. Olson, N. J., Air Force Materials Lab. Tech. Report AFML-TR-184, Part 1, June, 1968.
51. Olson, N. J., in *Thermal Analysis*, R. F. Schwenker and P. D. Garn, eds., Academic, New York, 1969, Vol. 1, p. 325.
52. Bremer, H., and G. Henrion, *Chem. Techn.*, **18**, 368 (1966).
53. Franck, R., and M. Harmelin, *Proceedings of the Third Analytical Chemistry Conference*, I. Buzas, ed., Akademiai Kiado, Budapest, 1970, Vol. 2, p. 219.
54. McBain, J. W., and A. M. Bakr, *J. Am. Chem. Soc.*, **48**, 690 (1926).
55. Sarasohn, I. M., and R. W. Tabeling, *Pittsburgh Conference of Analytical Chemistry and Applied Spectroscopy*, March 5, 1964.
56. Williams, H. W., *Thermochim. Acta*, **1**, 253 (1970).
57. Brown, H. A., E. C. Penski, and J. J. Callahan, *Thermochim. Acta*, **3**, 271 (1971).
58. Williams, J. R., E. L. Simmons, and W. W. Wendlandt, *Thermochim. Acta*, **5**, 101 (1972).
59. Chiu, J., *Anal. Chem.*, **39**, 861 (1967).
60. Paulik, J., F. Paulik and L. Erdey, *Anal. Chim. Acta*, **34**, 419 (1966).
61. Gulbransen, E. A., K. F. Andrew, and F. A. Brassart, in *Vacuum Microbalance Techniques*, Plenum, New York, 1965, Vol. 4, p. 127.
62. Feldman, R. F., and V. S. Ramachandran, *Thermochim. Acta*, **2**, 393 (1971).
63. Pedersen, E., *J. Sci. Instr.*, **1**, 1013 (1968).

64. Scott, K. T., and K. T. Harrison, *J. Nucl. Mater.*, **8**, 307 (1963).
65. Bradley, W. S., and W. W. Wendlandt, *Anal. Chem.*, **43**, 223 (1971).
66. Cahn Electrobalance Model RTL, Instructions No. 2006, Cahn Instrument Co.
67. Etter, D., and W. H. Smith, *J. Chem. Educ.*, **49**, 143 (1972).
68. Plant, A. F., *Industrial Res.*, July, 36 (1971).
69. Van Empel, F. J., E. C. Ballegooyen, F. Boersma, and J. A. Poulis, *Thermochim. Acta*, **5**, 129 (1972).
70. Wendlandt, W. W., *Chimia*, **26**, 1 (1972).
71. Ferguson, J. M., P. M. Livesey, and D. Mortimer, International Confederation of Thermal Analysis III, Davos, Switzerland, Aug., 1971, Paper I.
72. Wendlandt, W. W., *Anal. Chim. Acta*, **49**, 185 (1970).

CHAPTER IV

APPLICATIONS OF THERMOGRAVIMETRY

A. Introduction

The method of thermogravimetry is basically quantitative in nature in that the mass-change can be accurately determined. However, the temperature ranges in which the mass-changes occur are qualitative in that they depend upon the instrumental and sample characteristics. With the wide use of commercial thermobalances, TG data of a sample can be correlated from laboratory to laboratory if similar conditions of pyrolysis are employed.

Thermogravimetry is widely used in almost all of the areas of chemistry and allied fields. In the early 1950s it caused a revolution in inorganic gravimetric analysis, and if a similar analogy can be used in the 1960s, the revolution occurred in the field of polymer chemistry. Equally important has been the application of TG techniques to applied science problems such as the characterization of various materials used in road construction, determination of moisture contents in a wide variety of materials, and numerous others. As will be seen, TG is almost universally applied to a large number of analytical problems in the fields of metallurgy, paint and ink science, ceramics, mineralogy, food technology, inorganic and organic chemistry, polymer chemistry, biochemistry, geochemistry, and others.

The application of thermogravimetry to a particular problem is possible if a mass-change is observed on the application of heat. If no mass-change is observed, then other thermal techniques such as DTA, DSC, TMA, and so on, may have to be employed. If the mass-change is very small ($<1\%$), then perhaps other techniques such as evolved-gas analysis (EGA) may be more useful. Mass-changes (generally mass-losses) which can be detected by TG techniques are summarized in Figure IV.1.

Some of the many application of thermogravimetry are listed below:

(1) thermal decomposition of inorganic, organic, and polymeric substances
(2) corrosion of metals in various atmospheres at elevated temperatures
(3) solid-state reactions
(4) roasting and calcining of minerals
(5) distillation and evaporation of liquids
(6) pyrolysis of coal, petroleum, and wood
(7) determination of moisture, volatiles, and ash contents
(8) rates of evaporation and sublimation

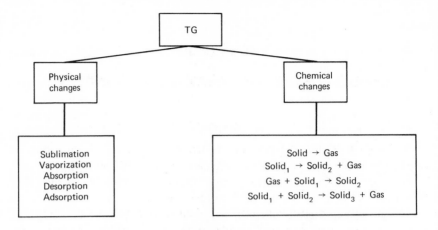

Figure IV. 1. Mass-changes detectable by TG.

(9) dehydration and hygroscopicity studies
(10) automatic thermogravimetric analysis
(11) thermal oxidative degradation of polymeric substances
(12) decomposition of explosive materials
(13) development of gravimetric analytical procedures
(14) reaction kinetics studies
(15) discovery of new chemical compounds
(16) vapor pressure determinations and heats of vaporization

The many applications of thermogravimetry have been reviewed or summarized in numerous reviews and books.   Mention should be made of the older review articles by Gordon and Campbell (1), Duval (2), Coats and Redfern (3), Lukaszewski and Redfern (4), and to the more recent biennial reviews by Murphy (5, 6) and Toursel (7).   Relevant books or book chapters include those by Duval (8, 9), Wendlandt and Smith (10), Wendlandt (11), Anderson (12), Doyle (13), Barrall (14), Liptay (15), Reich and Stivala (16), and others.   The book by Liptay (15) is a compilation of TG curves (and DTG and DTA) obtained by use of the Derivatograph.   Different sample sizes and heating rates for the same sample are also given by means of a clear plastic overlay sheet.

B. Applications to Analytical Chemistry

The principle applications discussed in this chapter are those for analytical determinations of various types.   Duval (2, 8, 9) has discussed these in detail, as have Palei et al. (17); the applications given by the latter are as follows:

(*1*) New weighing compositions in gravimetric analysis and the determination of their temperature stability ranges.

(*2*) For weighing substances which are unstable at ambient temperatures, such as those which absorb CO_2 and H_2O from the air.

(*3*) For studying the behavior of materials in atmospheres of various gases.

(*4*) For determination of the purity and thermal stability of analytical reagents, including primary and secondary standards.

(*5*) For determination of the composition of complex mixtures.

(*6*) For the systematic study of the properties of materials in relation to the methods used for their preparation.

(*7*) For automatic gravimetric analysis.

Duval (2) has also discussed the above topics in addition to others, such as the following:

(*1*) Various filtration techniques such as the ignition of filter paper.

(*2*) Should a precipitate be dried or ignited?

(*3*) Use of the thermobalance for discovery of new methods of separation and in gasometry.

(*4*) The study of the sublimation of various substances.

(*5*) Correction of errors in analytical chemistry.

(*6*) Use of thermogravimetry in functional organic analysis.

1. AUTOMATIC GRAVIMETRIC ANALYSIS

It was Duval (18) who envisioned the ability to determine the amount of a specific metallic ion in 15–20 min with the precision usually attainable in gravimetric procedures and independently of the skill of an operator. He also suggested the possibility of being able to determine simultaneously two or three different ions without having to carry out a preliminary separation. With these ideas as goals, Duval developed the technique known as *automatic gravimetric analysis*, in which a determination for a given metallic ion or mixture of ions could be rapidly carried out using the thermobalance. Selection of the gravimetric method to be used must be based upon the following requirements:

(*a*) Quantitative and immediate precipitation.

(*b*) Immediate filtration with no need to age the precipitate.

(*c*) Immediate drying.

(*d*) Production of a TG constant weight plateau at the lowest temperature possible.

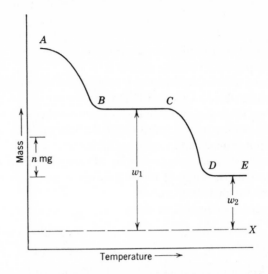

Figure IV. 2. Automatic gravimetric analysis for a single-component system (18).

The principle of this technique, for the single-component system shown in Figure IV.2, is based on the following description. Using a clean and dry crucible, the base line of the thermobalance is determined as indicated by the dashed line X. The crucible is removed from the balance, loaded with the wet precipitate, and then replaced on the balance. The precipitate is heated and the mass-loss curve recorded in the usual manner. From the horizontal mass plateau BC the mass w_1 can be obtained, and from DE the mass w_2 is taken. Since the mass levels indicate that a definite stoichiometry of the precipitate has been attained, multiplication of w_1 or w_2 by the appropriate gravimetric factor gives the mass of metal ion present. The metal-ion content obtained by calculation from w_1 will probably be the most accurate because of the greater accuracy in mass measurement and the smaller gravimetric factor of the precipitate.

For certain precipitates, the entire operation, filtration, drying, and recording on the thermobalance, takes only 12 min (18).

In the case of a binary mixture, the procedure is similar. Take the case of the mixture as illustrated in Figure IV.3. The mass-loss curves for the pure individual components, MX and NY, are given, as well as the curve for a mixture of $MX + NY$. Component MX decomposes from D to E, while NY decomposes from B to C. In the mixture curve, horizontal mass levels are formed at the same temperatures as were present on the two initial component curves. Thus, from the mixture curve, the amount of NY can be obtained by determining the value of BC; the amount of MX

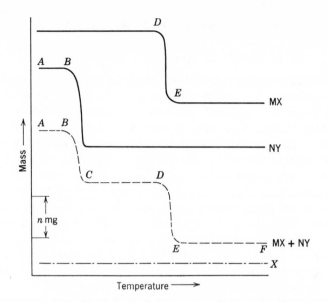

Figure IV. 3. Automatic gravimetric analysis of a binary mixture (18).

from the value of DE. Thus, in one simple operation, the analysis of certain binary or ternary mixtures can be obtained with reasonable accuracy.

Duval (18) used this technique for the analysis of a binary mixture containing calcium and magnesium ions. These two ions were precipitated as oxalates and the TG curve so obtained compared with curves for the individual metal oxalates. If x and y are the mass of calcium and magnesium, respectively, and m and n are the known masses (from the TG curve) of the mixtures present at 500° ($MgO + CaCO_3$) and 900° ($MgO + CaO$), then

$$\frac{100x}{40} + \frac{40.32y}{24.32} = m \tag{IV.1}$$

$$\frac{56x}{40} + \frac{40.32y}{24.32} = n \tag{IV.2}$$

hence,

$$x = \frac{m - n}{1.1} \tag{IV.3}$$

A synthetic mixture containing 0.1541 g of $CaC_2O_4 \cdot H_2O$ and 0.0453 g of $MgC_2O_4 \cdot 2H_2O$ gave $x = 0.0427$ g, compared with a theoretical value of 0.0422 g. A similar determination was carried out for the determination of a copper-silver alloy from a mixture of the metal nitrates.

C. Analytical Determinations Using TG

1. Drying of Analytical Precipitates

One of the first modern applications of the thermobalance to problems in analytical chemistry was the determination of the drying temperatures and weighing forms of analytical gravimetric precipitates. Duval (19) was impressed by the fact that authors were very specific about details concerning the conditions of precipitation, such as concentration of the reagents, volume of reagents, pH of the solution, time of aging precipitate, and other factors, but very vague about drying or pyrolysis temperatures. Such general statements as "ignite to constant weight," "heat not above a dull red," and so on, were entirely inexcusable when it came to gravimetric precipitates. With the aid of 17 collaborators, Duval prepared and heated about 1200 precipitates which had been prepared for use in inorganic gravimetric analysis. Only a small number of these were judged to be suitable for the gravimetric determination of various metal ions, based upon the ease of precipitation and the drying or ignition temperatures.

One of the early TG curves which was used in the gravimetric determination of calcium is that shown in Figure IV.4 (20). As is well known, the curve plateau at temperatures from 25 to 100°C corresponds to the composition for the initial compound; that from 226 to 346°C corresponds to CaC_2O_4; that from 420 to 660°C corresponds to $CaCO_3$; and that from 840 to 980°C corresponds to CaO. Thus, the drying and ignition temperatures and the composition of the compound at any temperature can be determined. The question of drying temperatures and the nonexistence of a TG curve plateau (horizontal) have been discussed by numerous investigators, especially Simons and Newkirk (21). For a multistage thermal decomposition reaction, which would include most of the analytical precipitates

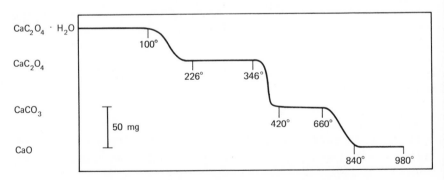

Figure IV.4. TG curve of $CaC_2O_4 \cdot H_2O$ by Peltier and Duval (20).

studied, the following general conclusions can be drawn (21):

(a) The appearance of a plateau for a compound on a TG curve does not necessarily imply that the compound is isothermally stable, either in a thermodynamic or practical sense, at all or any temperatures that lie on that plateau.

(b) If the curve obtained for a multistage reaction has no intermediate portion in which the mass remains constant with time over a range of temperature, one can make the reasonable inference that the reactions leading to the formation and to the subsequent decomposition of the intermediate are not independently sequential, but overlap at least partly.

(c) In the absence of a true plateau, one cannot determine from a curve for successive reactions exact value for either the initial or final temperatures of the plateau (T_i or T_f), or the stoichiometric mass level, although a reasonable inference as to the latter can often be made.

Thus, the transfer of drying or ignition temperatures from a TG curve plateau to isothermal measurements appears to be questionable, although it is a widely used practice.

2. The Drying and Decomposition of Sodium Carbonate

The drying of sodium carbonate is important in the standardization of acids for various types of acidimetric titrations. The recommended drying-temperature range is from 250 to 300°C, although Duval (22) stated that a horizontal mass level was obtained from 100 to 840°C. In a more recent study, Newkirk and Aliferis (23) found that the decomposition temperature of anhydrous sodium carbonate was dependent on the type of crucible container the sample was heated in. The results of this study are illustrated in Figure IV.5 and described in Table IV.1.

When the sodium carbonate was heated in platinum or gold sample holders, the mass-loss was much less rapid and was probably due to the decomposition of the sample to form sodium oxide and carbon dioxide. As seen in curve 6, the presence of a nitrogen gas stream resulted in a faster rate of mass-loss, while when water was present (curve 7) the observed mass-loss rate was less. A sample of sodium carbonate dried at 350°C showed no further mass-change on further heating for 12 h at 600°C and 4 h at 650°C in a platinum crucible in air. The reaction of sodium carbonate with coarse silica sand occurred rapidly at 800–850°C as shown by curve 9. On grinding the silica mixture, the first evidence of mass-loss was at about 500°C (curve 10) or somewhat less that the temperature in curve 9.

It was recommended that sodium carbonate for analytical use be dried by

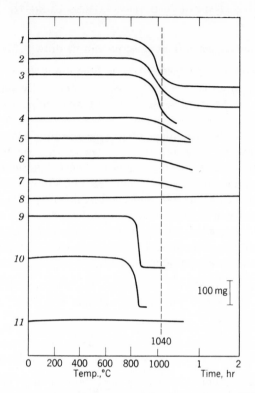

Figure IV.5. Mass-loss curves of sodium carbonate (23).

TABLE IV.1

Curve[a]	Crucible	Atmosphere	Sample
1	Porcelain	Air	Na_2CO_3
2	Porcelain	Air	Na_2CO_3[b]
3	Porcelain	Dry N_2[c]	Na_2CO_3
4	Alumina	Air	Na_2CO_3
5	Platinum	Air	Na_2CO_3
6	Platinum	Dry N_2[c]	Na_2CO_3
7	Platinum	Wet N_2[c]	Na_2CO_3
8	Platinum	CO_2[c]	Na_2CO_3
9	Platinum	Dry N_2[c]	$Na_2CO_3 + SiO_2$
10	Platinum	Dry N_2[c]	$Na_2CO_3 + SiO_2$[d]
11	Gold	Dry N_2[c]	Na_2CO_3[e]

[a] Heating rate $300°C/h$, except runs 9 and 10.
[b] Crucible covered.
[c] Gas flow rate 250 ml/min.
[d] Heating rate $300°C/h$ to $520°C$, then $50°C/h$.
[e] Maximum temperature $922°C$, but sample cooled and held 1 h at $915°C$ after reaching $922°C$.

heating in dry air or carbon dioxide using a platinum or other inert sample container in the temperature range of 250 to at least 700°C.

3. IGNITION TEMPERATURE OF ALUMINUM OXIDE

Although in 1949 Dupuis and Duval (24) studied the pyrolysis of hydrous alumina, $Al_2O_3 \cdot nH_2O$, prepared by using some 25 precipitating agents, more recent works by Erdey and Paulik (25) and Milner and Gordon (26) have raised questions concerning the low ignition temperatures obtained. The minimum temperatures for ignition to Al_2O_3, as found by Dupius and Duval (24), were from 280°C (for bromine) to 1031°C (for aqueous ammonia). Little agreement was found with the above results by Erdey and Paulik (25) in that most of the samples were still losing mass at 1000°C, the maximum temperature of the Derivatograph.   Milner and Gordon (26) recommended that a minimum temperature of 1200°C be used for aluminum oxide precipitates that are to be ignited and weighed by conventional techniques.   The latter conclusion is based partially on the results in Table IV.2.   The results show that the minimum *conventional* ignition temperature is 1 h at 1100°C, following charring of the filter paper.   If the sulfate ion is present, as in the basic sulfate method, an even higher temperature is indicated.

Duval (27) maintains that if the sample is to be heated, cooled and weighed outside of the thermobalance, it is necessary to ignite the aluminum oxide to a higher temperature such that it will not be hygroscopic while it is being cooled

TABLE IV.2

Effect of Ignition Temperatures on the Weights of Aluminum Oxide Precipitates Obtained by Different Methods (26)

Temp.,[a] °C	Percent excess mass over final reference value			
	Method A (urea-basic sulfate method), %	Method B (urea-basic succinate method), %	Method C (urea method),[b] %	Method D (ammonium hydroxide method), %
650	19.2	3.9	3.1	4.5
800	9.8	2.3	1.7	2.4
950	3.4	1.0	1.0	1.2
1100	0.6	0.0	0.0	0.2
1200	0.2	0.0	0.0	0.1
(2nd hour)	Ref. value	Ref. value	Ref. value	Ref. value

[a] After charring of the filter paper, the precipitates were ignited at 500°C for 8 h before being ignited for 1 h at each of the stated temperatures.

[b] Chloride, but not sulfate or succinate, was present.

and weighed on an analytical balance.   It was stated that, using automatic thermogravimetric analysis, this source of error was eliminated, and that the lower temperatures can be employed.   Duval (27), however, does not comment on the different curves presented by Erdey and Paulik (25).   The latter concluded that the internal structure of the hydrous aluminum oxide is determined by such variables as the rate and temperature of precipitation, and only to a small extent by the nature of the precipitant.   They also stated that the lower temperatures reported by Dupuis and Duval (24) were due to the variable precipitation conditions employed, since these are difficult to reproduce.   Even small variations would have a marked effect on the ignition temperatures.

4. Complex Mixtures of Hydrates

Griffith (28) has applied thermogravimetry to the determination of the water content of mixtures of hydrated and anhydrous salts up to six phases. The method is based upon the fact that when a proper rate of heating is employed, a selective decomposition of the phase with the highest dissociation pressure takes place.   When the phase of the highest dissociation pressure is completely decomposed, the compound with the highest dissociation pressure begins to decompose, and so on.   Thus, the water content in two different phases of a mixture of hydrates may be observed by the rate of loss of water from the mixture.

The thermal decomposition curve of a mixture of sodium carbonate

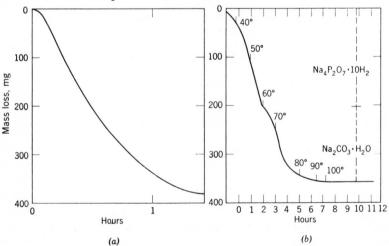

Figure IV.6. (a) Decomposition curve of $Na_4P_2O_7·10H_2O$ and $Na_2CO_3·H_2O$ at 100°C; (b) Same mixture with slowly rising temperature (28).

1-hydrate and sodium pyrophosphate 10-hydrate at a constant temperature of 100°C is given in Figure IV.6a. There is no indication from the curves as to where the $Na_4P_2O_7 \cdot 10H_2O$ was completely dehydrated or where the $Na_2CO_3 \cdot H_2O$ began to dehydrate. The dissociation pressures of these two compounds are very nearly the same at higher temperatures, but at lower temperatures ($\sim$60°C) the differences in dissociation pressures become appreciable. If this mixture is decomposed at 60°C, there is a distinct change of slope at the stoichiometric end point of the curve. In Figure IV.6b the same mixture heated at a slow rate, 8.7°C/h, of temperature rise from 30 to 100°C is illustrated. A pronounced curve break was observed in the curve, which demonstrates the need for a proper choice of rate-of-temperature-rise program.

Other examples presented by Griffith (28) were $Na_5P_3O_{10} \cdot 6H_2O$ and mixtures of $Na_5P_3O_{10}$, $Na_5P_3O_{10} \cdot 6H_2O$, $Na_4P_2O_7$, $Na_4P_2O_7$, $Na_4P_2O_7 \cdot 10H_2O$, Na_2CO_3, and $Na_2CO_3 \cdot H_2O$; $Na_2B_4O_7 \cdot 10H_2O$ and $Na_4P_2O_7 \cdot H_2O$ and $K_2CO_3 \cdot 1.5H_2O$. The average error found in the analysis of the above mixtures was $\pm 1.2\%$.

5. ANALYSIS OF CLAYS AND SOILS

Hoffman et al. (29, 30) studied the thermogravimetry of soils, relatively pure clays, crystalline carbonates, and soils to which known amounts of clays and carbonates were added. Sharp breaks were observed in the decomposition curves of relatively pure clay minerals at elevated temperatures, which suggested the use of thermogravimetry for the determination of pure clays and simple clay mixtures. The quantities that could be determined by this technique were water content, organic matter content, and inorganic carbonates.

The decomposition curves of most soils showed horizontal mass levels starting at 150–180°C and extending to 210–240°C, indicative of either hygroscopic moisture or hygroscopic moisture plus easily volatile organic compounds. In general, the mass-loss values fell between those obtained by the Karl Fischer method and oven drying at 105°C. The organic matter started to burn off between 210 and 240°C and was usually completely burned off at 500°C. In organic soils and those containing less than 15% clay, a relatively close estimate of the organic matter could be made from the mass-loss curve. When the clay content varied from 15 to 40%, the loss in mass at 500°C usually gave an estimate of the organic matter, which was in satisfactory agreement with dry-combustion and wet-oxidation data. When the clays contained more than 40% clay, it was not possible to distinguish between mass-losses due to decomposition of organic matter and those due to

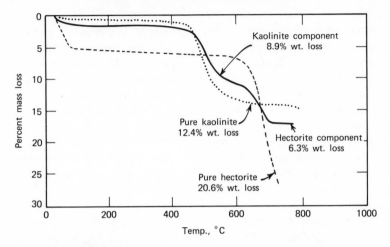

Figure IV.7. TG curves of kaolinite-hectorite mixture (53).

the elimination of the lattice water of clays. This work also suggests that the lattice water in the pure clay samples can be quantitatively determined. Because the lattice water came off at different temperatures with different clays, it may be possible to use these temperatures as an additional means of identification and characterization.

Thermogravimetry was used to determine the composition of a kaolinite-hectorite mixture, as shown in Figure IV.7 (58). The dashed lines are the curves for the pure clays; kaolinite evolved 20.6% water (by a dehydroxyl-ation reaction), while hectorite lost 12.4%. From the TG curve for the mixture, the composition was found to be 70% kaolinite and 30% hectorite. This agreed very well with the theoretical values of 70% and 30%, respectively.

Mulley and Cavendish (31) found that TG could be used to analyze mixtures of calcium hydrogen orthophosphate 2-hydrate (brushite) and anhydrous calcium hydrogen orthophosphate (monetite) based on their loss of water on heating. The water loss on the heating of brushite corresponds to 2.5 moles per mole of compound, while the monetite evolves 0.5 mole per mole of compound. The total mass-loss on heating a mixture of the two compounds is

$$\text{mass-loss due to}$$
$$CaHPO_4 \cdot 2H_2O = b = 0.0662a\% \qquad \text{(IV.4)}$$

and

$$\text{mass-loss due to}$$
$$CaHPO_4 = a = 0.2617b\% \qquad \text{(IV.5)}$$

The total mass-loss is

$$T\% = 0.0662a + 0.2617b \qquad \text{(IV.6)}$$

but $b = (100 - a)$; therefore the total mass-loss is

$$T\% = 26.17 - 0.1955a \qquad (IV.7)$$

From a plot of T versus a, which is a straight line, the composition of the mixture can be read directly off the curve.   By a similar series of calculations, the analysis of three component systems could be carried out.

High- and low-magnesium calcite minerals, in addition to aragonite and normal calcite, were analyzed by several different methods, including TG (32).   The results obtained by TG compared favorably with X-ray diffraction and wet chemical analyses.

Paulik et al. (33) developed a Derivatographic method for the determination of pyrites content in bauxite and clay minerals.   This method could be used to analyze the samples if they contained more than 0.2% pyrite.   Use was made of the DTG curve, rather than the TG curve, for the determination.

The analysis of lunar material by TG techniques is described in Chapter VIII.

6. ANALYSIS OF CALCIUM SILICATE HYDRATES FOR FREE LIME AND CARBONATE

The free lime [$Ca(OH)_2$] and carbonate ($CaCO_3$) contents of calcium silicate hydrates, ranging from 1.0 to 20%, were determined by a thermogravimetric method by Biffen (34).   The mass-loss curves for a series of calcium silicate hydrates, calcium silicate hydrates plus varying amounts of calcium hydroxide, and calcium silicate hydrates plus varying amounts of calcium carbonate are given in Figure IV.8.

The curves for the calcium silicate hydrates are all quite similar, show no sharp breaks, and exhibit a gradual slope for a straight line between 375 and 650°C.   The curve breaks above 600°C are due to the decomposition of carbonate content in the sample.

Using synthetic mixtures of calcium silicate hydrate and calcium hydroxide, the series of curves obtained all indicated curve breaks, at about 500°C, which were caused by calcium hydroxide decomposition as was shown by authentic mass-loss curves for the pure compounds.   By taking the vertical distance from the point at which the straight-line curve starts to change due to evolution of the combined water from the calcium silicate hydrate to the point where it resumes the calcium silicate decomposition drop, and calculating the calcium hydroxide from the loss in mass of water equivalent to this vertical distance, a good estimate of the amount of calcium hydroxide was obtained in all cases.

The decomposition of calcium silicate hydrate samples containing added amounts of calcium carbonate, and in some cases calcium hydroxide, is

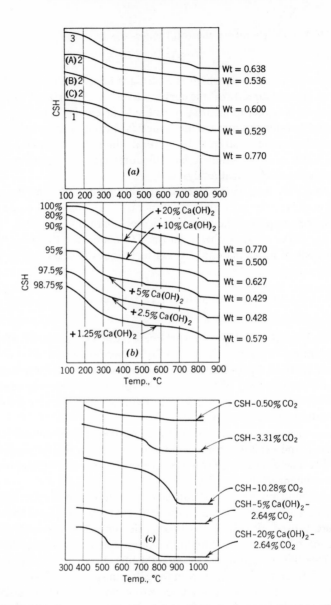

Figure IV.8. Mass-loss curve of: (a) calcium silicate hydrate preparation; (b) calcium silicate hydrate preparation plus added Ca(OH)$_2$; and (c) calcium silicate hydrate preparations plus added amounts of CaCO$_3$ (31).

112

given in Figure IV.8.   The presence of calcium carbonate is indicated by the curve break, due to the evolution of carbon dioxide, in the temperature range 700–900°C.   If a vertical distance is measured between the points where this straight line begins to drop and then becomes horizontal, the carbon dioxide content of the sample can be easily obtained.

Good agreement with other accepted methods was reported for the determination of water, free lime, and carbonate in calcium silicate hydrates by the thermogravimetric method.

Ramachandran (35) determined the $Ca(OH)_2$ content in calcium silicate mixtures by the water-loss between 450 and 550°C.

7. DETERMINATION OF MAGNESIUM, CALCIUM, STRONTIUM, AND BARIUM

The determination of calcium, strontium, and barium ions in the presence of one another has been carried out by thermogravimetry by Erdey et al. (36, 37).   The ions are precipitated in the form of mixed metal oxalate hydrates and decomposed on the thermobalance.   From the resulting mass-loss curves, the amounts of calcium, strontium, and barium can be determined.

The mass-loss curve and its first derivative (see Chap. II) of a mixture of calcium, strontium, and barium oxalate hydrates are shown in Figure IV.9.

From the curve, it can be seen that the decomposition processes are going on independently of one another.   Between 100 and 250°C, the water of hydration is evolved since each ion forms a metal oxalate 1-hydrate.   According to the curves of individual compounds, the water contents are lost in the following order: barium, strontium, and calcium.   However, under the conditions of mixed precipitates, the decomposition of strontium and calcium oxalates hydrates took place simultaneously.

After the loss of the water of hydration, the curve exhibited a horizontal mass level from 250 to 360°C, which corresponded to the composition for anhydrous metal oxalates.   Decomposition of the three oxalates then took place simultaneously, the process being completed at about 500°C.   The anhydrous metal carbonates were then stable from about 500 to 620°C, followed by strontium carbonate, which also began to decompose in this range and was completely decomposed at 1100°C, at which temperature barium carbonate began to decompose.

From the mass-loss curve, then, the following data are obtained: D, mass of dry precipitate at 100°C; E, mass of water of hydration; F, mass of carbon monoxide formed by the decomposition of the anhydrous metal oxalates; G, mass of carbon dioxide formed by the decomposition of calcium

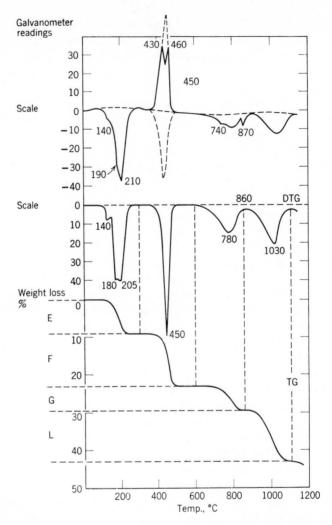

Figure IV.9. Mass-loss curves of calcium, strontium, and barium oxalate hydrates (32).

carbonate; and L, the mass of carbon dioxide formed by the decomposition of strontium carbonate. From these data, the amounts of calcium, C, strontium, S, and barium, B, can be calculated from

Amount of calcium, $C = 0.91068 \cdot G$
Amount of strontium, $S = 1.9911 \cdot L$
Amount of barium, $B = 0.58603 \cdot D - 1.9457 \cdot G - 2.5788 \cdot L$

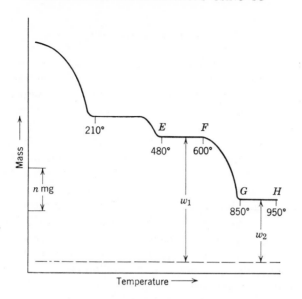

Figure IV.10. Mass-loss curve of a mixture of magnesium and calcium carbonate precipitate (wet) (34).

Assuming that the amounts of C, S, and B are unity, the error of the determination was calculated as

$$\frac{\Delta D}{D} = \frac{\Delta E}{E} = \frac{\Delta F}{F} = \frac{\Delta G}{G} = \frac{\Delta L}{L} = 0.1\% \qquad (IV.8)$$

The simultaneous determination of calcium and magnesium by thermogravimetry has been described by Dupuis and Duval (38). Using the mass-loss curve of a typical dolomite sample, as illustrated in Figure IV.10, the amounts of calcium and magnesium can be calculated. Using the principles previously discussed under automatic thermogravimetric analysis, EF corresponds to a mixture of MgO and $CaCO_3$, and GH corresponds to a mixture of MgO and CaO. The difference, $w_1 - w_2$, is equal to the mass of carbon dioxide evolved between 500 and 900°C by the decomposition of calcium carbonate. The amount of calcium oxide is then given by

$$w(CaO) = (w_1 - w_2) \cdot \tfrac{56}{44} = (w_1 - w_2) \cdot 1.272 \qquad (IV.9)$$

and the amount of magnesium oxide by the difference

$$w(MgO) = w_2 - w(CaO) \qquad (IV.10)$$

The thermal decomposition of pure $BaC_2O_4 \cdot H_2O$ in a nitrogen atmosphere was studied using the combined Derivatograph–thermo-gas-titrimetry technique (39) (see Chap. IX).

8. POTASSIUM HYDROGEN PHTHALATE

Although the use of the thermobalance was supposed to eliminate the confusion concerning drying and decomposition temperatures of analytical precipitates and reagents, in many cases it has only contributed to this confusion. In comparing four different investigations concerning the drying and decomposition temperatures of potassium hydrogen phthalate, four different results were obtained, not to mention the drying temperatures recommended by nonthermogravimetric methods. Dupuis and Duval (40) first reported that the decomposition of $KHC_8H_4O_4$ began at 172°C; Duval (41), in a later study, found a decomposition temperature of 240°C at a 150°C/h heating rate, and 236°C at a 300°C/h heating rate. Belcher et al. (42) reported that the compound began to decompose at 200°C and recommended a drying temperature of 100–150°C. Lastly, Newkirk and Laware (43) reported a procedural decomposition temperature of about 260°C. In view of the previous discussion on the limitations of thermogravimetry, these conclusions are perhaps not unusual. It is believed (43) that these studies have little value in determining the safe, long-term drying temperature for a primary standard substance.

The mass-loss curves of $KHC_8H_4O_4$, under various atmospheric conditions, are given in Figure IV.11.

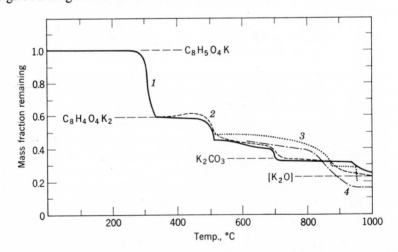

Figure IV.11. Thermal decomposition of 0.1 g samples of $KHC_8H_4O_4$ at a heating rate of 300°C/h, under various conditions: ——, air atmosphere, Pt crucible; – – – –, air atmosphere, porcelain crucible, without protective sleeve on crucible support rod; · · · ·, nitrogen atmosphere, porcelain crucible; — — — —, nitrogen atmosphere, preheated sample to 300°C (38).

There are four major decomposition reactions that take place during the pyrolysis: (1) water and phthalic anhydride volatize and a residue of di-potassium phthalate is formed, $K_2C_8H_4O_4$; (2) the latter compound decomposes to form potassium carbonate and carbonaceous material; (3) the carbonaceous material loses mass slowly and finally burns giving a residue of K_2CO_3; and (4) the potassium carbonate decomposes with the evolution of carbon dioxide, while the K_2O formed reacts with the porcelain crucible sample holder.

The various conditions under which the thermal decomposition takes place are illustrated by the curves in Figure IV.11. The slight mass-increase between 425 and 450°C, noted on the dashed curve, was caused by the evaporation of phthalic anhydride from the furnace walls with increasing temperature and its condensation on the crucible support rod.

Newkirk and Laware (43) found that the initial isothermal rate of decomposition of $KHC_8H_4O_4$ in carbon dioxide at 235°C was about 15 mg/g/h. This compares with an extrapolated value of 7 mg/g/h previously reported. However, Duval (41) found no observable mass-loss from isothermal runs at 150, 160, and 170°C, respectively. If a sample mass of 0.5 g is assumed for his experiments, Newkirk and Laware (43) calculated that the mass-changes expected would be 0.004, 0.011, and 0.030, respectively. Since these changes are too small to be detected by the Chevenard thermobalance, it is not surprising that Duval observed no mass-changes.

9. SOME AMINE MOLYBDOPHOSPHATES

The thermal decomposition of oxine (8-quinolinol), ammonium, and quinolinium phosphomolybdates have been studied by thermogravimetry. Much controversy has existed in the literature concerning the drying or ignition temperature of ammonium 12-molybdophosphate. Dupuis and Duval (44) stated that the precipitate, using Treadwell's method, has the composition $(NH_4)_3PO_4(MoO_3)_{12} \cdot 2HNO_3 \cdot H_2O$. The precipitate loses water and nitric acid on heating to 180°C and then gives a plateau to 410°C. However, if the precipitate is first dried in air, the mass-loss up to 180°C corresponds only to the loss of the 2 moles of nitric acid and not $2HNO_3 + H_2O$, the water apparently being lost on air drying. The horizontal mass level from 180 to 410°C corresponds to the formula $(NH_3)_3PO_4(MoO_3)_{12}$.

Wendlandt (45), using Stockdale's method of precipitation, obtained the mass-loss curves as shown in Figure IV.12. The air-dried precipitate began to evolve loosely held water at 60°C, giving a horizontal mass level from 160 to 415°C, which corresponded to the composition $(NH_4)_2HP(Mo_3O_{10})_4 \cdot H_2O$. Above 415°C, additional mass-loss occurred to give the oxide level,

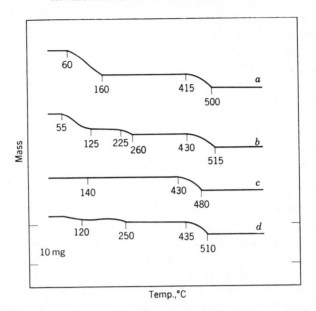

Figure IV.12. Mass-loss of ammonium 12-molybdophosphate (40); (*a*) HNO$_3$ washed, air-dried; (*b*) NH$_4$NO$_3$ washed, air-dried; (*c*) HNO$_3$ washed, oven-dried; (*d*) NH$_4$NO$_3$ washed, oven-dried.

P$_2$O$_5$·24MoO$_3$, beginning at 500°C. The 160°C oven-dried sample gave a similar TG curve except for the evolution of the loosely held water. The ammonium nitrate-washed precipitate, after being air-dried, began to evolve loosely held water at 55°C. From 225 to 260°C, NH$_4$NO$_3$ was evolved, giving a horizontal mass level from 260 to 430°C, which corresponded to the composition (NH$_4$)$_3$ [P(Mo$_3$O$_{10}$)$_4$]. Total decomposition began at 435°C, resulting in the P$_2$O$_5$·24MoO$_3$ mass level at 510°C.

Undoubtedly, the disagreement between various workers is due to the nature of the precipitation reaction, and not to the instrumentation involved. The drying temperatures of the nitric-acid-washed precipitate found by Wendlandt (45), 160–415°C, were similar to the 180–410°C temperatures reported by Dupuis and Duval (44), although the compositions were different. Even with the somewhat different compositions, the molecular weights of the two compounds would be almost identical, so that little error would result no matter which gravimetric factor was employed.

The same problems were encountered with the oxine phosphomolybdate as were found for the corresponding ammonium compound. Duval (46) first reported that the composition corresponding to that of the 2-hydrate was stable in the temperature range 176–225°C; he later reported (47) that the

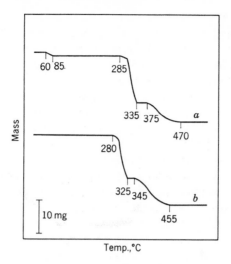

Figure IV.13. Thermal decomposition curves of oxine molybdophosphate; (a) air-dried; (b) oven-dried at 140°C (43).

2-hydrate was stable in the temperature range 236–268°C. Beyond 341°C the compound began to decompose, giving the $P_2O_5 \cdot 24MoO_3$ level beginning at 765°C. No reference was made, nor were supporting data given, for the apparent change in drying-temperature ranges.

Wendlandt and Brabson (48) studied the thermal decomposition of oxine molybdophosphate prepared by the method of Brabson and Edwards (49). The TG curves of this compound are given in Figure IV.13. The air-dried precipitate began to lose mass at 60°C, giving a horizontal mass level from 85 to 285°C, which corresponded to the composition for the anhydrous compound, $3C_9H_7ON \cdot H_3(PMo_{12}O_{40})$. This compound began to decompose at 375°C, resulting in the oxide level, $P_2O_5 \cdot 24MoO_3$, beginning at 470°C. The oven-dried sample decomposed in a manner similar to that of the above compound.

D. Polymers

1. INTRODUCTION

Perhaps the greatest number of applications of thermogravimetry during the past ten years has been in the characterization of polymeric materials. These studies have been useful not only in the applied areas but also in the theoretical aspects of high polymers. Applications of TG include comparisons of the relative thermal stability, the effect of additives on the thermal stability, moisture and additive contents, studies of degradation kinetics, direct quantitative analysis of various copolymer systems, oxidation stability, and many others.   In studies of thermal oxidative degradation (50), TG can reveal the molecular structure and arrangement of repeating units, existence of cross-links between chains, side groups in homopolymer and copolymer chains, and so on.   Rate constants, reaction orders, frequency factors, and activation energies of degradation can also be obtained (16).

A number of methods have been used to classify polymers according to their thermal stability.   As discussed by Fock (50), classification is difficult because of the wide variety of possible thermal reactions at elevated temperatures.   For example, the onset of decomposition may be the degradation of a side chain with the main polymer chain remaining intact; at some higher temperature, further decomposition could occur, resulting in drastic changes in the properties of the material.   One substance may degrade completely in a single step, while a second substance under identical furnace conditions may leave a residue at the upper temperature limit.

Since the usual decomposition temperatures obtained from TG are highly dependent on the experimental procedure employed, Doyle (51) has used the expression "procedural decomposition temperature" as a precaution against mistakenly regarding such trivial data as definitive.   Two types of procedural decomposition temperatures were defined by Doyle (51).   The first of these was called the "differential procedural decomposition temperature" (dpdt) which was used to define the location of "knees" in normalized TG curves. The second type was called "integral procedural decomposition temperature" (ipdt) and was a means of summing up the entire shape of the normalized mass-loss curve.

The ipdt values are determined from a mass-loss curve as follows: The curve, as shown in Figure IV.14, is divided into small squares.   The area under the curve is integrated by weighing a paper cutout of the curve on an analytical balance.   The mass of the crosshatched region in Figure IV.14 divided by the mass of the total rectangular plotting area is the total curve area, A^*, normalized with respect to both residual mass and temperature. The quantity A^* is converted to a temperature T_A^*, by

$$T_A^* = 875A^* + 25 \qquad \text{(IV.11)}$$

In T_A^*, it is presumed that all materials volatilize below 900°C and do so at a single temperature.   Thus, T_A^* represents a characteristic end-of-volatilization temperature, rather than an ipdt having practical significance.   However, it does serve as a measure of refractoriness, but is not very satisfactory.

To put all materials on an equal basis with respect to experimental temperature range, as in A^*, but also with respect to their individual refractory

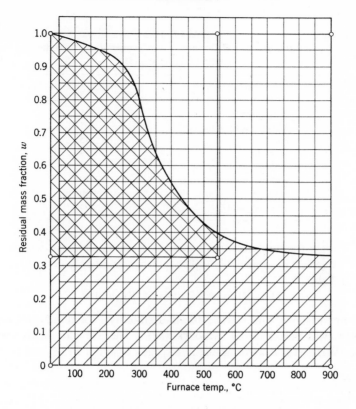

Figure IV.14. TG curve areas, A^* and $\mathbf{R}^*$ (46).

contents, a second curve area, K^*, is the ratio between the doubly cross-hatched area and the rectangular area bounded by the characteristic end-of-volatilization temperature, T_A^*, and the residual mass fraction at the fixed end-of-test temperature of 900°C.

Doyle (46) showed that the product A^*K^* represented a comprehensive index of intrinsic thermal stability for 54 polymers of widely different basic types. It was also shown that by substituting A^*K^* for A^* in equation (IV.11) the ipdt obtained had a practical meaning as a half-volatilization temperature. Unlike ordinary half-volatilization temperatures, defined as the temperature at which half the ultimate volatilization has occurred, the ipdt based on the residual mass fraction of 900°C was appropriate whether decomposition occurred in a single step or in several consecutive steps.

As a quantity derived from curve areas, the ipdt was highly reproducible and its value was only slightly affected by small vagaries or systematic errors in the data curve, especially as contrasted with indices derived on the basis of

TABLE IV.3
Integral Procedural Decomposition Temperatures of Some
Common Polymers (51)

Polymer	ipdt, °C
Polystyrene	395
Maleic-hardened epoxy	405
Plexiglass	345
66 Nylon	419
Teflon	555
Kel-F	410
Viton A	460
Silicone resin	505

residual mass fraction end points alone.   Even small variations in heating
rate do not affect it appreciably.   The ipdt of several polymeric materials are
given in Table IV.3.

2. RELATIVE THERMAL STABILITY

A comparison of the relative thermal stability of a number of different
polymers was described by Newkirk (52).   The TG curves, as shown in
Figure IV.15, were heated rapidly in nitrogen to about 340°C and then more

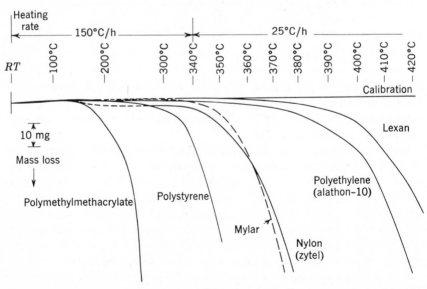

Figure IV.15. Comparison of thermal stabilities of various polymers according to Newkirk
(47).

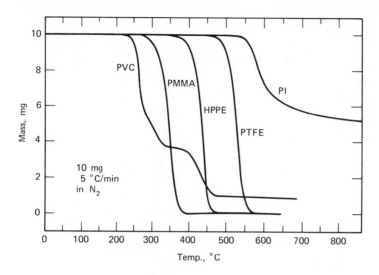

Figure IV.16. Relative thermal stability of polymers by TG (48).

slowly.   An order of stability is readily observable: poly(methyl methacrylate) < polystyrene < Mylar and nylon < polyethylene < Lexan.

In another TG study, Chiu (53) compared the relative thermal stabilities of five polymers, as shown in Figure IV.16.   The five polymers, poly (vinyl chloride) (PVC), poly(methyl methacrylate) (PMMA), high pressure polyethylene (HPPE), polytetrafluoroethylene (PTFE), and an aromatic polypyromellitimide (PI), were all heated under identical conditions in the same thermobalance.   Each polymer showed its characteristic mass-loss curve in a specific temperature region.   This type of information can be used as a guide for further studies on the decomposition mechanisms.

3. Additive Content

Besides its use in thermal stability studies, TG can be used for analysis of additives, either organic or inorganic, in a polymer system.   Analysis of a plasticizer in a polymer is usually tedious by conventional methods, but on many occasions TG provides a rapid determination of the amount of plasticizer present (53).   This is illustrated by the curves in Figure IV.17: the TG curve of unplasticized polyl(vinyl butyral) resin, a resin containing 31 % plasticizer, and the same resin with the plasticizer presumably extracted by n-hexane. Apparently, the first mass-loss is due to volatilization of the plasticizer, and can be used as a measure of its amount.   The precision of this determination can be improved by using a low heating rate or isothermal thermogravimetry.

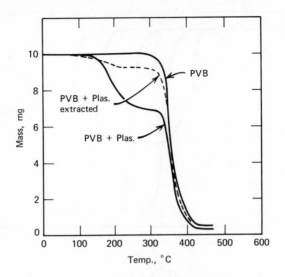

Figure IV.17. Estimation of plasticizer content in poly(vinylbutyrate) (PVB) resins by TG (48).

Light et al. (54) determined the polytetrafluoroethylene (PTFE) (Teflon) content of a silica-filled PTFE polymer by a TG technique.   The polymer was completely volatilized by heating to 600°C in air; in an inert atmosphere only the PTFE was volatilized at this temperature.

4. COMPOSITION OF POLYMER BLENDS AND COPOLYMERS

Numerous TG studies have been made on the characterization of copolymer systems.   In general, the thermal stability of a copolymer falls between those

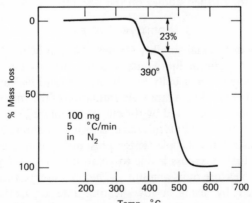

Figure IV.18. TG curve of ethylene-vinyl acetate copolymer (48).

TABLE IV.4
Correlation between TG and Chemical Analyses of Ethylene-Vinyl Acetate Copolymer
(53)

Vinyl acetate, % (chemical)	Mass-loss due to acetic acid, %	Vinyl acetate, % (TG)	Absolute deviation, %
4.3	3.2	4.6	0.3
8.3	5.8	8.3	0.0
11.2	7.6	10.9	0.3
14.9	10.2	14.6	0.3
27.1	18.9	27.1	0.0
31.1	21.7	31.1	0.0

of the two homopolymers, and changes in a regular fashion with the copolymer composition (53). In the case of ethylene-vinyl acetate copolymers, acetic acid is evolved rapidly and quantitatively during the initial stage of thermal decomposition. Only at higher temperatures (in inert atmosphere) do the residual hydrocarbon segments decompose. A typical TG curve of an ethylene-vinyl acetate copolymer is shown in Figure IV.18. The copolymer composition can be estimated from the initial mass-loss. Compared to chemical, infrared, and nmr methods, TG is both rapid and accurate. Typical results for six copolymer samples with vinyl acetate contents ranging from 4.3 to 31.1 % are given in Table IV.4. These results are compared with a chemical saponification method (53).

5. MISCELLANEOUS

The determination of the thermal life rating of a magnet wire enamel by the customary method, ASTM D-2307, is very time-consuming as it requires the heating of several wire samples at each of several temperatures and the periodic testing of each sample until electrical failure occurs. David (55) observed a correlation between the temperature of initial deflection (extrapolated to zero heating rate) of a TG curve of a magnet wire and the $T_{20,000}$ of the wire. Brown et al. (56) demonstrated correlations between TG (and DTA) data and $T_{20,000}$ for 15 commercial magnet wires. The wire samples were heated in a dynamic oxygen atmosphere, and the temperature at which 5% of the enamel mass had been lost correlates well with the $T_{20,000}$ value determined by the standard method. The kinetics of the decomposition was also determined (57).

E. Miscellaneous Inorganic Applications

The value and versatility of TG as an analytical technique when applied to problems of inorganic compounds and materials was demonstrated by

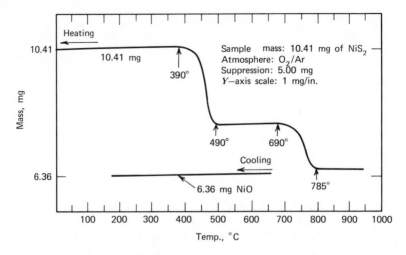

Figure IV.19. TG curve of NiS_2 to determine sulfur content (59).

Williams (59). TG is particularly useful when the sample of interest is in the form of single crystals that are too small to be analyzed by conventional methods. Such samples are quite common as a result of various low-yield reactions or are part of a low-volume multiphase product and are separated physically from the mixture. It was shown that the metal and sulfur contents of single crystals could be determined using about 10-mg samples. Oxygen contents were determined also by oxidation of $SrCrO_3$ to $SrCrO_4$ while $CrVO_4$ was reduced to $CrVO_3$.

As an example of the determination of sulfur, the TG curve of the oxidation of NiS_2 to NiO is shown in Figure IV.19. The sulfur was removed in two temperature regions, 390–490°C and 690–785°C. After cooling the system back to ~ 100°C, a hydrogen-argon atmosphere was introduced and the NiO

TABLE IV.5
Results of Oxygen Determinations by TG (59)

Starting compound	Atmosphere	Result, %	Calc., %	Reaction temperature, °C
$CrVO_4$ (ortho)	H_2–Ar	9.5	9.6	515–630
Cr_2O_5	H_2–Ar	O_2 loss 17.3	17.4	340–430
Fe_3O_4	H_2–Ar	O_2 loss 27.7	27.6	320–520
$SrCrO_3$	O_2–Ar	O_2 loss 8.3	8.5	375–750

TABLE IV.6
Conversion of Rare Earth Sulfides to Oxysulfides by TG (59)

		Net % mass loss[a]	
Starting material	Product	Calc.	Found
Nd_2S_3	Nd_2O_2S	8.4	8.3
Alpha Dy_2S_3	Dy_2O_2S	7.6	7.5
Er_2S_3	Er_2O_2S	7.5	7.3
Yb_2S_3	Yb_2O_3S	7.3	6.9
Pr_2S_3	Pr_2O_2S	8.5	8.5
Ho_2S_3	Ho_2O_2S	7.5	7.1
Dy_2S_3	Dy_2O_2S	7.6	7.4
Ce_2S_3	Ce_2O_2S	8.5	8.5
Y_2S_3	Y_2O_2S	11.7	11.3
La_2S_3	La_2O_2S	8.6	8.4
Tm_2S_3	Tm_2O_2S	7.4	6.9

[a] Loses 2 sulfurs, gains 2 oxygens.

reduced to nickel metal. Total sulfur loss found was 52.1%; the theoretical value was 52.2%.

The results of oxygen determination of various materials are shown in Table IV.5, while the transformations of rare earth sulfides to oxysulfides are summarized in Table IV.6.

A similar approach was used by Wiedemann (60) to determine the titanium content of a nonstoichiometric sample of titanium carbide. The sample is heated in a chlorine gas atmosphere to 975°C; titanium chloride is formed which vaporizes from the sample leaving a residue of carbon. The amount of carbon in the residue is then determined by heating the sample between 475° and 600°C in air. These processes are illustrated in the TG curve in Figure IV.20.

The thermal properties of analytical-grade reagents have been investigated by use of the Derivatograph. Compounds studied include sodium salts (61), rubidium salts (62), cesium salts (63), and ammonium salts (64). High-temperature fusion reactions were also investigated by this technique (65).

F. Applications to Vapor Pressure and Sublimation Determination

The vapor pressure or the sublimation behavior of organic compounds can be determined conveniently using thermogravimetry. Ashcroft (66), using the Langmuir equation

$$m = \alpha \left(\frac{M}{2\pi RT} \right)^{1/2} p \qquad \text{(IV.12)}$$

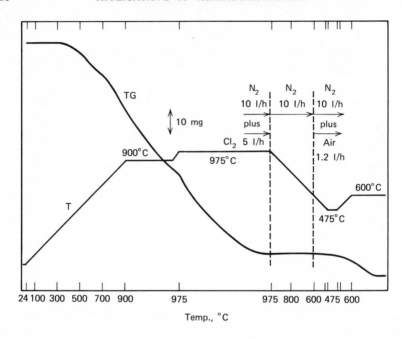

Figure IV.20. TG determination of titanium and carbon in a nonstoichiometric sample of titanium carbide (60).

where M is the molar mass of the gaseous substance, T is the Kelvin temperature, and α is the sublimation coefficient (assumed unity), determined the enthalpy of sublimation of a number of organic compounds and inorganic chelates.   Application of the Clausius-Clapeyron equation to a sublimation process during which the surface area of the sample is constant shows that a plot of log $[m(T)^{1/2}]$ against $10^3/T$ has a slope of $-0.0522\ \Delta H_{sub}$, from which ΔH_{sub} may be calculated in kJ.   Rates of mass-loss of powdered 50–100-mg samples, contained in a platinum boat, were recorded at a series of five or six temperatures over a 20–30° range.   By choosing the temperature to give low rates of mass-loss and low ($<2\%$) overall loss, good straight-line plots were obtained from which the slopes, as calculated by the least-squares method, were reproducible to about 5%.   Enthalpies of sublimation obtained by this method are shown in Table IV.7.   There was good agreement between the two series of values for the first five compounds, although some variations may be expected from the different temperature ranges of measurement.

The determination of the vapor pressure of various compounds was reviewed by Wiedemann (67).   He discussed the determination of vapor pressure by TG techniques based on the Knudsen effusion method.   The sample holder that was employed is illustrated in Chap. III (Figure III.6).

TABLE IV.7
Enthalpies of Sublimation of Various Compounds in kJ.mole^{-1} (66)

Compound[a]	ΔH_{sub}	Temp. range, °K	Literature ΔH_{sub}	Lit. temp. range, °K
Anthraquinone	105.9	335–356	112.0	298
			127	470–590
			106.2	428
1,4-Dihydroxyanthraquinone	94.5	324–351	103.6	408
1,8-Dihydroxyanthraquinone	96.5	335–356	105.9	405
1-Aminoanthraquinone	90.9	361–386	113.8	461
Benzoic acid	89.1	299–329	91.5	343–387
			100	420–480
Thymol	69.0	229–312	91.3	273–313
			67.0	420–480
Sc^{III} (acac)$_3$	97.2	335–361	169	445–555
			99.6	389
			49.8	377–387
Cr^{III} (acac)$_3$	85.9	335–356	125	490–595
			110.9	397
			27.8	389–397
Mn^{III} (acac)$_3$	117.3[b]	335–356	77.8	383–391
Fe^{III} (acac)$_3$	114.9	335–356	116	452–535
			99.0	391
			65.3	378–388
			23.4	393
Co^{III} (acac)$_3$	86.3	335–361	74.9	378–393
Cu^{II} (acac)$_2$	106.1	335–361	57.3	475–560

[a] acac refers to acetylacetonate.
[b] Probably a maximum value.

For some measurements, a Pyrex glass cell having a diameter of about 15 mm was used. Four organic compounds were studied: p-chlorophenyl-N', N'-dimethyl urea (Monuron, a herbicide), p-phenacetin, anthracene, and benzoic acid, in the temperature range of 250–400°K. The vapor-pressure curves of these compounds, in the range from 0 to 10^{-6} Torr, are shown in Figure IV.21. The ΔH_s values calculated were: Monuron, 27.4; p-phenacetin 27.6; anthracene, 20.1; and benzoic acid, 20.7 kcal/mole.

Adonyi (68) developed a method in which the Derivatograph could be used for vapor-pressure determinations. The method is similar to that discussed by Ashcroft (66).

G. Miscellaneous Applications

Probably the most important characteristic of military and commercial explosives and solid rocket propellants is performance as related to end use

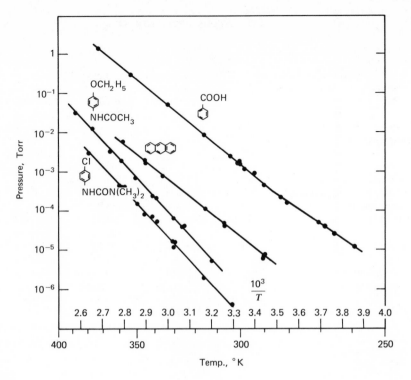

Figure IV.21. Vapor pressure curves obtained by TG using the Knudsen effusion method (67).

TABLE IV.8
Physical Processes and Method of Investigation (69)

Physical process	Indicated by TG Increase	Decrease	DTA Endothermic	Exothermic	Conventional thermal decomposition rate
Melting	—		+		
Sublimation		+	+		+
Boiling		+	+		+
Phase transformation	—		+	+	+
Desorption		+	+		+
Adsorption	+			+	+
Solid reaction	—		+	+	—
Decomposition		+	+	+	+
Explosion		+		+	+
Oxidation, reduction	+	+	+	+	+

Key: + process is indicated; — process is not indicated.

130

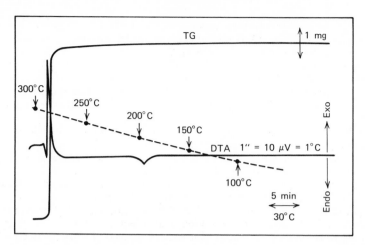

Figure IV.22. TG–DTA curves for 15 mg of β-HMX in a dynamic He atmosphere at a heating rate of 6°C/min (69).

and safety. Performance can be described by a variety of conventional properties such as thermal stability, shock sensitivity, friction sensitivity, explosive power, burning, or detonation rate, and so on. Thermal analysis methods, according to Maycock (69), show great promise for providing information on both these conventional properties and other parameters of explosive and propellant systems. The thermal properties have been determined mainly by TG and DTA techniques, and isothermal or adiabatic constant-volume decomposition. Physical processes in pseudostable materials, which may be observed by DTA, can often be interpreted only with a knowledge of TG or gas evolution data. A selection of the physical processes and the methods used for their investigation is shown in Table IV.8.

The correct interpretation of the large amount of experimental data obtained by these techniques is a problem not to be taken lightly, since many important decisions relative to the use of a particular material may be based on these interpretations. It is obvious that these techniques can be used both for quality control and as an approach to the basic science of pseudostable materials.

A typical TG (and DTA) curve of the explosive, β-HMX, is shown in Figure IV.22. The TG curve indicates that a meaningful range which can be used for isothermal mass studies is from 220 to 280°C. From the latter, kinetics data may easily be obtained.

References

1. Gordon, S., and C. Campbell, *Anal. Chem.*, **32**, 271R (1960).
2. Duval, C., *Anal. Chem.*, **23**, 1271 (1951).

3. Coats, A. W., and J. P. Redfern, *Analyst*, **88**, 906 (1963).
4. Lukaszewski, G. M., and J. P. Redfern, *Lab. Pract.*, **10**, 552 (1961).
5. Murphy, C. B., *Anal. Chem.*, **42**, 268R (1970).
6. Murphy, C. B., *Anal. Chem.*, **44**, 513R (1972).
7. Toursel, W., *Z. für Chemie*, **7**, 265 (1967).
8. Duval, C., *Inorganic Thermogravimetric Analysis*, Elsevier, Amsterdam, 1953.
9. Duval, C., *Inorganic Thermogravimetric Analysis*, Second Ed., Elsevier, Amsterdam, 1963.
10. Wendlandt, W. W., and J. P. Smith, *Thermal Properties of Transition Metal Ammine Complexes*, Elsevier, Amsterdam, 1967.
11. Wendlandt, W. W., in H. A. Flaschka and J. A. Barnard, *Chelates in Analytical Chemistry*, Marcel-Dekker, New York, 1967, Vol. 1, Chap. 5.
12. Anderson, H. C., in *Techniques and Methods of Polymer Evaluation*, P. E. Slade and L. T. Jenkins, eds., Marcel-Dekker, New York, 1966, Chap. 3.
13. Doyle, C. D., in *Techniques and Methods of Polymer Evaluation*, P. E. Slade and L. T. Jenkins, eds., Marcel-Dekker, New York, 1966, Chap. 4.
14. Barrall, E. M., in *Guide to Modern Methods of Instrumental Analysis*, T. H. Gouw, ed., Wiley-Interscience, New York, 1972, Chap. 12.
15. Liptay, G., ed., *Atlas of Thermoanalytical Curves*, Akademiai Kiado, Budapest, 1971, Vol. 1.
16. Reich, L., and S. S. Stivala, *Elements of Polymer Degradation*, McGraw-Hill, New York, 1971.
17. Palai, P. N., I. G. Sentyurin, and I. S. Sklyarenko, *Zh. Anal. Khim.*, **12**, 329 (1957).
18. Duval, C., Ref. 9, p. 84.
19. Duval, C., Ref. 9, Chap. VIII.
20. Peltier, S., and C. Duval, *Anal. Chim. Acta*, **1**, 345 (1947).
21. Simons, E. L., and A. E. Newkirk, *Talanta*, **11**, 549 (1964).
22. Duval, C., *Anal. Chim. Acta*, **13**, 32 (1955).
23. Newkirk, A. E., and I. Aliferis, *Anal. Chem.*, **30**, 982 (1958).
24. Dupuis, T., and C. Duval, *Anal. Chim. Acta*, **3**, 191 (1949).
25. Erdey, L., and F. Paulik, *Acta Chim. Acad. Sci. Hung.*, **7**, 45 (1955).
26. Milner, O. I., and L. Gordon, *Talanta*, **4**, 115 (1960).
27. Duval, C., Ref. 8, pp. 227–228.
28. Griffith, E. J., *Anal. Chem.*, **29**, 198 (1957).
29. Hoffman, I., M. Schnitzer, and J. R. Wright, *Chem. Ind.* (London), **1958**, 261.
30. Hoffman, I., M. Schnitzer, and J. R. Wright, *Anal. Chem.*, **31**, 440 (1959).
31. Mulley, V. J., and C. D. Cavendish, *Analyst*, **95**, 304 (1970).
32. McCaleb, S. B., *Quest* (Sun Oil Co.), 24 (1966).
33. Paulik, F., S. Gal, and L. Erdey, *Anal. Chim. Acta*, **29**, 381 (1963).
34. Biffen, F. M., *Anal. Chem.*, **28**, 1133 (1956).
35. Ramachandran, V. S., *Thermochim. Acta*, **2**, 41 (1971).
36. Erdey, L., F. Paulik, G. Svehla, and G. Liptay, *Z. Anal. Chem.*, **182**, 329 (1961).
37. Erdey, L., F. Paulik, G. Svehla, and G. Liptay, *Talanta*, **9**, 489 (1962).
38. Dupuis, T., and C. Duval, *Mikrochim. Acta*, **1958**, 186.
39. Paulik, F., J. Paulik, and L. Erdey, *Anal. Chim. Acta*, **44**, 153 (1969).
40. Dupuis, T., and C. Duval, *Chim. Anal.*, **33**, 189 (1951).
41. Duval, C., *Anal. Chim. Acta*, **13**, 32 (1955).
42. Belcher, R., L. Erdey, F. Paulik, and G. Liptay, *Talanta*, **5**, 53 (1960).
43. Newkirk, A. E., and R. Laware, *Talanta*, **9**, 169 (1962).

44. Dupuis, T., and C. Duval, *Anal. Chim. Acta*, **4**, 256 (1950).
45. Wendlandt, W. W., *Anal. Chim. Acta*, **20**, 267 (1959).
46. Duval, C., *Inorganic Thermogravimetric Analysis*, Elsevier, Amsterdam, 1953, pp. 130, 132.
47. Duval, C., Ref. 9, pp. 245–246.
48. Wendlandt, W. W., and J. A. Brabson, *Anal. Chem.*, **30**, 61 (1958).
49. Brabson, J. A., and O. W. Edwards, *Anal. Chem.*, **28**, 1485 (1956).
50. Fock, J., *Some Applications of Thermal Analysis*, Mettler Instrument Corp., 1968.
51. Doyle, C. D., *Anal. Chem.*, **33**, 77 (1961).
52. Newkirk, A. E., *Proceedings at the First Toronto Symposium on Thermal Analysis*, H. G. McAdie, ed., Chemical Institute of Canada, Toronto, 1965, p. 33.
53. Chiu, J., in *Thermoanalysis of Fiber and Fiber-Forming Polymers*, R. F. Schwenker, ed., Interscience, New York, 1966, p. 25.
54. Light, T. S., L. F. Fitzpatrick, and J. P. Phaneuf, *Anal. Chem.*, **37**, 79 (1965).
55. David, D. J., *Insulation*, **13**, 38 (1967).
56. Brown, G. P., D. T. Haarr, and M. Metlay, *Proceedings of the 9th Electronics Inlation Conference, IEEE* 32C3-23 (1965), p. 160.
57. Brown, G. P., D. T. Haarr, and M. Metlay, *Thermochim. Acta*, **1**, 441 (1970).
58. DuPont Application Brief, No. 18, Feb. 1, 1968.
59. Williams, H. W., *Thermochim. Acta*, **1**, 253 (1970).
60. Wiedemann, H. G., Mettler Technique Series, *Tech. Bull.* No. T-103.
61. Erdey, L., J. Simon, S. Gal, and G. Liptay, *Talanta*, **13**, 67 (1966).
62. Erdey, L., G. Liptay and S. Gal, *Talanta*, **12**, 883 (1965).
63. Erdey, L., G. Liptay, and S. Gal, *Talanta*, **12**, 257 (1965).
64. Erdey, L., G. Liptay, and S. Gal, *Talanta*, **11**, 913 (1964).
65. Erdey, L., and S. Gal, *Talanta*, **10**, 23 (1963).
66. Ashcroft, S. J., *Thermochim. Acta*, **2**, 512 (1971).
67. Wiedemann, H. G., *Thermochim. Acta*, **3**, 355 (1972).
68. Adonyi, Z., *Periodica Polytech.*, **10**, 325 (1966).
69. Maycock, J. N., *Thermochim. Acta*, **1**, 389 (1970).

CHAPTER V

DIFFERENTIAL THERMAL ANALYSIS AND DIFFERENTIAL SCANNING CALORIMETRY

A. Differential Thermal Analysis

1. INTRODUCTION

Differential thermal analysis (DTA) is a thermal technique in which the temperature of a sample, compared with the temperature of a thermally inert material, is recorded as a function of the sample, inert material, or furnace temperature as the sample is heated or cooled at a uniform rate. Temperature changes in the sample are due to endothermic or exothermic enthalpic transitions or reactions such as those caused by phase changes, fusion, crystalline structure inversions, boiling, sublimation, and vaporization, dehydration reactions, dissociation or decomposition reactions, oxidation and reduction reactions, destruction of crystalline lattice structure, and other chemical reactions. Generally speaking, phase transitions, dehydration, reduction, and some decomposition reactions produce endothermic effects, whereas crystallization, oxidation, and some decomposition reactions produce exothermic effects.

The temperature changes occurring during these chemical or physical changes are detected by a *differential* method, such as is illustrated in Figure V.1. If the sample and reference temperatures are T_s and T_r, respectively, then the difference in temperature, $T_s - T_r$, is the function recorded. Perhaps a better name for this technique would be *differential thermometry*; the term "differential thermal analysis" implies that it has something to do with analysis, which, as with any other analytical technique, may or may not be the case. In *thermal analysis* (another misnomer?), the temperature of the sample, T_s, is recorded as a function of time (see Chap. X), and a heating or cooling curve is recorded. Small temperature changes occurring in the sample are generally not detected by this method. In the *differential* technique, since the detection thermocouples are opposed to each other, small differences between T_s and T_r can be detected with the appropriate voltage amplification devices. Thus, small samples (down to several μg in mass) may be employed and are, as a matter of fact, more desirable.

A comparison between the two techniques is shown in Figure V.2. In parts (*a*) and (*b*), the sample temperature is recorded as a function of time as

134

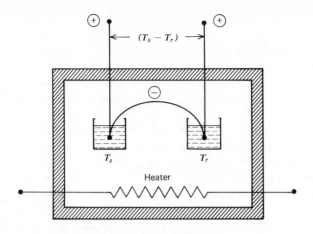

Figure V.1. Basic DTA system.

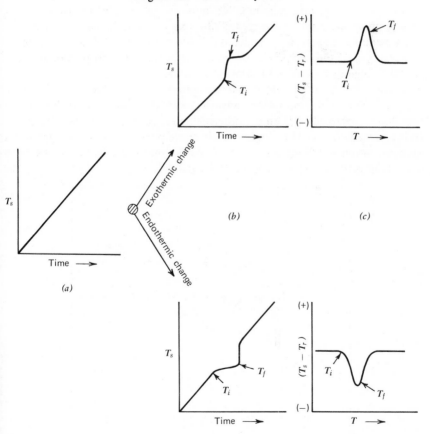

Figure V.2. Comparison between thermal analysis and differential thermal analysis.

the system temperature is increased at a linear rate.   However, the difference between the curves in (a) and (b) is that no enthalpic transition take place in the sample in (a), while in (b) exothermic and endothermic changes occur. Since no other temperature changes take place in the sample in (a), no deviation from the linear temperature rise is detected in the sample temperature.   However, in (b), deviations occur at the procedural initial reaction temperature, T_i, due to temperature changes caused by endothermic or exothermic changes.   These changes are essentially completed at T_f and the temperature of the sample returns to that of the system.   In the curves in (c), the difference in temperature, $T_s - T_r$, is recorded as a function of system temperature, T.   At T_i, the curve deviates from a horizontal position to form a peak in either the upward or the downward direction, depending upon the enthalpic change.   The completion of the reaction temperature, T_f, does not occur at the maximum or minimum of its curve but rather at the high-temperature side of the peak.   Its exact position depends upon the instrumental arrangement.   Thus, in the differential method, small temperature changes can be easily detected while the peak area is proportional to the enthalpic change ($\pm \Delta H$) and sample mass.

A typical DTA curve is illustrated in Figure V.3.   Four types of transitions are illustrated: (I) second-order transition in which a change in the horizontal base line is detected; (II) an endothermic curve peak caused by a fusion or melting transition; (III) an endothermic curve peak due to a decomposition or dissociation reaction; and (IV) an exothermic curve peak caused by a crystalline phase change.   The number, shape, and position of the various

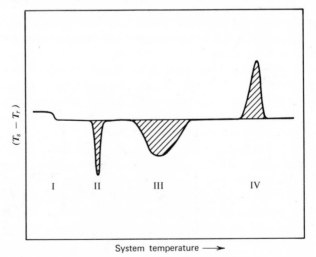

Figure V.3. Typical DTA curve.

endothermic and exothermic peaks with reference to the temperature may be used as a means for the qualitative identification of the substance under investigation. Also, since the area under the peak is proportional to the heat-change involved, the technique is useful for the semiquantitative or, in some cases, quantitative determination of the heat of reaction. As the heat of reaction is proportional to the amount of reacting substance, DTA can be used to evaluate quantitatively the amount of substance present if the heat of reaction is known. Thus, the technique finds much use in the qualitative and semiquantitative identification of organic and inorganic compounds, clays, metals, minerals, fats and oils, polymeric materials, coal and shales, wood, and other substances. It can also be used to determine the radiation damage of certain polymeric materials, the amount of radiation energy stored in various minerals, heats of adsorption, effectiveness of catalytic materials, heats of polymerization, and others. Quantitatively, it can be used for the determination of a reactive component in a mixture, or the heat of reaction involved in physical or chemical changes.

2. HISTORICAL ASPECTS

Some mention should be made of the historical aspects of DTA, which represents an interesting example of a multiscientific discipline approach to the development of a new experimental technique. Although Le Chatelier (1) has been credited as being the father of this technique, his early work in 1887 on clays and minerals consisted of thermal analysis determinations, employing a single thermocouple immersed in the sample. The thermocouple was embedded in the clay sample which was heated at about 100°C/ min. Response from a galvanometer was measured by reflection of flashes from an induction coil from the galvanometer mirror to a photographic plate. The temperature of the sample was then displayed on the developed photographic plate as a series of lines, each of which corresponded to a spark discharge which occurred at intervals of 2 sec. An endothermic reaction was indicated by closely spaced lines, while wider spacing indicated an exothermic reaction. Accordingly (2), Le Chatelier was the father of thermal analysis but not of differential thermal analysis.

Le Chatelier and other investigators, Ashley (3), Wholin (4), Rieke (5), Wallach (6), and Mellor and Holdcroft (7) studied the thermal changes which took place in a substance as it was heated by recording its temperature, as measured with a thermocouple, as a function of time. Breaks in the "heating curve" were thus obtained, indicative of dehydration, decomposition, phase transitions, and other reactions initiated by heating. This heating-curve method was not very sensitive to small heat effects and was adversely affected by changes in heating rate, recording equipment, and other factors.

In 1899, Roberts-Austen (8) suggested that a two-thermocouple system be employed. One of the thermocouples was placed in the sample and the other in a reference block in the furnace. Thus, the differential temperature reading, which was more sensitive to small temperature changes in the sample than the single-thermocouple method, was recorded or plotted as a function of time or temperature. In a much-neglected paper, Burgess (9) in 1908 discussed the merits of the single- and double-thermocouple systems, as applied to cooling-curve data. He discussed the representation of cooling-curve data by plots of sample temperature, T, versus time, t; difference in temperature between T and the temperature of a neutral body, T', or $T - T' = \Delta T$, in which ΔT was plotted against t; T versus dT/dt; and T versus dt/dT. These various representations were interpreted for three kinds of transformations: (a) those in which the substance remains at a constant temperature; (b) those in which the substance cools at a reduced rate, which may or may not be constant over a portion of the transformation; and (c) those in which the substance undergoes an increase in temperature during the first part of the transformation. Burgess also discussed the various types of experimental arrangements and recording systems known at that time. Equations were also presented for the calculation of the heats of transformation that were observed.

Perhaps the first application of the differential thermocouple method to the study of a chemical problem was made by Houldsworth and Cobb (11), although Fenner (12) had previously studied phase transitions in silicate minerals. The former investigators studied the behavior of fire clays and bauxite on heating, a field of investigations in which DTA was to become an important tool. Since this first paper, numerous investigations on the thermal decomposition of various clays and minerals have appeared in the literature; included among them are studies by Norton (17), Grim (18), Berkelhamer (13), Kerr and Kulp (14), Kauffman and Dilling (15), Foldvari-Vogl (16), and Mackenzie (19). DTA is now accepted as a routine tool for the qualitative and semiquantitative determination of a large number of clays and minerals.

The pioneering efforts of the clay mineralogists and metallurgists kept the technique of DTA alive both experimentally and theoretically. As far as chemistry is concerned, the renaissance of DTA development and application occurred during the early 1950s as a result of the work of Stone (20) and Borchardt and Daniels (21, 22). Stone is credited with the development of the first modern, high-quality commercial DTA instrumentation, which served as a stimulus to further developments in this badly neglected area. Borchardt, in his Ph.D. thesis at the University of Wisconsin, applied the DTA technique to problems in inorganic and physical chemistry. His work on the determination of homogeneous reaction kinetics by DTA is a classic investigation in differential thermal analysis.

In the 1960s, due to the availability of commercial instrumentation, the DTA technique was applied with great vigor to problems in polymer chemistry. This technique, aided by the newly developed technique (1963) of differential scanning calorimetry (DSC), was uniquely suited for the elucidation of the thermal properties of polymers. Indeed, many of the analytical uses of DTA and DSC at the present time are in the field of polymer chemistry.

As would be anticipated in a rapidly developing research technique, there are a number of reviews on DTA and its application to chemical problems. It is impossible to list all of the reviews here, but mention should be made of the biennial reviews since 1958 (25) by Murphy, the last one of which appeared in 1972 (26). Recent books and/or book chapters on DTA include those by Smothers and Chiang (27, 28), Wendlandt (23, 24), Garn (29), Mackenzie (30), Gordon and Campbell (32, 33), Kissinger and Newman (31), Barrall and Johnson (34), David (35), Barrall (36), Schultze (37), Ramachandran (38), Wunderlich (39), Porter and Johnson (40, 41), and Schwenker and Garn (42).

3. THEORETICAL ASPECTS

There have been a number of different theories concerning the theoretical interpretation of the DTA curve. All of the theories relate, in some manner, the area of the differential curve peak to the various parameters of the sample and apparatus. The equations representing these parameters were developed through the use of conventional heat-transfer relationships and the geometry of the sample and sample holder. The derivation of each of these theories is beyond the scope of this discussion, so only the final mathematical expressions will be presented.

In the theory developed by Speil et al. (10) and modified by Kerr and Kulp (14), the area enclosed by the differential curve is

$$\frac{m(\Delta H)}{gk} = \int_{t_1}^{t_2} \Delta T \, dt \qquad (V.1)$$

where m is the mass of reactive sample, ΔH is the heat of reaction, g is a geometrical shape constant for the apparatus, k is the thermal conductivity of the sample, ΔT is the differential temperature, and t_1 and t_2 are the integration limits of the differential curve. This expression is perhaps one of the simplest and relates the heat of reaction of the sample to the peak area through use of the proportionality constants or near-constants, g and k. It neglects the differential terms and the temperature gradients in the sample and also considers the peak area to be independent of the specific heat of the sample.   It is basically only an approximate relationship.

Vold (43) derived the expression

$$\frac{\Delta H}{C_s}\left(\frac{df}{dt}\right) = \left(\frac{dy}{dt}\right) + A(y - y_s) \tag{V.2}$$

where C_s is the heat capacity of the cell plus its contents, f is the fraction of the sample transformed at any time t, y is the differential temperature, y_s is the steady-state value of the differential temperature achieved a sufficiently long time after the initial condition $y = y_1$ at $t = t_1$, and A is a constant.

The inherent limitations of this theory are: (a) the assumption of a constant value of the heat capacity of the sample, and (b) the assumption that the sample temperature is uniform throughout at each time instant.   The heat capacity of the sample is that of the cell plus that of the transformed amount of the sample plus that of the untransformed amount.   These amounts change during the course of the reaction.   Thus, in practice, if the heat capacity of the cell is made large, this fluctuation is considered minor, although sensitivity is reduced.   The nonuniformity of the sample temperature is not considered important enough to vitiate the method, although it does affect the transformation temperature rather than the calculation of the heat effects.   Reduction of the heating rate, measurement of the sample temperature at its outside surface nearest the furnace wall, and various extrapolation procedures all reduce the error but do not eliminate it entirely (43).

Using a sample block constructed from an infinitely-high-thermal-conductivity metal such as nickel, in which the sample holder geometry is a cylinder, Boersma (44) found that the peak area was equal to

$$\int_{t_1}^{t_2} \Delta T \, dt = \frac{qa^2}{4\lambda} \tag{V.3}$$

where t_1 and t_2 are the times at the beginning and end of the peak, q is the heat of transformation per unit volume, ΔT is the differential temperature, a is the radius of the cavity filled with sample, and λ is the thermal conductivity of the sample material.

For a spherical metal sample container

$$\int_{t_1}^{t_2} \Delta T \, dt = \frac{qa^2}{6\lambda} \tag{V.4}$$

and for a one-dimensional case of a flat plate

$$\int_{t_1}^{t_2} \Delta T \, dt = \frac{qa^2}{2\lambda} \tag{V.5}$$

Lastly, for an infinitely large ceramic block, there are no finite solutions for the one- and two-dimensional cases; however, there is a solution for a spherical holder:

$$\int_{t_1}^{t_2} \Delta T \, dt = \frac{qa^2}{6}\left(\frac{2}{\lambda_c} + \frac{1}{\lambda_s}\right) \tag{V.6}$$

where λ_c is the thermal conductivity of the ceramic material and λ_s is the thermal conductivity of the sample.

In the above equations as applied to a conventional DTA apparatus, the sample is used for two entirely different purposes: (a) as a producer of heat, and (b) as a heat measuring resistance in which the flow of heat develops a temperature difference to be measured. To separate these two functions, Boersma (44) recommended the use of metal sample and reference cups in which the temperature difference was measured from outside the sample and reference materials. The peak area then depended on the heat of reaction by

$$\int_{t_1}^{t_2} \Delta T \, dt = \frac{mq}{G} \tag{V.7}$$

where m is the mass of the sample, and G is the heat-transfer coefficient between the nickel cup and the surrounding nickel shield. Although no data were presented, Boersma (44) claimed that measurements on the dehydration of $CuSO_4 \cdot 5H_2O$ confirmed equation (V.7) quantitatively. The samples were said to differ widely in packing density.

Lukaszewski, in a series of eleven papers, discussed the complex heat transfer problem in various types of DTA systems (45–55). These problems were simplified into three categories (53):

(a) Heat transfer between the heat source (furnace wall or heater) and the block calorimeter by conductive, convective, and radiative mechanisms.

(b) Heat conduction between the block calorimeter and some medium within it (reference or sample materials).

(c) The active sample in the system may periodically undergo heat-absorbing (endothermic) or heat-generating (exothermic) phenomena as functions of time, temperature, and position in the medium. These involve complex heat transfer between the sample and the calorimeter under conditions where the physical properties of the sample are undergoing rapid change.

The main problems, those of (b) and (c), can be represented mathematically as

$$C_s \varphi_s \left(\frac{\partial T}{\partial t}\right)_s = \text{div } k_s \text{ grad } T \pm A_s(P, t) \qquad (V.8)$$

where C_s and φ_s are the specific heat and density of sample, respectively; the heat absorption or generation term can be represented as

$$A(P, t) = Qb(1 - \alpha)^n \qquad (V.9)$$

where Q is the heat of reaction, b is the velocity constant $(Z \exp (-E/RT)$, α is the fraction of sample transformed, and n is the order of reaction. Imposing the condition that k_s is position- and temperature-independent, equation (V.8) reduces to

$$\left(\frac{\partial T}{\partial t}\right)_s = d_s \nabla^2 T \pm \frac{A_s(P, t)}{C_s} \qquad (V.10)$$

where d_s is the thermal diffusivity of the sample and C_s is the heat capacity of the sample per unit volume. Similarly, for the reference material, which exhibits no heat absorption or generation effects [so that $A(P, t) = 0$], equation (V.10) becomes

$$\left(\frac{\partial T}{\partial t}\right)_r = d_r \nabla^2 T \qquad (V.11)$$

where d_r is the thermal diffusivity of the reference material.

The heat transfer problem for a DTA system containing ring thermocouples has been treated by David (56, 57). In order to obtain a mathematical expression for C_p, the heat capacity of the sample (or reference), two factors must be considered: (a) the effects of the system upon the differential thermocouple, and (b) the effects of the system plus sample upon the differential thermocouple. The heat capacity of the sample holder containing a sample which is undergoing exothermic or endothermic changes can be expressed as

$$C_{p,s} \, dT_2 = K_s(T_3 - T_2) \, dt + dH \qquad (V.12)$$

and for the reference side

$$C_{p,r} \, dT_1 = K_r(T_3 - T_1) \, dt \qquad (V.13)$$

where T_1, T_2, and T_3 are the temperatures of the reference, sample, and furnace thermocouples, respectively, and $C_{p,s}$ and $C_{p,r}$ are the total heat capacities of the sample holder and sample and the reference holder and reference material, respectively.

Pacor (58) derived an expression for the relationship between the area of a DTA curve peak and the total amount of heat produced or absorbed for the DuPont DTA block-type sample holder. For the reference, which is not subject to chemical reaction,

$$\frac{1}{a_r}\left(\frac{\partial T_r}{\partial t}\right) = \nabla^2 T_r \qquad (V.14)$$

where T_r is the reference temperature and a_r is the thermal diffusivity of the reference. Likewise, for the sample, in which chemical reaction (or transition) can occur,

$$\frac{1}{a_s}\left(\frac{\partial T_s}{\partial T}\right) = \nabla^2 T_s + \frac{1}{\lambda}\dot{Q} \qquad (V.15)$$

where T_s is the sample temperature, a_s is the thermal diffusivity of the sample, λ is the thermal conductivity, and Q is transition heat per unit volume. Introducing the differential temperature, ΔT, and multiplying both sides by dt and integrating over a time interval large enough to make all the transient terms negligible, the following expression is obtained:

$$\int_{t_1}^{t_2}\left(\frac{1}{a_s}\frac{\partial T_s}{\partial t} - \frac{1}{a_r}\frac{\partial T_r}{\partial t}\right) dt = \nabla^2 \int_{t_1}^{t_2}\Delta T \, dt + \frac{Q}{\lambda} \qquad (V.16)$$

The peak area in DTA is the area enclosed by the curve line and the base line, so the area between the zero line and the base line must be subtracted; thus

$$\int_{t_1}^{t_2}\Delta T^1 \, dt = \int_{t_1}^{t_2}\left(\frac{1}{a_s}\frac{\partial T}{\partial t} - \frac{1}{a_r}\frac{\partial T_r}{\partial t}\right) dt \qquad (V.17)$$

Subtracting equation (V.17) from equation (V.16) and defining the heating rate, $\alpha = dT_r/dt$,

$$\nabla^2\int_{t_1}^{t_2}(\Delta T - \Delta T^1) \, dt = \frac{Q}{\lambda} = \nabla^2\frac{S}{\alpha} \qquad (V.18)$$

where S is the surface of the peak. In the case of a cylinder of infinite length, with the boundary condition $\nabla^2 T = 0$ for $r = R$ (thus $S = 0$), the solution with the thermocouple in the center is

$$S = \frac{\alpha R^2}{4}\frac{Q}{\lambda} = \frac{\alpha R^2}{4}\frac{Q_w\rho}{\lambda} \qquad (V.19)$$

where R is the radius of the sample holder, Q_w is the transition heat per unit mass, and ρ is the density.

The temperature distribution within the sample and the influence of the sample parameters on the DTA curve has been discussed in detail by Melling et al. (59). The temperature distribution in the sample obeys the well-known diffusion equation

$$\frac{\partial T}{\partial t} - \partial \nabla^2 T = \frac{\alpha}{K} \frac{\partial q}{\partial t} \qquad \text{(V.20)}$$

where $\alpha = K/\rho c$, and ρ is the density, c the specific heat, K the thermal conductivity, and $\partial q/\partial t$ the rate of internal heat generation per unit volume. If the sample has the form of a cylinder of radius r and length l, then

$$\nabla^2 T = \frac{\partial^2 T}{\partial r^2} + \frac{1}{r} \frac{\partial T}{\partial r} + \frac{1}{r^2} \frac{\partial^2 T}{\partial \theta^2} + \frac{\partial^2 T}{\partial t^2} \qquad \text{(V.21)}$$

If it is assumed that the distribution of temperature of the outer surface of the sample is independent of position, that is, far from the ends of the cylinder, then equation (V.16) reduces to

$$\nabla^2 T = \frac{1}{r} \left[\frac{\partial}{\partial r} \left(r \frac{\partial T}{\partial r} \right) \right] \qquad \text{(V.22)}$$

Further use is made of these and other equations in a later section of this chapter.

A general theory for describing DTA curves (and DSC and TG) was developed by Gray (60) which employs the same initial equations and assumptions as previously discussed (43, 44) and others. The essential components of a thermal analysis cell are shown schematically in Figure V.4.

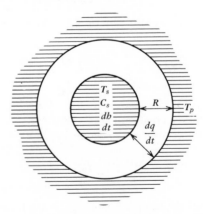

Figure V.4. Schematic diagram of a thermal analysis cell (60).

They consist of the sample and its container, at temperature T_s; a source of heat energy, at temperature T_p; and a path having a certain thermal resistance, R, through which the heat energy flows to or from the sample at a rate of dq/dt. It is assumed that (a) the sample temperature, T_s, is uniform and equal to that of the container; (b) the total heat capacity of the sample plus container, C_s, and the controlling thermal resistance, R, are constant over the temperature range of interest; and (c) heat generated by the sample per unit time, dH/dt, is positive, and heat absorbed is negative.

At any instant, the sample is generating heat at a rate dH/dt. Heat generated by the sample can either increase the sample temperature or be lost to the surroundings. Since heat must be conserved, the sum of these two effects must equal dH/dt. Therefore,

$$\frac{dH}{dt} = C_s\left(\frac{dT_s}{dt}\right) - \frac{dq}{dt} \qquad (V.23)$$

The rate of heat-loss to the surroundings is controlled by the thermal resistance and the temperature difference between the sample and surroundings. According to Newton's Law,

$$\frac{dq}{dt} = \frac{T_p - T_s}{R} \qquad (V.24)$$

and substituting into equation (V.23)

$$\frac{dH}{dt} = C_s\left(\frac{dT_s}{dt}\right) + \frac{T_p - T_s}{R} \qquad (V.25)$$

In a DTA apparatus, two cells as used as illustrated except one of them is the reference, where $dH/dt = 0$. Writing an equation similar to equation (V.25), the expression for the instantaneous rate of heat generation by the sample is

$$R\left(\frac{dH}{dt}\right) = (T_s - T_r) + R(C_s - C_r)\left(\frac{dT_r}{dt}\right) + RC_s d\frac{T_s - T_r}{dt} \qquad (V.26)$$

It is assumed that R for the reference cell is the same as that for the sample cell. The heat capacity of the reference, C_r, will not be equal to that of the sample, C_s. The heating rate for the reference, dT_r/dt, is the same as that for the sample, dT_s/dt, and is therefore a constant.

From equation (V.26), at any time, RdH/dt can be considered as the sum of three terms in units of temperature (see Figure V.5):

Part (I): $T_s - T_r$, which is the differential temperature of the recorded curve

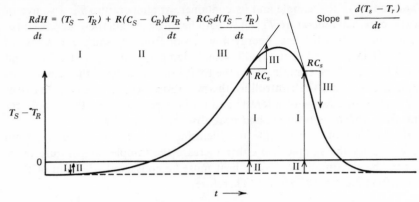

Figure V.5. Determination of RdH/dt from a DTA curve (60).

Part (II): $R(C_s - C_r)(dT_r/dt)$, which is the baseline displacement from the zero level.

Part (III): $RC_s\, d[(T_s - T_r)/dt]$, which is the slope of the curve at any point multiplied by a constant, RC_s. The term RC_s is called the time constant of the system and has the units of time.

At any point on the curve,

$$R\left(\frac{dH}{dt}\right) = \text{I} + \text{II} + \text{III} \qquad (V.27)$$

or if the tangent has a negative slope,

$$R\left(\frac{dH}{dt}\right) = \text{I} + \text{II} - \text{III} \qquad (V.28)$$

Thus, knowing RC_s, a curve can be graphically constructed which directly reflects the instantaneous thermal behavior of the sample.

4. FACTORS AFFECTING THE DTA CURVE

Differential thermal analysis, since it is a dynamic temperature technique, has a large number of factors which can affect the resulting experimental curves. These factors, which are similar to those discussed in thermogravimetry (Chap. II), are more numerous in DTA and can have a more pronounced effect on the curve. If the DTA curve is used for qualitative purposes, the shape, position, and number of endothermic and exothermic curve peaks is important. By a simple change of conditions, say

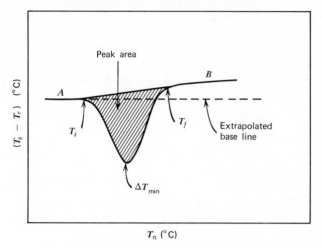

Figure V.6. Generalized DTA curve.

heating rate or furnace atmosphere, the positions (with reference to the T axis) will be changed, and perhaps the number of curve peaks as well. Changing from a nitrogen to an oxygen atmosphere can create additional exothermic peaks. For quantitative studies, the area enclosed by the curve peak is of great interest, so the effect of the experimental parameters on the area must be known. When DTA is used for specific heat measurements, the baseline deviations become important and such conditions as particle size of sample and diluent, system symmetry, sample packing, and so on must be taken into account if accurate and reproducible results are to be obtained.

A generalized DTA curve which will be used for purposes of this discussion is shown in Figure V.6. An endothermic peak is illustrated in which A is the pretransition (or prereaction) base line, B is the post-transition base line, T_i is the procedural initial deviation temperature which the instrument can detect, $\Delta T_{\min}$ is the procedural peak minimum temperature, and T_f is the procedural final temperature of the curve peak. For the temperature axis, T_n is the temperature of the reference (T_r), sample (T_s), or furnace (external) (T_e). The Y axis is that of the differential temperature, $T_s - T_r$ or ΔT.

As with the technique of thermogravimetry, the DTA curve is dependent upon two general categories of variables: (a) instrumental factors and (b) sample characteristics. The former category includes

(a) Instrumental factors
    (1) furnace atmosphere,
    (2) furnace size and shape,

(3) sample-holder material,
(4) sample-holder geometry,
(5) wire and bead size of thermocouple junction,
(6) heating rate,
(7) speed and response of recording instrument,
(8) thermocouple location in sample,

while the latter consists of

(b) Sample Characteristics
(1) particle size,
(2) thermal conductivity,
(3) heat capacity,
(4) packing density,
(5) swelling or shrinkage of sample,
(6) amount of sample,
(7) effect of diluent,
(8) degree of crystallinity

a.   Heating Rate

Although the effects of heating rate on the $\Delta T_{\min}$ temperatures and peak areas have been known for a number of years, only recently have these changes been explained in detail.   In general, an increase in heating rate, say from 2 to 20°C/min, will increase the T_i, $\Delta T_{\min}$, and T_f temperatures. As for peak area, the effect of heating rate depends upon the T_n temperature used.   If $T_s - T_r$ is plotted against T_s, the peak area will be proportional to the heating rate if the latter remains constant during the reaction (59).   If the peak area is measured as $T_s - T_r$ versus time, then it is independent of heating rate.   The conclusion reached by Garn (61) that the area of the peak increases with heating rate because of "the problem of heat transfer" was not substantiated by Melling et al. (59).   A higher heating rate will also decrease the resolution of two adjacent peaks, thereby obscuring one of the peaks.   At very low heating rates, the peak areas become very small or nonexistent on certain types of instruments, depending on the type of sample holder.

Kissinger (62, 63) has shown that the $\Delta T_{\min}$ temperature is dependent on the heating rate, according to

$$\frac{d[\ln (\beta/\Delta T_{\min}^2)]}{d(1/\Delta T_{\min})} = -\frac{E}{R} \tag{V.29}$$

where β is the heating rate, $\Delta T_{\min}$ is the peak minimum temperature, E is the activation energy, and R is the gas-law constant.   A plot of $\ln \beta/\Delta T_{\min}^1$

versus $1/\Delta T_{min}$ should yield a curve whose slope is E/R. The dependency of ΔT_{min} on heating rate was confirmed by Melling et al. (59), but the determination of E/R was found to be invalid in a practical experiment.

Speil et al. (10) first pointed out that the actual peak temperature is the point at which the differential heat input equals the rate of heat absorption and therefore

$$\Delta T_{min} = \left(\frac{dH}{dt}\right)_{max} \frac{m}{gk} \qquad (V.30)$$

as given in equation (V.1). A high rate of heating will cause dH/dt to increase because more of the reaction will take place in the same interval of time, and therefore the height or the apex or the differential temperature, ΔT_{min}, will be greater. As the return to the base line is a time function, as well as a temperature-difference function, the return will occur at a higher actual temperature with more rapid heating. This is illustrated for kaolin in Figure V.7. The peak areas were reported to be equal to within $\pm 3\%$, although there seemed to be a slight tendency toward smaller areas with low heating rates.

The effect of heating rate on T_i, ΔT_{min}, and T_f temperatures of sodium hydrogen carbonate, as determined by Barrall and Rogers (64), is shown in Figure V.8. This compound showed two endothermic peaks between 110 and 140°C. According to the least-squares extrapolations to zero heating rate, $NaHCO_3$ began to dissociate at $111.8 \pm 0.2°C$, and attained the first ΔT_{min} at $123 \pm 1°C$ and the second ΔT_{min} at $131.7 \pm 0.2°C$.

The variation in ΔT_{min} is fairly small if the differential temperature is measured against the sample temperature rather than the reference. In a careful investigation, Vassallo and Harden (66) studied the variation of ΔT_{min} for the fusion of benzoic acid and Marlex 50. The results are shown in Table V.1. The sample temperature, C_{sample} (C_s), was essentially constant over the entire range of heating rates studied. Using the reference temperature, A_{ref} (A_r), the ΔT_{min} values for benzoic acid varied about 4.0°C over the heating-rate ranges. Similar results were obtained for Marlex 50.

The effect on the resolution of two adjacent peaks in a DTA curve with heating rate is shown in Figure V.9. At heating rates of 2.5–10°C/min, Johnson and Miller (67) showed that cholesteryl propionate gave a curve with three transitions: crystal → smectic at 99°C; smectic → cholesteric at 110°C; and cholesteric → isotropic at 110°C. However, on changing the heating rate to 30°C/min, the 110°C peak was no longer detected, but a new, small endothermic peak was detected at about 64°C.

They suggested the use of high heating rates to detect small transitions which would not be detected at lower heating rates. Another effect of

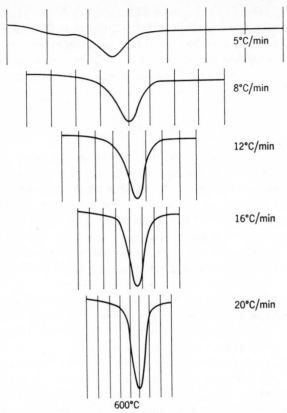

Figure V.7. Variation of peak temperature with heating rate (10).

heating-rate increases on the curve peaks observed in this study was to increase the peak amplitude.   This is illustrated in Figure V.10.

b. Furnace Atmosphere

In the case of a reaction which involves the evolution or absorption of a gaseous component, the peak temperature and the shape of the peak will be affected by the gas pressure of the system.   If the gaseous environment is identical to the evolved or absorbed gas, the changes will be more pronounced, as can be shown thermodynamically.   The relationship between transition temperature and pressure is expressed by the well-known Clapeyron equation which gives the rate of change of vapor pressure with temperature,

$$\frac{dp}{dT} = \frac{\Delta H}{T \, \Delta V} \tag{V.31}$$

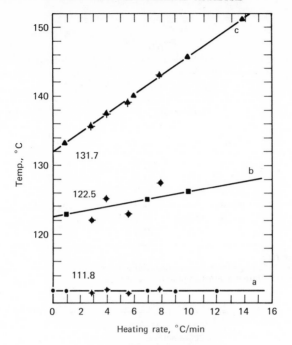

Figure V.8. Least-squares extrapolation of peak temperatures to zero heating rate (62).

where p is the vapor pressure, ΔH is the heat of transition, and ΔV is the change in volume of the system due to the transition. For reversible volatilization processes, several assumptions may be made leading to the familiar Clausius-Clapeyron equation, the integrated form of which is

$$\ln p = -\frac{\Delta H_V}{RT} + C \qquad (V.32)$$

where p is the vapor pressure in atm, ΔH_V is the heat of vaporization in cal/mole, R is the gas constant, T is the temperature, and C is a constant related to the entropy of transition.

TABLE V.1
Variation of ΔT_{max} for Some Fusion Reactions with Heating Rate (66)

ϕ,[a] ($^{\circ}$C/min)	ΔT_{max}, $^{\circ}$C			
	Benzoic acid		Marlex 50	
	A_{ref}[b]	C_{sample}[c]	A_{ref}[b]	C_{sample}[c]
5	121.5	121.8	136.0	134.2
10	121.6	121.7	138.0	134.4
15	122.1	121.9	138.7	134.2
25	124.0	121.9	139.5	134.2
40	125.5	121.9		134.2
80		121.8		134.4

[a] ϕ = heating rate.
[b] $A_{ref} = \Delta T$ plotted against reference material temperature.
[c] $C_{sample} = \Delta T$ plotted against sample temperature.

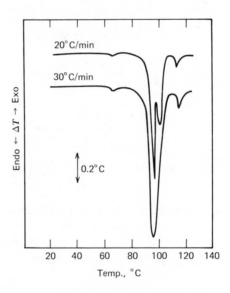

Figure V.9. Effect of heating rate on curve peak resolution (65). Compound used was cholesterol proprionate.

152

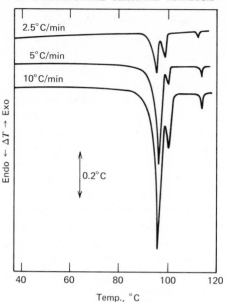

Figure V.10. Effect of heating rate on the peak amplitude (65).   Compound used was
cholesterol proprionate.

The Clausius-Clapeyron equation may be considered a special case of the
more general van 't Hoff equation

$$\frac{d \ln k_p}{dT} = \frac{\Delta H}{RT^2} \tag{V.33}$$

where K_p is the equilibrium constant for a reversible and ideal equilibrium
process.   For a *solid-gas*-type process, which is frequently studied by DTA,

$$A_{\text{solid}} \rightleftharpoons B_{\text{solid}} + C_{\text{gas}}$$

an approximate form of the van 't Hoff equation can be used,

$$\ln \frac{(K_p)_2}{(K_p)_1} = \ln \frac{(P_c)_2}{(P_c)_1} = \frac{\Delta H(T_2 - T_1)}{R(T_2 T_1)} \tag{V.34}$$

where $(P_c)_2$ and $(P_c)_1$ are the partial pressures of C_{gas} in the system at temperatures T_2 and T_1.

This relationship not only provides a convenient method for determining
the heat of volatilization processes, but also explains the response of volatilization peaks in DTA curves to changes in purge-gas flow rates exhibited in
dynamic-gas DTA systems.   Under dynamic flow conditions, increasing the

flow rate generally reduces the peak temperature at which volatilization occurs when the purge gas is noninteracting with the sample (and is different from the effluent gas). When the purge and effluent gases are identical, the volatilization peak either shifts to higher temperatures or remains unchanged relative to zero flow rate, depending upon the operating conditions. This behavior is a reflection of the changes in partial pressure of the effluent gas in the immediate vicinity of the sample surface. When the purge gas dilutes the effluent gas, the partial pressure of the effluent diminishes with increasing flow rate, and the peak moves to lower temperatures. Thus, the peak position can be shifted merely by controlling the partial pressure at the sample surface using the purge gas as a diluent (68).

In an atmosphere containing a fixed partial pressure of the evolved gas, a substance will not begin to dissociate to an appreciable extent until the dissociation pressure of the decomposition reaction equals or exceeds the partial pressure of the gaseous component in the surrounding atmosphere. The higher the partial pressure of the surrounding gas, the higher the dissociation temperature of the substance. Thus, the surrounding gaseous environment has a pronounced effect on the DTA curves so obtained. Furthermore, the reaction of the gaseous atmosphere with the sample can also produce peaks in the curve; for example, oxygen in the air causing an oxidation reaction and hence an exothermic peak.

Generally, two types of gaseous-atmosphere are employed: (*a*) a static gaseous atmosphere, usually in an enclosed system; and (*b*) a dynamic gaseous atmosphere in which a gas flow is either maintained through the furnace or through the sample and reference materials. The first type is the most difficult to reproduce since the atmosphere surrounding the sample is continually changing in concentration due to gas evolution by the sample and by furnace convection currents. Under controlled conditions, the dynamic atmosphere is the simplest to maintain and reproduce.

In a comprehensive study, Stone (69) compared the results obtained in static and dynamic gas atmospheres, as well as the effect of various gas atmospheres at different pressures, on certain decomposition reactions. A comparison between static and dynamic gas atmospheres on the thermal decomposition of illitic shale is given in Figure V.11. As can be seen, the peak minimum temperature, ΔT_{min}, is shifted to lower temperatures in the dynamic-gas technique. The shapes of the curves are similar in the two techniques.

The effect of two different gaseous atmospheres on the curve obtained for lignite is illustrated in Figure V.12. In the dynamic nitrogen atmosphere, the lignite pyrolyzes and distills off volatile matter; in oxygen, the lignite oxidizes, giving rise to exothermic instead of endothermic peaks.

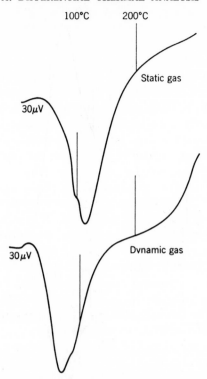

Figure V.11. Difference between DTA curves in static and dynamic atmospheres (108). Sample is an illitic shale.

The effect of the introduction of an atmosphere of the evolved gas on the DTA curves is shown in Figure V.13. In an oxygen atmosphere, the rhombic → hexagonal transition of $SrCO_3$ and the decomposition peak overlap each other.   On introducing an atmosphere of carbon dioxide, the transition peak remains at a ΔT_{min} of 927°C but the decomposition peak

$$SrCO_3 \text{ (s)} \rightleftarrows SrO \text{ (s)} + CO_2 \text{ (g)}$$

is shifted to much higher temperatures (70).

The choice of sample-holder shape will affect the interaction of the gas atmosphere with the sample.   If a glass capillary-tube-type holder is used, the changing of the furnace atmosphere will have little effect on the DTA curve due to the long gas diffusion path between the sample and the furnace atmosphere.   A flat-dish-type holder is perhaps ideal for control of the gas-solid reaction but may cause loss of ΔT sensitivity due to radiant heat loss.

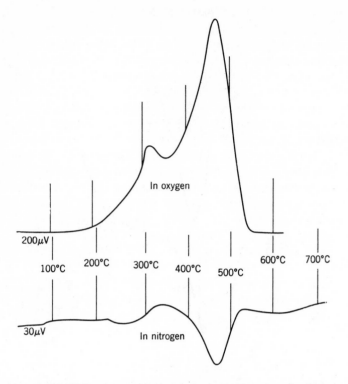

Figure V.12. Effect of O_2 and N_2 atmospheres on the DTA curve of a mixture of 2.5%
lignite in Al_2O_3 (108).

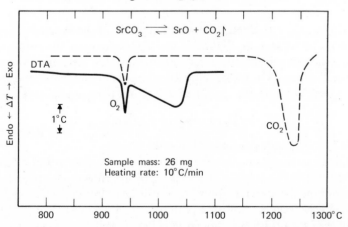

Figure V.13. Effect of atmosphere on the thermal decomposition of $SrCO_3$ (67). The
$solid_1 \rightarrow solid_2$ transition at 927°C is not affected.

For reactions in which the gas atmosphere plays no part, a spherical sample holder might be ideal, but would cause difficulty in introducing the sample. The effect of pressure changes on the DTA curve has been studied by numerous investigators. Increasing the pressure in the system, even with an inert gas, increases the transition temperatures, T_i, $\Delta T_{\min}$, and T_f. At low pressures, <1 Torr, the product gases are removed rapidly; hence the transition temperatures are shifted to lower temperatures and there is also a decrease in peak resolution. The effect of pressure changes from 5×10^{-6} Torr to 2300 p.s.i.g. on the DTA curves of $MgSO_4 \cdot 7H_2O$ are shown in Figure V.14 (71). Resolution of the curve peaks is very poor for the low-pressure curve in that only a single endothermic peak is observed plus a small exothermic recrystallization peak. On increasing the pressure, new endothermic peaks appear in the dehydration process. Unfortunately, the origin of the peaks was not discussed. Similar effects were shown by Locke (72), Garn (73), Levy et al. (74), David (75), and others.

c. Sample Holders

Since the shape of a DTA curve is influenced by the transfer of heat from the source to the sample and by the rate of internal generation or absorption of heat by the reactive sample, the sample holder plays an extremely important part in the DTA experiment. The specific effects of the sample-holder design on the DTA curve have been discussed in detail by Wilburn et al. (76) and Melling et al. (59). Using an analog computer, the latter generated DTA curves using various time constants, RC, for the sample holder. These results are shown in Figure V.15. As the sample-holder time constant is increased, the shape of the curve becomes distorted; this is the same as saying that as the holder diffusivity decreases and/or the heat capacity increases, and the shape of the curve will be markedly changed. Low-diffusivity-block-type holders cause the DTA peak to overshoot the zero baseline and so produce what appears to be an exothermic peak after an endothermic peak. This could lead to erroneous conclusions in the interpretation of the DTA curve, but is due, in fact to the use of a low-diffusivity block.

At high temperatures, transfer of heat will be mainly by radiation,

$$\frac{dH}{dt} = f(T_1 - T_2)^4 \qquad (V.35)$$

so that heating will be more rapid. This is equivalent to reducing the diffusivity of the holder.

Wilburn et al. (76) state that to prevent the adverse distortion of DTA curves, the sample size should be small, heat leakage between the samples

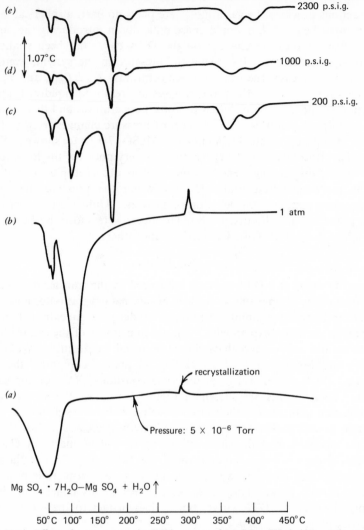

Figure V.14. Effect of pressure on the DTA curves of $MgSO_4 \cdot 7H_2O$ (68).   Nitrogen gas pressure;   heating rate of 10°C/min.

should be kept to a minimum, and heat transfer to the samples must be rapid.

DTA curves for identical samples show a change in shape with increasing sample radius (59) because of the development of a difference temperature at the beginning while the return to the base line at the end of the reaction is delayed.   With increasing radius, the "S" form of the leading edge becomes

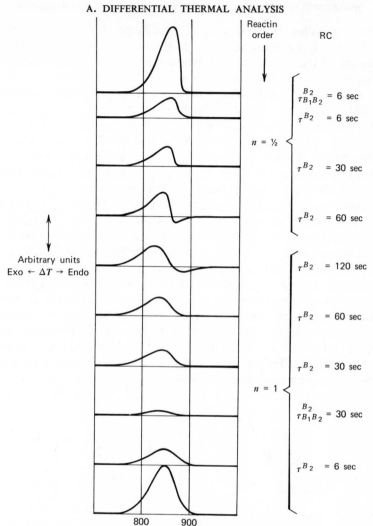

Figure V.15. Effect of sample holder diffusivity on the shape of the DTA peak (74).

more pronounced, and distortion of peak shape is more noticeable. If the peak shape is to be used for the determination of any reaction parameter, samples of small radius must be used.

As the sample radius is increased, the peak temperature increases for both reference and sample materials but the increase in sample temperature is much less than that occurring in the reference material. Hence, for minimum variation between samples of different radii, curves are best plotted using sample temperature as the abscissa. Also, to ensure that the heating rate in the sample remains reasonably constant during the reaction, it is imperative to use samples having small radii.

The time constant of the sample holder has previously been discussed by Gray (60) in equation (V.26). It is obviously an advantage in any sample holder to make RC_s as small as possible and ideally so small that the distance III in Figure V.5 is negligible compared to I + II. The smaller and more constant RC_s is, the more accurately will the instrument record the instantaneous thermal behavior of the sample. Obviously, an R value equal to zero in a DTA sample holder would result in no curve peaks at all. For high sensitivity, a large R is required, which is therefore incompatible with the requirement for fast response or high resolution that RC_s be small. In an instrument where the sample itself makes a major contribution to the thermal resistance, R, so that R is not constant during the transition, the curve peak area will not be proportional to the heat of transition.

Dosch (77) developed a simple electrical technique which could be used to measure both heat sensitivity and response time of a sample holder without interaction. Using the electrical analog for an isolated sample holder, as shown in Figure V.16, a small electrical heater was placed in the sample container and used in the measurements. The heater sensitivity was calculated by relating the temperature change at equilibrium to the input power, and the response time was determined by measuring the time constant. The

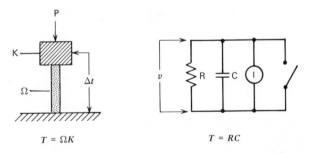

$$T = \Omega K \qquad\qquad T = RC$$

Figure V.16. Hypothetical isolated DTA sample holder and its approximate electrical analog circuits; in both cases the time, T, is the time required for the dependent variable (t or v) to reach $1/e$ or 63.2% of its final value (73). Equivalent quantities:

Time, Seconds	T————T	Time, Seconds
Thermal Resistance	Ω————R	Resistance, Ohms
Heat Capacity	K————C	Capacitance, Farads
Power	P————I	Current, Amperes
Temperature	t————v	Potential, Volts

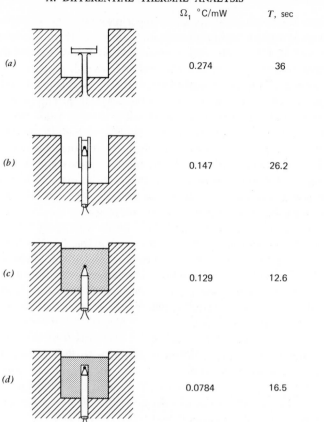

	Ω, °C/mW	T, sec
(a)	0.274	36
(b)	0.147	26.2
(c)	0.129	12.6
(d)	0.0784	16.5

Figure V.17. Sample holder configurations and the resulting values of Ω and T (73).

latter is the time required for the exponentially changing signal to reach 63.2% ($1/e$) of its equilibrium value.

The thermal resistances and time constants for a number of different sample holders are shown in Figure V.17. The resistance values obtained in C and D were not very reproducible due to the positioning of the heater in the sample area. The time-constant values, however, were much less affected. It is well known that the heat sensitivity of an isolated-container-type holder decreases with increasing temperature. This effect, which is due to the increase in thermal conductivity, k, results in a reduced value for the thermal resistance.

The geometry of the sample holder has a large effect on the intensity of the peak and the peak areas obtained. In Boersma's (44) equations for peak area, the value of $qa^2/4\lambda$ was obtained for a cylindrical metal sample holder, and $qa^2/6\lambda$ for a spherical holder. Experimentally, the spherical holder is difficult to construct, although an approach to a sphere has been made by Lehmann et al. (78), who used a cylindrical holder with a rounded bottom. In a spherical arrangement, the thermal effects from all parts of the sample arrive "in phase" at the centrally located thermocouple junction (79).

In general, a sample holder, either of the block or the cup type, constructed of a low-thermal-conductivity material will give better peak resolution for an endothermic reaction than one constructed of a high-thermal-conductivity material (79–81); for exothermic reactions, the resolution is worse for a low-thermal-conductivity material. Since most reactions studied by DTA are endothermic, it would appear that the low-thermal-conductivity sample holders would be preferred. In actual practice, however, the metal sample holders (high thermal conductivity) are more widely used, perhaps because of the ease of their fabrication and their durability.

A comparison between ceramic and metal sample holders has been made by Webb (82), Mackenzie (83), Arens (79), and Gerard-Hirne and Lamy (81). Webb found that for reactions involving the evolution of a gas, the ceramic sample holders gave peaks with maxima shifted to lower temperatures. However, for a crystalline phase transition such as the $\alpha \rightarrow \beta$ quartz transformation, the results were identical for metal and ceramic containers. Thus, it was concluded that the difference was caused by the ceramic holder allowing the gaseous decomposition products to diffuse into the furnace atmosphere, lowering the concentration of the gas in the sample, and so leading to a more rapid completion of the reaction. By using silica liners in the ceramic holder, this gaseous diffusion did not take place and hence the results were identical for ceramic and metal containers.

A comparison of the two types of sample holders (82) for endothermic reactions, such as the decomposition of $Ca(OH)_2$, $CaCO_3$, and $MgCO_3$, showed that the nickel block was only slightly less sensitive than a ceramic sample holder. At temperatures of about 500°C, it gave peak areas which were about 80% of those for the ceramic, and at about 900°C, about 70%. The metal holder yielded peaks which were sharper and also increased the resolution of adjacent peaks. Webb (82) explained this as follows: An endothermic reaction begins in the portion of the sample nearest the walls of the sample well, and in the case of the metal (nickel, in this case) holder, heat is readily available from the large mass of metal of high thermal conductivity in contact with the cooler decomposing material. Rapid heat flow into this superficial layer masks the early part of the reaction by neutralizing the endothermic effect before it can affect the thermocouple junction. It is for this reason that the endothermic reaction appears to start at a higher temperature. When the temperature reaches a value at which the rate of decomposition

TABLE V.2
Comparison of Block and Isolated Container Sample Holders (84)

Advantages	Disadvantages
Block type	
1. Good temperature uniformity	1. Poor exchange with atmosphere
2. Good thermal equilibration	2. Poor calorimetric precision
3. Good resolution	3. Difficult sample manipulation
4. Good for b.p. determinations	4. Sensitive to sample density change
Isolated container type	
1. Good exchange with atmosphere	1. Poor resolution
2. Good calorimetric precision	
3. Good for high temperature use	

becomes so rapid that the heat from the metal can no longer penetrate the increasingly thick layer of decomposed material of low thermal conductivity sufficiently rapidly enough to neutralize the endothermic effect, the reaction quickly manifests itself, reaching, for the rest of the reaction period, a rate comparable with that prevailing in the ceramic holder.

Comparing a ceramic (porous alumina) and a metal (nickel) sample holder, Mackenzie (83) found that for the endothermic peak in kaolinite, the peak was smaller in the metal holder (about 75% that of the ceramic) and was shifted about 6% higher in temperature.

The two basic types of sample holders, the block and isolated-container types, are compared in Table V.2 (84).

d. Thermocouple Location

The temperature distribution in reference and sample materials has been calculated by Smyth (85) and is illustrated in Figure V.18. The distribution curves for the reference material are at all instants of time identical parabolas, as would be deduced from the equation

$$T = T_c + \alpha t + \frac{1}{2}\left(\frac{1}{a}\right)\alpha x^2 \qquad (V.36)$$

where T is the temperature, t is the time, x is the distance in the direction of heat flow, α is the heating rate, a is the diffusivity of the material, and T_c is a constant having the dimensions of a temperature.

The sample-material curve starts out as a parabola; as the outer layers reach the inversion temperature, so much heat is required to change them from the low to the high form that the heat supply is interrupted. The rate of heating at the center of the sample material slows up before the material at

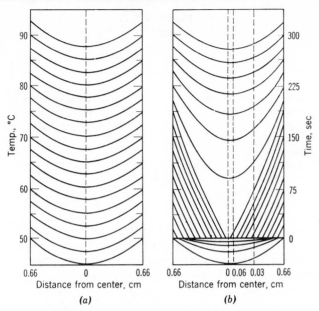

Figure V.18. Temperature distribution, at 18.75-sec intervals, of (a) reference materials; (b) sample material (114).

the center has reached the inversion temperature. As soon as the rate of heating at the center starts decreasing, the differential temperature curve will start to deviate from its base line. At this point, neither the center of the sample material nor the center of the reference material is at the inversion temperature. Smyth (85) states that not too much importance can be attached to this point of initial deviation.

When the sample-material inversion is completed, the distribution curve comes to a sharp point at its center. Since such a sharp point corresponds to a very high value of the second derivative of the equation

$$\frac{\partial^2 T}{\partial x^2} = \frac{1}{a}\frac{\partial T}{\partial t}$$
(V.37)

it would be expected that there would be a rapid rise in temperature at the center. This rapid rise gradually slows down, until after more than $300s$ the sample material has caught up with the reference material, causing the differential temperature to again become zero.

To illustrate more clearly the applications of Smyth's (114) calculations, the manner in which the differential temperature curve deviates as a function of reference temperature is presented. In DTA, the shape of the curve and

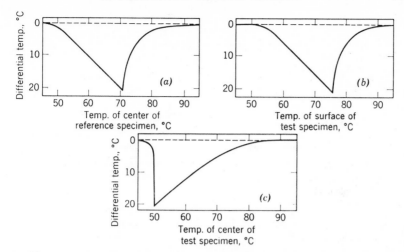

Figure V.19. Differential temperature plotted against the temperature (a) at the center of the reference material; (b) of the surface of the sample material; and (c) at the center of the sample material (114).

also the peak maxima temperatures are variable depending upon the source of the reference temperature. This is illustrated by the curves in Figure V.19 in which the differential temperature is plotted against the temperature of the center of the reference material, the surface of the sample material, and center of the sample material.

Curve (a) departs from zero some distance below the 50°C inversion temperature and reaches its peak some 20°C above the inversion temperature. In curve (b), the sample-material surface temperature would correspond to the temperature of the metal block in which the sample and reference cavities are located. This curve starts deviating from the base line at the inversion temperature, which, if this temperature could be accurately determined, would have useful significance on such a curve. If the differential temperature is plotted against the temperature at the center of the sample material, as shown in curve (c), the peak maximum temperature would be equal to the inversion temperature.

Smyth (85) illustrated the type of DTA curve obtained if the sample and reference thermocouples are not symmetrically located in their respective chambers. This asymmetry effect is shown in Figure V.20.

Curve (a) shows the differential temperature curve if the thermocouple is 0.06 cm from the center, instead of at the center, of the sample material. In curve (b), a rather extreme case is presented in which the sample thermocouple is 0.30 cm from the sample center. In both curves, the peak maxima

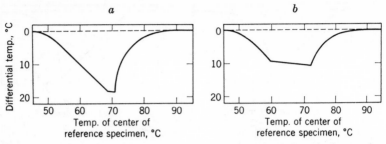

Figure V.20. Effect of having an asymmetric arrangement of sample and reference thermocouples. (a) thermocouple 0.06 cm from center of sample; (b) thermocouple 0.30 cm from center of sample (114).

temperatures are completely different from those of the symmetrically centered thermocouples.

Barrall and Rogers (86) also studied the effect of thermocouple asymmetry and the deviation of the differential temperature base line. The effect of the location of the thermocouple whose output is recorded against the differential temperature is illustrated in Figure V.21.

In case (1), the system or X axis thermocouple was located directly in the middle of the sample-holder block. As can be seen from curves a, b, and c, the ΔT_{min} temperature increases rapidly with an increase in the heating rate. For case (2), the thermocouple was located in the center of the reference sample, and the resulting curves, d and e, showed a less pronounced change ΔT_{min} with heating rate. In case (3), in contrast to the first two cases, there was no change in heating rate. It was also stated that the endothermic peak was more symmetrical and the recorded melting range was narrower for case (3).

Barrall and Rogers (86) also determined the effect on the base line of the differential curve if the thermocouples were not located symmetrically in the sample and reference chambers. The irregularities observed were even more pronounced when the sample packing was altered. These effects are illustrated in Figure V.22.

In curve (1) is illustrated the type of curve obtained with careful packing of the sample such that the thermocouples were symmetrically located not only with respect to the walls of the sample holder block, but also with respect to the top and bottom of the material in the sample and reference chambers. If one thermocouple was displaced approximately 2 mm toward the side, top, or bottom of the sample chamber, curve (2) was obtained. It can be seen that the displaced thermocouple heats up more quickly than the symmetrically located thermocouple, thus causing the deviation of the

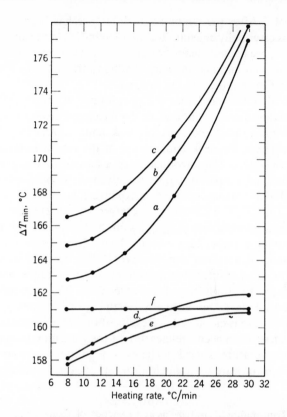

Figure V.21. Effect of location of system thermocouple and heating rate on temperature at ΔT_{min} using successive runs on different sample weights of 8.3% salicylic acid on carborundum (115). Case 1: (a) 0.0952 g; (b) 0.1125 g; (c) 0.1555 g. Case 2: (d) 0.0952 g; (e) 0.0822 g. Case 3: (f) 0.0952 g.

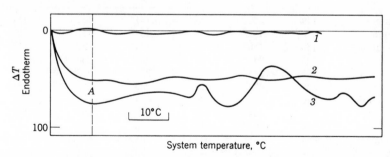

Figure V.22. System temperature versus ΔT for carborundum in both cells (115). Curve 1: symmetrically located thermocouples; curve 2: unsymmetrically located thermocouples in uniformly packed samples; curve 3: unsymmetrical thermocouples in a nonuniformly packed sample.

167

curve up to *A*.   After the temperature difference has been established, the curve remains essentially constant for the remainder of the run.   In curve (*3*), it is seen that loose packing leads not only to a displacement of the curve, but also to maxima and minima, presumably due to the shift of support particles as they expand during the heating process.

Melling et al. (59) found that the positioning of the thermocouple in a sample of small radius does not affect the measured peak temperatures.   In samples of large radius, the shift of peak temperatures for the sample is significant, whereas the peak temperature of the reference material remains reasonably constant.   The introduction of a measuring thermocouple does, however, affect the temperature distribution in the sample.

David et al. (138) made a comprehensive investigation of the use of furnace temperature (mode 1) or sample temperature (mode 2) as the temperature axis.   A comparison of these two modes on the DTA curves of polyethylene, a material having a rather large specific heat, showed that the ΔT_{min} temperatures occurred at lower temperatures when mode 2 was used.   The effect of the mode of temperature measurement on the various peak temperatures of pure metals is illustrated in Table V.3.   The extrapolated leading edge of the peak is taken normally as the transition temperature when furnace temperature is employed.   Results of the three methods showed that peak temperature is an accurate measure of the transition temperature when the sample temperature is used.   The extrapolated temperatures are low in this case, and if they are used a calibration procedure is necessary.

TABLE V.3

Peak Temperature Measurements as a Function of Temperature Axis (87)

Sample[a]	Furnace temp., °C			Sample temp., °C		
	1st deviation	Peak	Extrapolated	1st deviation	Peak	Extrapolated
Indium	149°	152°	157°	151°	157°	151°
Indium	150°	152°	156°	152°	157°	153°
Indium	150°	150°	157°	151°	158°	151°
Tin	226°	234°	228°	226°	234°	229°
Tin	225°	234°	227°	229°	232°	231°
Tin	225°	233°	228°	229°	232°	231°
Aluminum	646°	664°	649°	644°	661°	647°
Aluminum	644°	664°	648°	643°	660°	646°
Aluminum	644°	662°	648°	643°	660°	646°
Aluminum	644°	665°	647°	645.6°[b]	661.6°[b]	—
Silver	—	—	—	958°	961°	—

[a] Known values: indium, 156.6°C; tin, 231.8°C; aluminum, 660.1°C; and silver, 960.8°C.

[b] Determined with higher chart speed.

e. Thermocouples

Boersma (44) has shown by theoretical considerations that the heat-loss by conduction along the thermocouple wires is fairly large and can have a considerable effect on the area of the peak obtained. Since the temperature in the sample center is measured by means of a thermocouple, part of the heat produced in the sample is carried away by the thermocouple wires and therefore too low a sample temperature is measured.

In Figure V.23 is illustrated a spherical sample-filled cavity in a nickel block containing a thermocouple junction of radius r_0. During a reaction, a temperature gradient exists in the wires over length l. It is assumed that at the distance l, which is slightly larger than a, the radius of the sample cavity, the wires have attained the temperature of the nickel block. If the area of the thermocouple wires is A and θ_0 is the thermojunction temperature, the amount of heat carried away by the thermocouple leads is

$$Q = \int_{t_1}^{t_2} \frac{A\lambda_p}{l} \theta_0 \, dt \qquad (V.38)$$

where λ_p is the thermal conductivity of the wires. By various mathematical manipulations, the peak area for a spherical sample holder becomes

$$\int_{t_1}^{t_2} \theta_0 \, dt = \frac{qa^2}{6\lambda} \cdot \frac{\alpha}{1 + (\Lambda/\lambda)} \qquad (V.39)$$

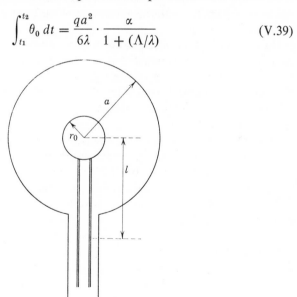

Figure V.23. Heat leakage through thermocouple wires (44).

where α, which is very nearly unity, comes from the altered geometry of the holder and is equal to

$$\alpha = 1 - \frac{r_0^2}{a^2}\left(3 - 2\frac{r_0}{a}\right) \tag{V.40}$$

and Λ is the heat leakage through the wires, or

$$\Lambda = \lambda_p \frac{r_0}{1}\frac{A}{4\pi r_0^2}\left(1 - \frac{r_0}{a}\right) \tag{V.41}$$

For low-thermal-conductivity samples, $(\lambda/\Lambda) \ll 1$, the peak area will become independent of the sample conductivity λ, whereas a high sample conductivity will cause an inversely proportional relationship.

For cylindrical holders, the expression is

$$\int_{t_1}^{t_2} \theta \, dt = \frac{qa^2}{4\lambda} \cdot \frac{1 - (r_0^2/a^2)[1 + 2\ln(a/r_0)]}{1 + (A/l)(\lambda_p/\lambda)[\ln(a/r_0)/2\pi h]} \tag{V.42}$$

For most samples, the thermal conductivities are about $0.3 J/ms$ °C, which makes λ/Λ about equal to unity. Therefore, according to equation (V.39), the heat-leakage through the wires reduces the peak area to less than 50% of its theoretical value.

Hauser (88) also studied the effect of wire and thermojunction size on the peak shapes. He concluded that the larger wire size (No. 22 compared to No. 28 gauge) and the larger bead size (1.43 mm compared to 0.8 mm) gave the more pronounced peaks on the thermal decomposition curves. In considering the electrical resistance characteristics of the thermocouple wires, he found that No. 28 gauge wires have too high a resistance for proper electrical properties without too great a thermal conductivity. A 1.43-mm thermojunction diameter was much more efficient in maintaining the emf than the smaller junctions, but still the mass of the junction was not great enough to absorb an excessive amount of heat.

The effect of thermocouple wire size on the transition temperature of benzoic acid and toluene was studied by Vassallo and Harden (66). For No. 28 gauge wire, at a heating rate of 10°C/min, the melting point (mp) of benzoic acid was 121.7°C. The effect was also observed for a heating rate of 40°C/min. For the No. 40 gauge wire, the peak heights were about 15% less than those obtained for the larger wire. The difference in peak shapes and intensities for the sample and reference temperatures using the various wire gauges is illustrated in Figure V.24. The shapes of the curves are entirely different for the different size wires.

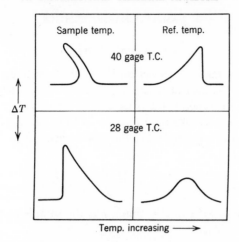

Figure V.24. Character of boiling endotherm using sample and reference temperature measurements at a heating rate of 10°C/min (64).

Heat leakage due to thermocouple leads can be expressed by the equation (59)

$$\frac{dh}{dt} = K_H(T_S - T_A) \tag{V.43}$$

where dh/dt is the rate of heat-loss, T_s is the temperature at the center of the sample, T_A is the ambient temperature, and

$$K_H = \frac{k_w a_w}{l_w} \tag{V.44}$$

where K_H is the heat-loss factor, k_w is the thermal conductivity of the wire, a_w is the cross-sectional area of the wire, and l_w is the length of wire from thermojunction to ambient temperature. As would be expected, increasing heat-loss along the thermocouple wire causes the area under a typical DTA peak to decrease, but it is also interesting to note that the peak temperature also decreases. The base line of the curve will only be zero if the reference and sample parameters are identical and if heat leakage from both thermocouples is identical. If heat-leakage differs, the differential temperature will increase linearly with temperature. A nonlinear base line change may be due to the temperature dependence of the heat-leakage factor. Heat-leakage along the thermocouple leads reduces the time for the quasisteady state to be reached and reduces the rate of heating.

The calibration of the temperature-axis thermocouples is discussed in Chap. VI.

5. SAMPLE CHARACTERISTICS

a. Sample Mass

According to the various theories on DTA, the area under the curve peak is proportional to the heat of reaction or transition, and hence, the mass of reactive sample.   In general, the peak area is inversely proportional to the thermal conductivity, k; directly proportional to the density, φ, and the heat of reaction, ΔH; and independent of the specific heat (59).   The relationship can be expressed as

$$A \propto \frac{\varphi r^2 \Delta H l}{k} \tag{V.45}$$

where A is the peak area ($\Delta T \times$ time), r is the sample radius, and l is the length of sample.   Then

$$A = \frac{\varphi V \Delta H}{k} \tag{V.46}$$

where V is the volume of sample, and hence

$$A = \frac{Gm \Delta H}{k} \tag{V.47}$$

where G is a calibration factor and m the sample mass.

The effect of sample mass on the peak area will be discussed in greater detail in the sections on quantitative DTA and DSC.

b. Sample-Particle Size and Packing

There are a number of conflicting studies concerning the effect of particle size and particle-size distribution of the sample on the peak areas and ΔT_{min} values.   Speil et al. (10) found that the peak areas under the kaolin dehydration peak varied from 725 to 2080 mm^2 over the particle-size range of 0.05–0.1 μ to 5–20 μ.   It was also found that the ΔT_{min} values varied from 580 to 625°C.   However, Norton (89) found that the ΔT_{min} values remained essentially constant, but that the temperature at which the dehydration reaction was completed varied from 610 to 670°C over a particle-size range of <0.1 to 20–44 μ.   Grimshaw et al. (90) agreed with the latter study in that, with particle sizes down to 1 μ, the thermal characteristics of the kaolin samples were independent of particle size.   This effect is illustrated in Table V.4.

Carthew (91), who also studied the decomposition of kaolinite, in the particle size range of >2 to 0.25–0.1 μ, agreed with the work of Norton (118)

TABLE V.4
Effect of Particle Size on the Thermal Characteristics of Kaolin (90)

| Average particle size, μ | Endothermic reaction | | Exothermic |
	ΔT_{min}, °C	Finishing temp., °C	ΔT_{min}, °C
10–44	600	670	980
0.5–1.0	605	650	980
0.25–0.5	605	630	980
0.10–0.25	600	615	980
<0.10	600	610	945, 990[a]
<0.10 dialyzed	605	610	955, 986[a]

[a] Two peaks.

and Grimshaw et al. (90).   For particle sizes from 2–1 μ to 0.25–0.1 μ, the peak areas were essentially constant, as were the ΔT_{min} values.   The disagreement with Spiel et al. (10) was attributed to the fact that they obtained their particle-size fractions by a grinding process, which could reduce the degree of crystallinity of the kaolin.

Barrall and Rogers (92) found that in a blank run of small glass beads against large beads, the base line displacement indicated that the large beads did not transmit heat as well as the small beads.   The base line gradually decreased with increasing temperature and a large fraction of the displacement remained at 200°C.

Langer and Kerr (65) found that an increase in the particle size of kaolinite produced a peak temperature increase for the dehydration reaction.   There was no significant shift in the phase-transition exothermic peak.   An increase in sample packing density was said to increase heat transfer through the sample and cause changes in the baseline slope.

Rather dramatic changes in the DTA curves of silver nitrate were indicated (93) on change in particle size.   These curves, as shown in Figure V.25, indicate a change in the peak shape as well as the peak minimum temperature.   The ΔT_{min} for the original sample was 161°C, which shifted to 166.5°C in the finely ground sample.   After melting, the three samples all gave identical DTA curves.   According to Negishi and Ozawa (95), the effect of grinding on the transition is a kinetic one, namely, the formation of a barrier to the fusion transition.

A pronounced effect on the DTA curves of CdO due to grinding was also observed by Wada et al. (94).   The origin of the DTA peaks, which became more pronounced as the time of grinding increased, was due to the presence of $Cd(OH)_2$ contamination.

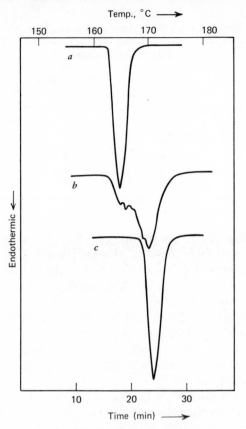

Figure V.25. DTA curves of silver nitrate (94). (*a*) the original sample; (*b*) the slightly ground sample; (*c*) the finely ground sample.

As discussed previously, the area of the DTA peak is inversely proportional to the thermal conductivity of the sample, which in turn is dependent on the particle-size distribution and packing of the sample (59).

The effect of sample packing on the DTA curve has been illustrated by Gruver (95) as shown in Figure V.26.   In curve (*a*), the kaolin sample was placed in the sample crucible and settled by a slight tapping action; in curve (*b*), the sample was tamped in place by use of a small glass rod.   The curves obtained turned out to be identical.   Admittedly, this is rather a crude method of testing this effect.

Bollin and Bauman (96) described a simple sample-packing technique which is capable of accurate reproduction of sample-packing density.   The

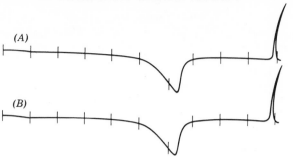

Figure V.26. Effect of packing of the sample (kaolin): (*A*) tapping; (*b*) tamping (122).

sample is packed around the thermocouple by means of a machined loading die. The result of this packing procedure is an equidimensional sample placed immediately around the thermocouple and surrounded by 60-mesh-size aluminum oxide. This type of sample-loading procedure was said to produce a more efficient geometry than the planar "sandwich" method of Barshad (97).

The effect of crystallinity of the sample is rather difficult to evaluate because of the definition of the term "degree of crystallinity." Carthew (91) defined the latter, in the case of kaolin samples, as the perfection of crystal orientation and not the size of the crystal. Using five different samples of kaolin, he found that the area of the endothermic dehydration peak decreased with decrease in sample crystallinity. The peaks appeared to be sharper as the degree of crystallinity of the sample increased. This effect of crystallinity was said to be similar to that of change in particle size, and could probably be explained in a similar manner.

c. Effect of Diluent

If samples are sufficiently diluted with a diluent (inert material), the physical properties of all tested materials will be more nearly the same, so that the peak area will be directly proportional to the heat of reaction or transition (59). Dilution will, however, reduce the heat effect, which in turn will reduce the peak area. The diluent must not, of course, react with the sample during the heating process.

The effect on the peak ΔT_{min} values for various concentrations of kaolinite and halloysite diluted with alumina has been studied by Dean (94a). For the former substance, the ΔT_{min} values ranged from 525°C for a 10% mixture to 570°C for a 60% mixture. Similar results were found for the halloysite mixtures. De Jong (98) found that there was a linear relationship between peak area and weight fraction of kaolinite diluted with alumina, provided

TABLE V.5

Peak Area Obtained by 0.01 g Salicylic Acid Diluted With Various Materials[a] (92)

Diluent	Salicylic acid, %	Area/0.01 g of acid (mm^2)
Carborundum	6.87	306
Iron metal	8.82	710
Iron(III) oxide	3.40	280
Glass beads, 0.029 mm	4.57	322
Glass beads, 0.29 mm	5.58	289
Alumina	8.60	313
Nujol	20.00	92

[a] ΔT sensitivity of 67 μV per in.; heating rate of 7.9°C min.

that the density of the mixtures did not change drastically. For illite-alumina mixtures, however, with a tendency for greater peak areas for higher illite concentrations, a linear relationship was not found.

In a comprehensive study, Barrall and Rogers (92) determined the effect of diluents such as carborundum, iron metal, and iron(III) oxide on the peak caused by the fusion of salicylic acid. This effect is illustrated in Table V.5. The variation in thermal conductivity of the diluent is probably the main cause of the peak area variation. Higher conductivity allows the thermal effect to be more efficiently conducted to the thermojunction in the center of the sample. It should be noted, however, that the diluent high thermal conductivity may decrease the peak area when the diluted sample is in direct contact with a metal sample block.

A "masking" effect has been noted (92) for certain peaks when the diluent reacts with the sample. This is illustrated in the case of 8-quinolinol diluted with carborundum and alumina, as shown in Figure V.27. When carborundum is used, the ΔT_{min} obtained was 76.3°C; with an alumina diluent, no endothermic peak was observed in this temperature region. Apparently, a complex had been formed between the alumina and 8-quinolinol.

6. Critique of Operational Parameters

The key operational parameters for DTA (or DSC), as given by Sarasohn (84), are shown in Table V.6. If a large sample size is chosen, lower heating rates are required. This in turn decreases the ΔT sensitivity and peak resolution. Small samples are perhaps the most convenient to use, especially with present-day commercial instruments. They also permit higher heating rates and give better peak resolution.

A summary of the operational parameters is given in Table V.7 (137).

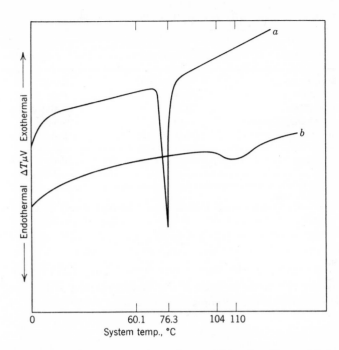

Figure V.27. Masking effects of sample peaks caused by the diluent: (*a*) 8-quinolinol diluted to 6.9% with carborundum; (*b*) 8-quinolinol diluted to 5.9% with alumina (121).

TABLE V.6
Key Operational Parameters (84)

Sample size

Large: Useful for detecting low-level transitions. Useful for nonhomogeneous samples. Curve peaks are broad; low resolution and temperature accuracy. Requires slow heating rate.

Small: Good resolution of curve peaks. Peaks are sharp, transition temperatures need equilibrium values for zero order reactions. Permits fast heating rate.

Heating rate

Fast: Increases sensitivity.
Decreases resolution.
Decreases temperature accuracy.

Atmosphere

Can react with sample.
Dynamic preferred over static.

177

TABLE V.7
Summary of Operational Parameters (84)

Parameter	Maximum resolution	Maximum sensitivity
Sample size	Small	Large
Heating rate	Slow	Fast
Sample holder	Block	Isolated container
Surface/volume of sample	Large	Small
Atmosphere	High k^a (He, H_2)	Low k (vacuum)

[a] Competing reversible volatilization reactions can frequently be resolved using dynamic gas atmosphere.

B. Quantitative Differential Thermal Analysis

1. INTRODUCTION

The determination of the heat of transition (or reaction) or the mass of the reactive sample from the area of the curve peak is a widely used procedure in DTA (and DSC). Expressed very simply,

$$\Delta Hm = KA \qquad (V.48)$$

where ΔH is the heat of transition (or reaction), m is the mass of reactive sample, K is the calibration coefficient, and A is the curve peak area. The calibration constant is related to the geometry and thermal conductivity of the sample holder and is usually determined by calibration of the system with compounds having known heats of transition (or reaction).

The use of DTA for determining the heat of transition has been reviewed by Bohon (99), Ozawa (100), and many others. The primary advantages of DTA (or DSC) techniques over classical calorimetry have been given as the following:

(*a*) Rapidity of the determination; a wide temperature range can be investigated in minutes or hours.

(*b*) Small sample masses; sample size may range from several mg to several hundred mg.

(*c*) Versatility; samples may be either liquids or solids.

(*d*) Simplicity and ease of procedure and analysis of data.

(*e*) Applicable to cooling processes and to measurements under high pressure.

(*f*) Can be used to study many different types of chemical reactions.

Disadvantages of the method include the following:

(a) Relative low accuracy and precision of the method, 5–10% in most cases.

(b) Cannot be used very conveniently to determine the ΔH of overlapping reactions.

(c) The need for calibration over the entire temperature range of interest because K is a function of temperature (DTA only).

(d) Inaccuracies in determining peak areas due to baseline change during the transition or reaction.

In many cases, the investigators have been perhaps somewhat optimistic about the results obtained and have not taken into account all of the variables of the method. No great accuracy can be obtained unless all of these variables are rigidly controlled, which in many cases is extremely difficult.

One of the first to determine the ΔH of fusion of several organic acids was Vold (43), who used equation (V.2). The value of A was obtained by plotting $\log (y - y_s)$ versus time, t. Thus, no calibration with standards was required.

In another very early investigation, Wittels (101) found that the area enclosed by the curve peak was proportional to the heat absorbed in the decomposition of calcite, $CaCO_3$. A linear relationship was found, using sample weights from 0.30 to 3.00 mg. In another study, Wittels (102) elucidated the effect of heating rate and sample mass on the peak areas obtained by the thermal decomposition of tremolite, $Ca_2Mg_5Si_8O_{22}(OH)_2$. The relationship

$$\Delta H = \frac{A}{\tan [\ln R(R - c)/m]} \tag{V.49}$$

was derived in which R is the heating rate, A is the peak area, and m and c are the constants. The best response of the instrument was obtained at a heating rate of 30°C/min, and fell off rapidly below 15°C/m.

The more recent investigations will be discussed later in this section.

2. CALIBRATION

The calibration coefficient, K, is determined by use of compounds having known heats of transition. Most of the standards used involve the heat of fusion, ΔH_f, or a $solid_1 \rightarrow solid_2$ heat of transition. The standards, obviously, must meet certain qualifications such as chemical stability during the transition, low vapor pressure so the heat of vaporization does not contribute to the heat effect, and so on.

The expression used for K depends on the type of instrument used and the method of recording the DTA or DSC curves. If the variables of ΔT sensitivity and recorder chart speed are included, the expression for K can be written as (103)

$$K = \frac{\Delta H m C}{A \, \Delta T_s} \qquad (V.50)$$

where ΔH is the heat of transition in cal/g, m is the sample mass in mg, C is the chart speed in in./min, A is the peak area in in.^{2}, and ΔT_s is the differential temperature sensitivity in deg/in. With these units, K is expressed in mcal/min °C. A similar expression was used by David (104). For systems having a large heat capacity, Bohon (105) calculated K from the expression

$$K = t_h C_h \qquad (V.51)$$

where t_h is the apparent heat transfer coefficient and C_h is the heat capacity of the sample holder, where $C_h \cong \bar{c}_h w_c$ ($\bar{c}$ is the specific heat of sample container and w_c its mass). The heat-transfer coefficient can be obtained from

$$\log(\Delta T) = -\frac{t_h}{2.303} t + I \qquad (V.52)$$

In many of the methods of quantitative differential thermal analysis, the calibration coefficient can be mathematically determined and no experimental procedures are necessary. For example, Kronig and Snoodjik (104) calculated K for a cylindrical sample holder as

$$K = \varphi \frac{a^2}{4\lambda} = \frac{M}{4\pi h \lambda} \qquad (V.53)$$

and for spherical symmetry

$$K = \varphi \frac{a^2}{6\lambda} = \frac{M}{8\pi a \lambda} \qquad (V.54)$$

where φ is the sample density, a is the radius of the sample holder, and h is the height of the sample. These equations describe the temperature difference between the center of the reference material and the sample material, assuming very small thermocouple junctions and leads. The quantity, $4\pi h$ or $8\pi a$, is the geometrical shape factor g in Speil's theory. These results were similar to those obtained by Boersma (44) in whose work the influence of the heat-loss through the thermocouple wires was taken into account. Vold (43) obtained K by an analysis of the exponential decay of

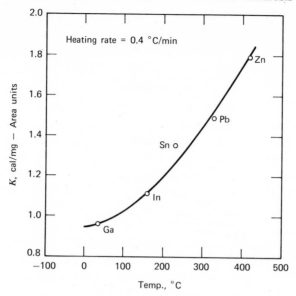

Figure V.28. Temperature dependence of K (82).

the curve after the reaction had ceased.   The relaxation time must have such a large value that after the reaction or transition is completed, the temperature relaxes over a range large enough to be interpretable.   Other attempts at calculating K were made by Ozawa (100), Pacor (58), and others.

The temperature dependence of K for the DuPont DTA sample holder is illustrated in Figure V.28 (106).   As can be seen, the calorimetric sensitivity of the apparatus decreases with temperature; that is, more heat is required per unit area.   In differential scanning calorimetry, such as with the Perkin-Elmer instrument, K is independent of temperature; hence, only a one-temperature calibration is required.   The problem of multitemperature calibration in DTA is also eliminated in the technique of constant-sensitivity DTA proposed by Wendlandt and Williams (107).

The calibration coefficient is also dependent on other instrumental variables such as furnace atmosphere.   David (104) determined K as a function of temperature while varying the composition of the furnace atmosphere (air, N_2, or He) and also the pressure (5×10^{-6} Torr to 147 atm).   The effect of different gaseous atmospheres on the value of K is shown in Figure V.29 (104).   The difference between the curves can be related to the different thermal conductivities of the gases studied.   It was found, as expected, that K was independent of heating rate in the range from 2 to 40°C/min.

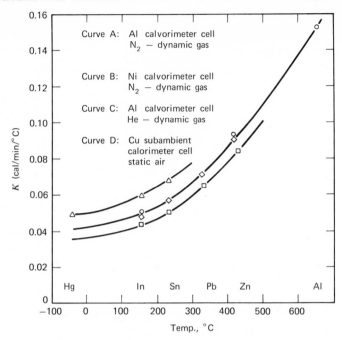

Figure V.29. Effect of different furnace atmospheres on K (80).

In many cases, the curve for K was expressed in equation form.   Currell (108) expressed the values for K as

$$K = -1.3 \times 10^{-4}T + 0.2200 \tag{V.55}$$

while Ozawa (100) found that the smoothed curve was fitted by the expression

$$K = 1.507259 + 8.782709 \times 10^{-3}T - 1.808468 \times 10^{-5}T^2 + 6.324056 \times 10^{-14}T^4 \tag{V.56}$$

The use of a DTA apparatus in which a substance having a known heat of fusion is used as the reference material, rather than an inert substance, was described by Wiedemann and van Tets (109, 110).

Ramachandran and Sereda (111) described the calibration of a DTA system using an internal standard of $AgNO_3$.   The peak height was used rather than peak area for all calculations.   The peak height was also used in the determination of quartz in a mixture by Davis and Holdridge (112), for calcium silicate (113) and other determinations.

3. CALIBRATION STANDARDS

Numerous compounds have been proposed to calibrate the DTA and DSC sample holders for quantitative determinations.   Most of the standards used are pure metals, although many organic compounds of high purity have also been employed.   The heat of fusion is the thermal transition normally used, although dehydration and decomposition reactions have also been recommended by numerous investigators.   A list of standard materials used for calibration purposes is given in Table V.8.

Most of the ΔH values given in Table V.8 are those obtained at constant pressure.   If the calibration is carried out at constant volume, such as was described by Bohon (105), the ΔH values must be corrected by the approximate relation:

$$\Delta H(v) \cong \Delta H(p) + RT\Delta n \qquad (V.57)$$

where $\Delta H(v)$ is the ΔH at constant volume, $\Delta H(p)$ is the ΔH at constant pressure, and Δn is the change in number of moles of gas formed in the reaction.   Corrected $\Delta H(v)$ values for several reactions are given in Table V.9.

4. CALCULATION OF ΔH

Once the sample cell is calibrated and the calibration coefficient determined in the temperature range of interest, the ΔH of the sample can be calculated using equation (V.8) or similar expressions.   The sample curve peak can be integrated using a planimeter or other means and the area determined.   If overlapping peaks are present in the curve, the curve may be integrated in parts, as shown in Figure V.30 (106).   For each of the various peak areas, a different K value must be used to calculate the ΔH (if DTA is used); the total ΔH is thus the sum of each area.   In DSC, since K is invariant with temperature, only one K value is needed.   This illustrates one of the advantages of DSC over DTA for quantitative measurements.

If the base line undergoes a large displacement during the reaction or transition, the integration of the curve is difficult and, if carried out, leads to large errors in the ΔH calculations.   Generally, small samples are used to prevent large base line displacements.   Also, the thermoelectric-disk-type sample holders have better base line stability than many other types of instruments.

5. PRECISION AND ACCURACY OF ΔH MEASUREMENTS

Despite the present advanced development of statistical methods, the great majority of investigators still present numerical data without qualification as to precision and accuracy.   Generally, when a new method or instrument appears, measurements are made on an arbitrary number of samples

TABLE V.8
Standards Used for DTA and DSC Calibration

Temperature °C	Standard	ΔH_f, cal/g	ΔH_t cal/g
34.6	Azoxybenzene	21.6	
44.9	C_2H_6		2.59
47	CBr_4		4.81
48.2	Benzophenone	23.5	
62.5	Palmitic acid	51.2	
69	Stearic acid	47.5	
69.8	Biphenyl	28.7	
99.3	Phenanthrene	25.0	
114	o-Dinitrobenzene	32.3	
121.8	Benzoic acid	33.9	
125	NH_4NO_3		12.6
128	KNO_3		12.86
130	$BaCl_2 \cdot 2H_2O$	116.6 ($-2H_2O$)	
137.2	NH_4Br		(882 cal/mole)
150	$CuSO_4 \cdot 5H_2O$	228.5 ($-4H_2O$)	
156.4	In	6.79	
177.0	KCNS	25.72	
179	Ag_2S		4.08
180	$CaSO_4 \cdot 2H_2O$	157.2 ($-2H_2O$)	
183.1	NH_4Cl		(1873 cal/mole)
187.8	Pentaerythritol	77.1	
212	$AgNO_3$	17.7	
231.9	Sn	14.4	
252	$LiNO_3$	88.5	
299.8	$KClO_4$		23.7
306.2	$NaNO_3$	44.2	
327.4	Pb	5.50	
337	KNO_3	28.1	
350	$CdCO_3$	134.5 ($-CO_2$)	
398	$K_2Cr_2O_7$		28.9
419.5	Zn	24.4	
498	$PbCl_2$	20.9	
553	LiBr	36	
575	$LiSO_4$	62	
588.8	Na_2WO_4		28.57
850	$CaCO_3$	427.1 ($-CO_2$)	

with data points obtained by an arithmetic mean approach.   The resulting numbers, according to Schwenker and Whitwell (114), take on a disconcerting absolutism and are frequently cited as definitive values.   Using the technique of DSC, these authors evaluated the resulting data using statistical methods.

The variables of heating rate, sample size, instrument sensitivity level, and

TABLE V.9
Heat of Reaction Values Corrected for Constant Volume (105)

Compound	$\Delta H(p)$	$\Delta H(v)$	$T,$
	cal/g		°C
$MgSO_4 \cdot 7H_2O$	402	398	130
$AgNO_3$	210	122	400
$NaNO_3$	776	537	677
$CaCO_3$	400	468	787

metal standard used were incorporated into a factorial, 3×2^3 design.   To keep the number of runs within reasonable limits, only three metals, indium, tin, and lead, were used as standards.

The predicted influence of the number of peak-area measurements on calibration coefficient precision is shown in Table V.10 (114).   These data indicate that the best procedure is to replicate samples rather than planimeter area measurements.   The standard errors show that no significant improvement in precision results from making four area measurements instead of two, whereas precision is markedly improved by increasing the number of samples.

The effect of the number of samples used to determine K and the number of samples for determining the ΔH of an unknown transition on precision, at the 95% confidence limits, is shown in Table V.11.   These results indicate that for the higher levels of precision, several samples are required for calibration.

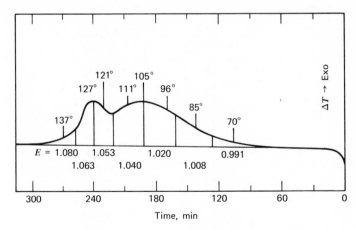

Figure V.30. Integration of DTA curve by parts (82).

TABLE V.10
Predicted Precision of Calibration Constants (114)

No. of measurements		Std. error	Estimated C.L., % (95% C.L.)
Samples	Area		
1	1	0.0286	—
	2	0.0277	±5.5
2	1	0.0202	—
	2	0.0196	±3.9
4	1	0.0168	—
	2	0.0139	±2.8
	4	0.0137	—
8	1	0.0101	—
	2	0.0098	±2.0
	4	0.0096	—

From a practical viewpoint, very few calorimetric studies are reported in which 30 samples are used to calibrate K and to determine ΔH values. Hence, the precision expected is certainly greater than 1%, as indicated in Table V.11. Since most of the measurements are of the first two types, the precision expected is from 4 to 7%, if not greater.

Sturm (115) has described a systematic error in quantitative DTA which is caused by the change in the apparent heat-transfer coefficient and the

TABLE V.11
Predicted Confidence Limits on ΔH (114)

No. of samples		Est. 95% C.L. in percent
Calibration	Unknown	
1	2	±6.52
	4	±6.22
	8	±5.90
2	2	±5.56
	4	±4.80
	8	±4.36
4	2	±4.86
	4	±3.96
	8	±3.42
8	2	±4.40
	4	±3.46
	8	±2.80
30	30	±1.00

apparent heat capacity of the sample and sample holder. The logarithm of the peak area furnished an approximate measure of these changes; the ratio of the logarithms of the areas of the standard and sample provided a correction factor for K.

C. Reaction Kinetics

Almost all of the kinetic methods used in DTA and DSC are based on the equation (59).

$$\frac{d\alpha}{dt} = f(\alpha, T) \tag{V.58}$$

where $d\alpha/dt$ is the rate of reaction, $f(\alpha, T)$ is a function of the amount reacted, and T the absolute temperature at time t. It is also assumed that the rate of heat generation, H, is directly proportional to $d\alpha/dt$, or

$$H = B\rho\left(\frac{d\alpha}{dt}\right) \tag{V.59}$$

where B is the heat per unit volume and ρ is the sample density. Although the kinetic equation used depends on the reaction mechanism, one expression which excludes diffusion-controlled, fusion, and inversion-type reactions is

$$\frac{d\alpha}{dt} = A(1 - \alpha)^n e^{-E/RT} \tag{V.60}$$

where A is the preexponential factor, α the fraction of sample reacted ($0 \leqslant \alpha \leqslant 1$), and n the "order of reaction." This type of equation has been criticized by Clarke et al. (116) as being unusable for solid-state decomposition reactions. They stated that it is misleading, if not meaningless, to use the concept of reaction order; also, the activation energy is not very well defined because it is not known to which of the many processes it applies.

Although the results of kinetic studies by DTA may be questionable, much effort has been expended to derive expressions from which E and n can be calculated using DTA or DSC data. Reviews of the determination of kinetics by DTA have been given by Friedman (117), Sestak and Berggren (118), Bohon (99), Murphy (1), and others.

One of the early methods used to obtain kinetic data from a DTA curve was that of Murray and White (119). They developed theoretical DTG curves and found that: (a) the shapes of DTA and DTG curves were similar, and (b) the maximum temperature difference, ΔT_{max}, occurred near $(d\alpha/dt)_{max}$. Using $n = 1$ for a series of clay samples and by taking the second derivative of the temperature form of equation (V.60),

$$\frac{d\alpha}{dT} = \left(\frac{A}{\beta}\right) e^{-E/RT}(1 - \alpha) \tag{V.61}$$

and setting $d^2\alpha/dT^2 = 0$, and substituting, they obtained

$$\frac{E}{RT_{max}^2} = \left(\frac{A}{\beta}\right) e^{-E/RT_{max}} \tag{V.62}$$

They also integrated equation (V.60) using the approximation

$$\int_{T_0}^{T} e^{-E/RT} \, dT \approx \left(\frac{RT^2}{E}\right) e^{-E/RT} \left(1 - \frac{2RT}{E}\right) \tag{V.63}$$

and equation (V.61) to give

$$-\ln (1 - \alpha_{max}) \approx 1 - \frac{2RT_{max}}{E} \tag{V.64}$$

Kissinger (62) differentiated equation (V.61) and obtained

$$\frac{d[\ln (\beta/T_{max}^2)]}{d(1/T_{max})} = -\frac{E}{R} \tag{V.65}$$

where β is the heating rate. For any value of n (63)

$$\frac{E}{RT_{max}^2} = \left(\frac{An}{\beta}\right)(1 - \alpha_{max})^{n-1} e^{-E/RT_{max}} \tag{V.66}$$

which was developed instead of equation (V.61), and for $n \neq 0$ or 1

$$n(1 - \alpha_{max})^{n-1} \approx 1 + (n - 1)\left(\frac{2RT_{max}}{E}\right) \tag{V.67}$$

Since $(n - 1)(2RT_{max}/E) \leqslant 1$, equation (V.65) may be further approximated by

$$n(1 - \alpha_{max})^{n-1} \approx 1 \tag{V.68}$$

Substitution of equation (V.68) into equation (V.67) gives an approximate equation that is the same as equation (V.61). Thus, Kissinger concluded that equation (V.65) was independent of order. The order of reaction, n, was obtained from the shape index, S, defined as the absolute value of the ratio of the slopes of the tangents to the curve at the inflection points,

$$n \approx 1.26S^{1/2} \tag{V.69}$$

Reed et al. (120) used Kissinger's method in their investigation of the kinetics of the decomposition of benzenediazonium chloride and found that the E value so obtained differed from other methods by about 42%. By this criterion, they judged the method unacceptable. Similar conclusions were made by Melling et al. (59) because the slopes they calculated from the computed DTA curves were all low when using the peak sample temperature. Hence, the method by which Kissinger evaluated E/R was invalid in a practical experiment. Piloyan et al. (121) cited the disadvantages of this method: (a) It requires the determination of several DTA curves at different heating rates, and (b) it was necessary to use a special programming device to control the temperature. However, provided that the appropriate experimental conditions are used, Akita and Kase (122) concluded that the peak minimum of the DTA curve did agree with the maximum rate of reaction, in agreement with Kissinger.

Piloyan et al. (121) developed a kinetic method which was also based on equation (V.60). By substitution of $\Delta T = S(d\alpha/dt)$, where S is the peak area, they obtained

$$\ln \Delta T = C - \ln f(\alpha) - \frac{E}{RT} \qquad (V.70)$$

where C is a constant. If α lies between 0.05 and 0.8 (about up to the peak minimum), $\ln f(\alpha)$ can be neglected and equation (V.70) can be approximated to

$$\ln \Delta T = C_1 - \frac{E}{RT} \qquad (V.71)$$

Estimated errors of the values of E obtained were between 15 and 20%.

Perhaps the most widely used kinetic method in DTA has been that derived by Borchardt and Daniels (22) in 1957. The method is based on the following assumptions: (a) The temperature in the sample and reference materials is uniform; obviously, this only can be applied to stirred liquids and not to solids; (b) heat is transferred by conduction only, a condition easily met with liquids and the temperature ranges usually employed (the heat transfer through the thermocouple is neglected); (c) the heat-transfer coefficients are identical for the sample and reference materials; and (d) the heat capacities of the sample and reference materials must also be identical, a condition approached if dilute sample solutions are investigated.

The actual rate of reaction at any temperature in terms of the slope of the curve ($d\Delta T/dt$) and the height ΔT is

$$-\frac{dN}{dt} = \frac{N_0}{KA}\left[C_p \frac{d\,\Delta T}{dt} + K\,\Delta T\right] \qquad (V.72)$$

where N is the number of moles of reactant present at any time and is equal to the initial number of moles, N_0, minus the number of moles that have reacted, or

$$N = N_0 - \int_0^t - \frac{dm}{dt}\, dt \qquad (V.73)$$

By various manipulations, the rate constant, k, was shown to be equal to

$$k = \left[\frac{KAV}{N_0}\right]^{n-1} \frac{C_p(d\,\Delta T/dt) + K\,\Delta T}{[K(A - a) - C_p\,\Delta T]^n} \qquad (V.74)$$

where V is the volume, A is the total peak area, a is the area up to time t, and n is the order of reaction.   In the case of a first-order reaction, $n = 1$,

$$k = \frac{C_p(d\,\Delta T/dt) + K\,\Delta T}{K(A - a) - C_p\,\Delta T} \qquad (V.75)$$

For differential scanning calorimetry, where $\Delta T = 0$, the total heat is equal to the peak area, A,

$$\Delta H = A \qquad (V.76)$$

It it is assumed that the heat evolved is directly proportional to the number of moles reacted, it follows that

$$-\frac{dN}{dt} = -\frac{N_0}{A}\left(\frac{dH}{dt}\right) \qquad (V.77)$$

and the rate constant is given by

$$k = \frac{(AV/N_0)^{n-1}(dH/dt)}{(A - a)^n} \qquad (V.78)$$

It was stated that this equation was not limited to DSC but could apply to any procedure where the rate of change of any physical property is measured as a function of temperature and time under conditions where the temperature is changing.   The physical property should be nearly independent of temperature.

Equation (V.75) has been simplified by numerous investigations; since the quantities $C_p(d\Delta T/dt)$ and $C_p\Delta T$ are an order of magnitude smaller than the quantities to which they are added and subtracted, they may be neglected to obtain

$$k = \frac{\Delta T}{(A - a)} \qquad (V.79)$$

for a first-order reaction, or

$$k = \frac{(AV/N_0)^{n-1} \Delta T}{(A - a)^n} \qquad (V.80)$$

for the general case.

Padmanabhan et al. (123) and Agarwala and Naik (124) have used the simplified expression, as shown by equation (V.79), to determine the kinetics of a thermal decomposition reaction involving a powdered solid. The use of this expression for solid-state reactions does not appear to be valid in view of the original assumptions made in the derivation of the original equation. Also neglecting $C_p(d\Delta T/dt)$, Borchardt (125) made the approximation

$$-\frac{d(N/N_0)}{dt} = -\frac{d\alpha}{dt} \approx \frac{\Delta T}{A} \qquad (V.81)$$

The kinetic method of Borchardt and Daniels (22) was subjected to an exhaustive examination by Reed et al. (120) in 1965. The equations were integrated numerically, producing theoretical DTA curves which agreed well with the corresponding experimental curves. The effects of the various parameters, such as heating rate, reaction order, E, and so on, on the DTA curves were also established by numerical integration.

Using the decomposition of benzenediazonium chloride, Reed et al. (120) compared the kinetic results obtained by several different methods; these comparisons are shown in Table V.12. They concluded that the Borchardt and Daniels method can be used for the quantitative determination of kinetic parameters if the experimental conditions closely approximate the assumptions of the theory, namely, reaction order, n, with respect to only one

TABLE V.12
Kinetic Constants for the Decomposition of Benzenediazonium Chloride (120)

Method	E, kcal/mole	$\log A$, min^{-1}
DTA		
Reed et al. (120), using Borchardt and Daniel's method	28.7	18.4
Reed et al. (120) using Kissinger's method	16.7	10.8
Borchardt and Daniels (22)	28.3	18.1
	29.1	
	28.5	
Wada (127)	30.6	
Conventional		
Crossley et al. (128)	27.2	17.3
Moelwyn-Hughes and Johnson (129)	27.025	17.26

component, and the absence of temperature gradients and overlapping peaks. Reich (126) modified equation (V.81) to give

$$k = \left(\frac{A}{W_0}\right)^{n-1} \frac{\Delta T}{\bar{a}^n} \tag{V.82}$$

where k is the rate constant, ΔT is the peak height, W_0 is a function of the initial sample mass, and

$$A = \int_0^\infty \Delta T \, dt;$$

$$\bar{a} = \int_0^\infty \Delta T \, dt - \int_0^\infty \Delta T \, dt \tag{V.83}$$

A difference method was developed in which the values from two DTA curves, obtained at two different heating rates, were used to calculate E and n.

Maycock (130), using the DSC curves shown in Figure V.31, and assuming a linear Arrhenius plot, found that E could be calculated by the expression

$$-E = \frac{R \ln d_1 - \ln d_2}{1/T_1 - 1/T_2} = \frac{4.58 \log (\alpha_1/\alpha_2)}{1/T_1 - 1/T_2} \tag{V.84}$$

where d_1 and d_2 are rates of heat evolution at T_1 and T_2, respectively.

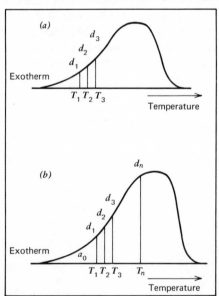

Figure V.31. DSC curves used in kinetic analysis by Maycock (106).

A number of other methods have been discussed, many of them based on the equations described here. For polymer degradation, the book by Reich and Stivala (131) discusses numerous kinetic methods.

D. Differential Scanning Calorimetry

1. INTRODUCTION

The term "differential scanning calorimetry" has become a source of confusion in thermal analysis the past several years. This confusion is understandable because at the present time there are two entirely different types of instruments which use the same name. These two instruments are based on different designs, of which one is actually a differential thermal analysis apparatus. The instruments are:

(a) differential scanning calorimeters which are heat-flow-recording instruments $[(dq/dt)_{\Delta T=0}]$ (Perkin-Elmer, Deltatherm, and others).

(b) "Differential scanning calorimeters" which are differential-temperature-recording or DTA instruments ($\Delta T \neq 0$) (DuPont, Stone, Fisher, Linseis, and others).

The term differential scanning calorimetry (DSC) was apparently first used by Watson et al. (132) to describe the instrumental technique recently developed (1963) by the Perkin–Elmer Corporation. This technique maintained the sample and reference materials isothermal to each other by proper application of electrical energy, as they were heated or cooled at a linear rate. The curve obtained is a recording of heat flow, dH/dt, in mcal/sec, as a function of temperature. A typical DSC curve is shown in Figure V.32. In the true thermodynamics sense, an endothermic curve peak is indicated by a peak in the upward direction (increase in enthalpy), while an exothermic peak is recorded in the opposite direction. In all appearances, the DSC curve looks very similar to that of a DTA curve except for the ordinate axis units. As in DTA, the area enclosed by the DSC curve peak is directly proportional to the enthalpy change,

$$\text{area} = K\Delta Hm \qquad (V.85)$$

except that K is independent of temperature.

The basic difference in the DSC instruments is schematically illustrated in Figure V.33. The DSC instrument shown is that described by O'Neill (133) while the "DSC" diagram, which is a DTA system, is that described by Baxter (134) or David (135). The basic difference between the two units is easily seen although the curves are similar in appearance. Much the same

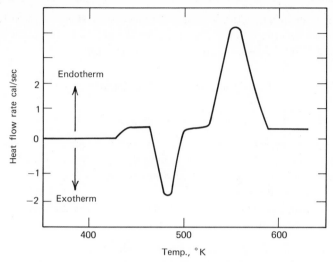

Figure V.32. Recommended presentation of a DSC curve.

type of information is available from the curves; however, until recently, the DSC sample holder has been limited to about 500°C, while DTA holders have been capable of much higher operation. For quantitative measurements, ΔH for example, the DSC technique is easier to use since the calibration coefficient, K, does not change with temperature. For overlapping curve peaks, there is only one value for K, which simplifies calculations.

Other types of calorimeters, such as the Calvert, Deltatherm, and others have been reviewed by Wilhoit (136). A bibliography of DSC application from 1964 to 1970 is available (137).

2. THEORY

The theory of the Perkin–Elmer calorimeter has been presented by O'Neil (133), Gray (60), and Flynn (135), while that of the DuPont and Stone instruments has been discussed by Baxter (134) and David (135), respectively. The theory of the latter two instruments has been discussed previously in this chapter.

Using the DSC curve in Figure V.34, Gray (60) developed the basic equation relating dH/dt to the measured quantities, as in the case of DTA,

$$\frac{dH}{dt} = -\frac{dq}{dt} + (C_s - C_r)\frac{dT_p}{dt} - RC_s\frac{d^2q}{dt^2} \qquad \text{(V.86)}$$

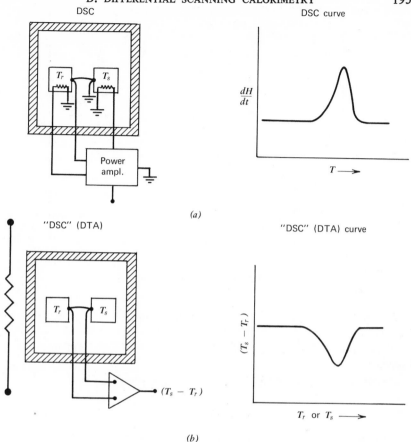

Figure V.33. Difference between (a) DSC and (b) "DSC" techniques (130).

The expression for dH/dt again involves the sum of three terms, as was described in the DTA theory. Two differences are noted between DSC and DTA: (a) the thermal resistance, R, occurs only in the third term in the equation; and (b) the area under the curve peak is $\Delta q = -\Delta H$. A calibration coefficient is still required, but it is used to convert area to calories and is an electrical conversion factor rather than the thermal constant used in DTA.

A somewhat more sophisticated treatment is that given by Flynn (138) which concerns three critical parameters of the DSC curve:

(a) The steady slope of the curve caused by a linearly increasing temperature increment and proportional to an interfacial conductivity term.

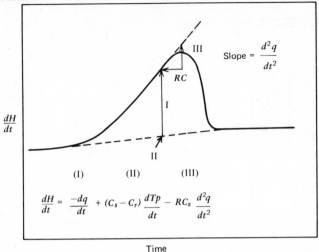

Figure V.34. Application of equation (V.86) to a DSC curve (60).

(b) An onset temperature of the transition, T_i, obtained from the intersection of this slope with the baseline.

(c) A decay constant, k_i, back to the baseline at the completion of the transition.

The slope of the curve can be described by Newton's law of cooling,

$$\frac{dq}{dt} = Ah(T_2 - T_1) \qquad \text{(V.87)}$$

where h is the interfacial thermal conductivity, A is the area of the interface, and $T_2 - T_1$ is the temperature difference between the sample and the holder. The temperature of the sample, T_1, is described by

$$T_1 = T_0 + \frac{1}{mc} \int \frac{dq}{dt}\, dt \qquad \text{(V.88)}$$

where T_0 is the initial temperature, m is the sample mass, and c is the heat capacity of the sample. At a transition, T_1 will remain at the transition temperature, T_i, for a residence time, t_i, determined from

$$\int_0^{t+t_i} \frac{dq}{dt}\, dt = m\, \Delta H_i \qquad \text{(V.89)}$$

where ΔH_i is the heat of transition.

From these equations, for a constant heat capacity, the steady slope during the melting transition is

$$\text{slope} = \beta hA = k_m\beta cm \qquad (V.90)$$

The onset temperature, T_i, is

$$T_i = T_m + \frac{\beta cm}{hA} = T_m + \frac{\beta}{k_m} \qquad (V.91)$$

and the decay constant at the termination of the transition is

$$\text{decay constant} = \frac{hA}{cm} = k_m \qquad (V.92)$$

In DSC curves it is usually assumed that the rate of change of the reference material temperature is equal to the programmed temperature. Brennan et al. (139) have considered this problem mathematically and found that the rate of change is controlled by the time constant of the reference material. If the time constant is about 1 sec, the assumption is valid after a few seconds. However, if it is much larger, the assumption will not apply.

References

1. Le Chatelier, H., *Bull. Soc. Franc. Mineral*, **10**, 203 (1887).
2. Murphy, C. B., *Minerals Sci. Eng.*, **2**, Oct., 51 (1970).
3. Ashley, H. E., *Ind. Eng. Chem.*, **3**, 91 (1911).
4. Wholin, R., *Sprechsaal*, **46**, 749, 767, 781 (1913).
5. Rieke, R., *Sprechsaal*, **44**, 637 (1911).
6. Wallach, H., *Compt. Rend.*, **157**, 48 (1913).
7. Mellor, J. W., and A. D. Holdcraft, *Trans. Brit. Ceram. Soc.*, **10**, 94 (1910–1911).
8. Roberts-Austen, W. C., *Proc. Inst. Mech. Engrs.*, (London), (1899); *Metallographist*, **2**, 186 (1899).
9. Burgess, G. K., *Nat . Bur. Std. (U.S.) Bull.*, **5**, 199 (1908–09).
10. Speil, S., L. H. Berkelhamer, J. A. Pask, and B. Davis, *U.S. Bur. Mines, Tech. Papers*, **664** (1945).
11. Houldsworth, H. S., and J. W. Cobb, *Trans. Brit. Ceram. Soc.*, **22**, 111 (1922–1923).
12. Fenner, C. N., *Am. J. Sci.*, **36**, 331 (1913).
13. Berkelhamer, L. H., *U.S. Bur. Mines, Rept. Invest.*, **R13762** (1944).
14. Kerr, P. F., and J. L. Kulp, *Am. Mineralogist*, **33**, 387 (1948).
15. Kauffman, A. J., and E. D. Dilling, *Econ. Geol.*, **45**, 222 (1950).
16. Foldvari-Vogl, M., *Acta Geol. Acad. Sci. Hung.*, **5**, 1 (1958).
17. Norton, F. H., *J. Am. Ceram. Soc.*, **22**, 54 (1939).
18. Grim, R. E., *Ann. N. Y. Acad. Sci.*, **53**, 1031 (1951).
19. Mackenzie, R. C., *Differential Thermal Analysis of Clays*, Central Press, Aberdeen, Scotland, 1957.
20. Stone, R. L., *J. Am. Ceram. Soc.*, **35**, 76 (1952).
21. Borchardt, H. J., and F. Daniels, *J. Phys. Chem.*, **61**, 917 (1957).

22. Borchardt, H. J., and F. Daniels, *J. Am. Chem. Soc.*, **79**, 41 (1957).
23. Wendlandt, W. W., in *Technique of Inorganic Chemistry*, Vol. I, H. B. Jonassen and A. Weissberger, eds., Interscience, New York, 1963, p. 209.
24. Wendlandt, W. W., *J. Chem. Educ.*, **49**, A623 (1972).
25. Murphy, C. B., *Anal. Chem.*, **30**, 867 (1958).
26. Murphy, C. B., *Anal. Chem.*, **44**, 513R (1972).
27. Smothers, W. J., and Y. Chiang, *Differential Thermal Analysis: Theory and Practice*, Chemical Publishing Co., New York, 1958.
28. Smothers, W. J., and Y. Chiang, Ref. 27, Second Ed., 1966.
29. Garn, P. D., *Thermoanalytical Methods of Investigation*, Academic, New York, 1965.
30. Mackenzie, R. C., ed., *Differential Thermal Analysis*, Academic, London, Vol. 1, 1970.
31. Kissinger, H. E., and S. B. Newman, in *Differential Thermal Analysis in Analytical Chemistry of Polymers*, Vol. XII, Part II, G. M. Kline, ed., Interscience, New York, 1962.
32. Gordon, S., in *Encyclopedia of Science and Technology*, Vol. 13, McGraw-Hill, New York, 1960, pp. 556–559.
33. Gordon, S., and C. Campbell, in *Handbook of Analytical Chemistry*, L. Meites, ed., McGraw-Hill, New York, 1963.
34. Barrall, E. M., and J. F. Johnson, in *Techniques and Methods of Polymer Evaluation*, P. E. Slade and L. T. Jenkins, eds., Vol. 1, Marcel-Dekker, New York, 1966, Chap. 1.
35. David, D. J., Ref. 34, Chap. 2.
36. Barrall, E. M., in *Guide to Modern Methods of Instrumental Analysis*, T. H. Gouw, ed., Wiley-Interscience, New York, 1972, Chap. 12.
37. Schultze, D., *Differentialthermoanalyze*, Deutscher Verlag der Wissenschaften, Berlin, 1969.
38. Ramachandran, V. S., *Differential Thermal Analysis in Cement Chemistry*, Chemical Publishing Co., New York, 1969.
39. Wunderlich, B., in *Physical Methods of Chemistry*, A. Weissberger and B. W. Rossiter, eds., Vol. 1, Part V, Wiley-Interscience, New York, 1971, Chap. VIII.
40. R. S. Porter and J. F. Johnson, eds., *Analytical Calorimetry*, Vol. 1, Plenum, New York, 1968.
41. R. S. Porter and J. F. Johnson, eds., Ref. 40, Vol. 2, 1970.
42. R. F. Schwenker and P. D. Garn, eds., *Thermal Analysis*, Academic, New York, Vols. 1 and 2, 1969.
43. Vold, M. J., *Anal. Chem.*, **21**, 683 (1949).
44. Boersma, S. L., *J. Am. Ceram. Soc.*, **38**, 281 (1955).
45. Lukaszewski, G. M., *Lab. Practice*, **14**, 1277 (1965).
46. Ref. 45, p. 40.
47. Lukaszewski, G. M., *Lab. Practice*, **15**, 75 (1966).
48. Ref. 47, p. 82.
49. Ref. 47, p. 187.
50. Ref. 47, p. 302.
51. Ref. 47, p. 431.
52. Ref. 47, p. 551.
53. Ref. 47, p. 664.
54. Ref. 47, p. 762.
55. Ref. 47, p. 861.
56. David, D. J., *Anal. Chem.*, **36**, 2162 (1964).

57. David, D. J., *Lab. Equip. Digest*, June–Aug., 110 (1968).
58. Pacor, P., *Anal. Chim. Acta*, **37**, 200 (1967).
59. Melling, R., F. W. Wilburn, and R. M. McIntosh, *Anal. Chem.*, **41**, 1275 (1969).
60. Gray, A. P., in *Analytical Calorimetry*, R. F. Porter and J. M. Johnson, eds., Plenum, New York, 1968, p. 209.
61. Garn, P. D., Ref. 29, p. 60.
62. Kissinger, H. E., *J. Res. Nat. Bur. Standards*, **57**, 217 (1956).
63. Kissinger, H. E., *Anal. Chem.*, **29**, 1702 (1957).
64. Barrall, E. M., and L. B. Rogers, *J. Inorg. Nucl. Chem.*, **28**, 41 (1966).
65. Langer, A. M., and P. F. Kerr, *DuPont Thermogram*, **3**, No. 1, 1 (1968).
66. Vassallo, D. A., and J. C. Harden, *Anal. Chem.*, **34**, 132 (1962).
67. Johnson, J. F., and G. W. Miller, *Thermochim. Acta*, **1**, 373 (1970).
68. Sarasohn, I. M., *DuPont Thermogram*, **2**, No. 1, 1 (1965).
69. Stone, R. L., *Anal. Chem.*, **32**, 1582 (1960).
70. *Mettler Thermal Technique Series, Tech. Bull. No T-106.*
71. David, D. J., *Am. Lab.*, Jan., 35 (1970).
72. Locke, C. E., in *Proceedings of the Third Toronto Symposium on Thermal Analysis*, H. G. McAdie, ed., Chemical Institute of Canada, Toronto, 1969, p. 251.
73. Garn, P. D., *Anal. Chem.*, **37**, 77 (1965).
74. Levy, P., G. Nieuweboer, and L. C. Semanski, *Thermochim. Acta*, **1**, 429 (1960).
75. David, D. J., *Anal. Chem.*, **37**, 82 (1965).
76. Wilburn, F. W., J. R. Hesford, and J. R. Flowers, *Anal. Chem.*, **40**, 777 (1968).
77. Dosch, E. L., *Thermochim. Acta*, **1**, 367 (1970).
78. Lehmann, H., S. S. Das, and H. H. Paetsch, *Tenind.-Ztg. u Keram. Rundschau*, (1954), 1.
79. Bayliss, P., and S. St. J. Warne, *Am. Mineralogist*, **47**, 775 (1962).
80. Arens, P. L., *A Study of the Differential Thermal Analysis of Clays and Clay Minerals*, Excelsiors Foto-Ottset, The Hague, 1951.
81. Gerard-Hirne, J., and C. Lamy, *Bull. Soc. Franc. Ceram.*, **1951**, 26.
82. Webb, T. T., *Nature*, **174**, 686 (1954).
83. Mackenzie, R. C., *Nature*, **174**, 688 (1954).
84. Sarasohn, I. M., ACS Short Course, American Chemical Society, Washington, D.C., 1969.
85. Smyth, H. T., *J. Am. Ceram. Soc.*, **34**, 221 (1951).
86. Barrall, E. M., and L. B. Rogers, *Anal. Chem.*, **34**, 1101 (1962).
87. David, D. J., D. A. Ninke, and B. Duncan, *Am. Lab.*, Jan., 31 (1971).
88. Hauser, R. E., B. S. thesis, N.Y. State College of Ceramics, Alfred, N.Y., 1953.
89. Norton, F. H., *J. Am. Ceram. Soc.*, **22**, 54 (1939).
90. Grimshaw, R. W., E. Heaton, and A. L. Roberts, *Trans. Brit. Ceram. Soc.*, **44**, 76 (1945).
91. Carthew, A. R., *Am. Mineralogist*, **40**, 107 (1955).
92. Barrall, E. M., and L. B. Rogers, *Anal. Chem.*, **34**, 1106 (1962).
93. Negishi, A., and T. Ozawa, *Thermochim. Acta*, **2**, 89 (1971).
94. Wada, M., Y. Iida, and S. Ozaki, *Japan J. Appl. Phys.*, **8**, 1569 (1969).
94a. Dean, L. A., *Soil Sci.*, **63**, 95 (1947).
95. Gruver, R. M., *J. Am. Ceram. Soc.*, **31**, 323 (1948).
96. Bollin, E. M., and A. J. Bauman, in *Analytical Calorimetry*, R. F. Porter and J. M. Johnson, eds., Vol. 2, Plenum, New York, 1970, p. 339.
97. Barshad, I., *Am. Mineralogist*, **37**, 667 (1952).

200 DIFFERENTIAL THERMAL ANALYSIS AND SCANNING CALORIMETRY

98. deJong, G. J., J. Am. Ceram. Soc., 40, 42 (1957).
99. Bohon, R. L., Proceedings of the First Toronto Symposium on Thermal Analysis, H. G. McAdie, ed., Chemical Institute of Canada, Toronto, 1965, p. 63.
100. Ozawa, T., Bull. Chem. Soc. Japan, 39, 2071 (1966).
101. Wittels, M., Am. Mineralogist, 36, 615 (1951).
102. Ref. 101, p. 760.
103. Collins, W. E., in Analytical Calorimetry, R. S. Porter and J. M. Johnson, eds., Vol. 2, Plenum, New York, 1970, p. 353.
104. David, D. J., Ref. 103, p. 369.
105. Bohon, R. L., Anal. Chem., 35, 1845 (1963).
106. Chiu, J., in Analytical Calorimetry, R. F. Porter and J. M. Johnson, eds., Vol. 2, Plenum, New York, 1970, p. 171.
107. Wendlandt, W. W., and J. R. Williams, International Confederation of Thermal Analysis III, Davos, Switzerland, Aug. 1971, paper I.
108. Currell, B. R., in Thermal Analysis, R. F. Schwenker and P. D. Garn, eds., Vol. 2, Academic, New York, 1969, p. 1185.
109. Wiedemann, H. G., and A. van Tets, Z. Anal. Chem., 233, 161 (1968).
110. Wiedemann, H. G., and A. van Tets, Thermochim. Acta, 1, 159 (1970).
111. Ramachandran, V. S., and P. J. Sereda, Nature, physical Science, 233, 134 (1971).
112. Davis, C. E., and D. A. Holdridge, Clay Minerals, 8, 193 (1969).
113. Ramachandran, V. S., J. Thermal. Anal., 3, 181 (1971).
114. Schwenker, R. F., and J. C. Whitwell, in Analytical Calorimetry, R. F. Porter and J. M. Johnson, eds., Vol. 1, Plenum, New York, 1968, p. 249.
115. Sturm, E., Thermochim. Acta, 4, 461 (1972).
116. Clarke, T. A., E. L. Evans, K. G. Robbins, and J. M. Thomas, Chem. Comm., 1969, 266.
117. Friedman, H. L., Proceedings of the Third Toronto Symposium on Thermal Analysis, H. G. McAdie, ed., Chemical Institute of Canada, Toronto, 1969, p. 127.
118. Sestak, J., and G. Berggren, Svazek, 64, 695 (1970).
119. Murray, P., and J. White, Trans. Brit. Ceram. Soc., 54, 204 (1955).
120. Reed, R. L., L. Weber, and B. S. Gottfried, I & EC Fundamentals, 4, 38 (1965).
121. Piloyan, G. O., I. D. Ryabchikov, and O. S. Novikova, Nature, 212, 1229 (1966).
122. Akita, K., and M. Kase, J. Phys. Chem., 72, 906 (1968).
123. Padmanablan, V. M., S. C. Saraiya, and A. K. Sundaram, J. Inorg. Nucl. Chem., 12, 356 (1960).
124. Agarwala, R. P., and M. C. Naik, Anal. Chim. Acta, 24, 128 (1960).
125. Borchardt, H. J., J. Inorg. Nucl. Chem., 12, 252 (1960).
126. Reich, L., J. Inorg. Nucl. Chem., 28, 1329 (1966).
127. Wada, G., Nippon Kagaku Zasshi, 1960, 1956.
128. Crossley, M. L., R. H. Kienle, and C. H. Benbrook, J. Am. Chem. Soc., 62, 1400 (1940).
129. Moelwyn-Hughes, E. A., and P. Johnson, Trans. Faraday Soc., 36, 948 (1940).
130. Maycock, J. N., Thermochim. Acta, 1, 389 (1970).
131. Reich, L., and S. S. Stivala, Elements of Polymer Degradation, McGraw-Hill, New York, 1971.
132. Watson, E. S., M. J. O'Neill, J. Justin, and N. Brenner, Anal. Chem., 36, 1233 (1964).
133. O'Neill, M. J., Anal. Chem., 36, 1238 (1964).
134. Baxter, R. A., in Thermal Analysis, R. F. Schwenker and P. D. Garn, eds., Academic, New York, 1969, p. 65.

135. David, D. J., *J. Thermal Anal.*, **3**, 247 (1971).
136. Wilhoit, R. C., *J. Chem. Educ.*, **44**, A571 (1967).
137. Perkin–Elmer Corp., Norwalk, Conn., May, 1970.
138. Flynn, J. H., in *Status of Thermal Analysis*, O. Menis, ed., NBS Special Publication 338, U.S. Gov't. Printing Office, Washington, D.C., Oct. 1970, p. 119.
139. Brennan, W. P., B. Miller, and J. C. Whitnell, *Thermochim. Acta*, **2**, 354 (1971).

CHAPTER VI

DIFFERENTIAL THERMAL ANALYSIS AND DIFFERENTIAL SCANNING CALORIMETRY INSTRUMENTATION

A. Differential Thermal Analysis

1. INTRODUCTION

As for thermogravimetry, there are a large number of different types of instruments for DTA. Perhaps the reason for this multiplicity is that before the advent of reliable commercial DTA instruments, each investigator designed and built his own apparatus. Needless to say, each apparatus varied widely in the type of components employed. Various types of sample holders, furnaces, ΔT amplifying devices, temperature programmers, and recorders were described. Because of this, there was little agreement in the DTA curves obtained for identical materials from instrument to instrument or from laboratory to laboratory. In fact, it was almost impossible to duplicate exactly the results obtained by another instrument in another laboratory, and this, at times, led to much controversy. Since the advent of good commercial DTA instruments in the late 1950's, some standardization has taken place, so that it is now possible to duplicate DTA data, providing, of course, that identical pyrolysis conditions are employed.

A typical DTA apparatus is illustrated schematically in Figure VI.1. The apparatus generally consists of (*a*) a furnace or heating device, (*b*) a sample holder, (*c*) a low-level dc amplifier, (*d*) a differential temperature detector, (*e*) a furnace temperature programmer, (*f*) a recorder, and (*g*) control equipment for maintaining a suitable atmosphere in the furnace and sample holder. Many modifications have been made of this basic design, but all instruments measure the differential temperature of the sample as a function of temperature or time (assuming that the temperature rise is linear with respect to time).

2. SAMPLE HOLDERS

One of the most important components of a DTA apparatus is the type of sample holder (and identical reference material holder) employed. There are a wide variety of sample holders available, in commercial instruments or that have been described. The type of sample holder (1) used depends, of

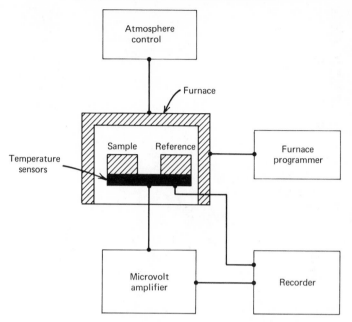

Figure VI.1. Schematic diagram of a typical DTA apparatus.

course, on the nature and quantity of the sample and also on the maximum temperature to be investigated. Sample holders have been constructed from alumina, zirconia (ZrO_2), borosilicate glass, Vycor glass, fused quartz, beryllia, boron nitride, graphite, stainless steel, nickel, aluminum, platinum or platinum alloys, silver, copper, tungsten, the sample itself, and numerous other materials. Some typical sample holders used in DTA are shown in Figure VI.2. In (a), the sample (~100 mg) is pressed into a closed-end tube and the tube is placed over the ceramic insulator tube containing the thermo-junction (2). This type of sample holder will permit the sample to dissociate in a self-generated atmosphere. It cannot be used with samples that fuse on heating, however. For determining heats of explosion for a number of explosive materials, the isochoric sample holder in (b) was used by Bohon (3). It consisted of a stainless-steel body and cap that was sealed with a screw cap and a copper gasket. The internal volume was about 0.085 ml and contained about 25 mg of sample. By means of a loading chamber, the sample holder could be charged with a gas at pressures up to 1000 p.s.i.g. Mazieres (4) developed the microsample holder in (c) for use with samples from 1 to 200 μg in mass. The sample was contained in a chamber drilled in the thermojunction itself. A similar cup-type holder was also described

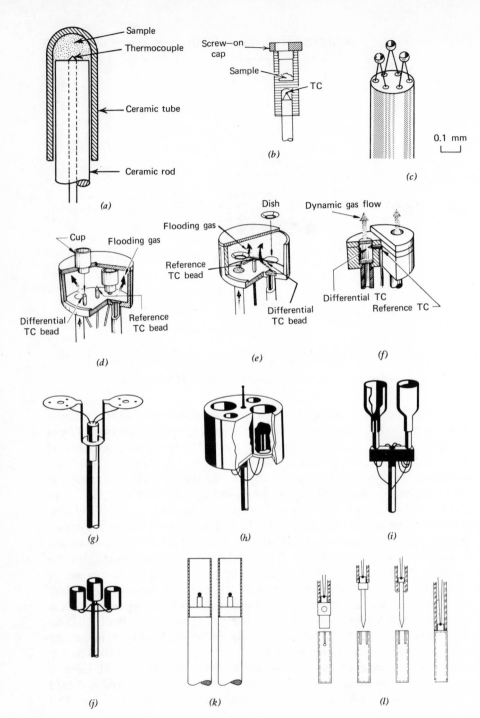

Figure VI.2. DTA sample holders.

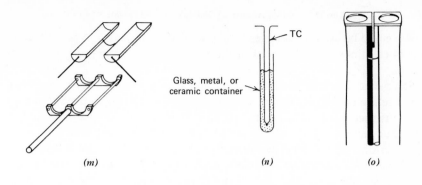

(m)

TC

Glass, metal, or ceramic container

(n)

(o)

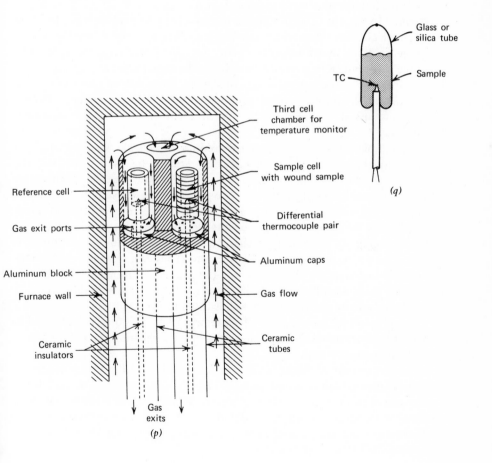

Glass or silica tube

TC

Sample

(q)

Third cell chamber for temperature monitor

Sample cell with wound sample

Reference cell

Differential thermocouple pair

Gas exit ports

Aluminum caps

Aluminum block

Furnace wall

Gas flow

Ceramic insulators

Ceramic tubes

Gas exits

(p)

which could be used for samples from 0.1 to 10 mg in mass. Some difficulty would certainly be experienced in the handling of microgram quantities of sample.

Sample holders used in the Stone instruments are shown in (d)–(f). Small cups are used in (d) to contain samples from 10 to 200 mg in mass; the cups are constructed from aluminum, stainless steel, nickel, or platinum and/or palladium alloys. For smaller samples, 0.1 to 20 mg, the highly sensitive ring thermocouple holder, as shown in (E), is used. The sample dishes can be made from aluminum, stainless steel, or platinum by the investigator using a simple press and die. True dynamic gas atmosphere control is featured in the sample holder in (f). The gas flow is through the sample and reference materials; it cannot be used with samples that fuse, however.

Sample holders illustrated in (g)–(j) are used in the Mettler thermoanalyzer system. In (g), the sample is placed in a small cup or crucible and placed on the small circular discs which contain the thermojunction. A block-type sample holder is shown in (h) in which an alumina block is employed. The sample is contained in a crucible which may be constructed from platinum or other metals. For macro amounts of sample, the holder in (i) may be used. The containers are constructed of alumina or of different metals. If only small amounts of sample are to be studied, the micro crucible sample holder in (j) may be used.

The sample holders in (k)–(m) are used in Linseis DTA equipment. In (k), removable sleeves made of metal or ceramic are used to contain the sample. This type of sample holder is convenient for cleaning purposes as the sleeve may be easily removed, leaving the exposed sample. A disadvantage of this type of sample holder is that the thermocouple is in direct contact with the sample and may be attacked by corrosive sample materials, thus changing its EMF output characteristics. A similar disadvantage is present for the probe-type sample holders in (l) and (n), in which the thermocouple is immersed in the interior of the sample. The glass sample container in (n) is usually a disposable capillary tube 1–2 mm in diameter. A sample holder for horizontal use is illustrated in (m).

A sample holder for use at very high temperatures (2200°C) is illustrated in (o). It is constructed of tungsten-tungsten, 26% rhenium, and is for use in the Mettler instruments.

The unique sample holder in (p) was developed by Miller (5) for use in the study of textile filaments and yarns. The filament sample is wound in the grooves cut in the outer surface of an aluminum cylinder. Three identical cylinders are mounted symmetrically in the center of a vertical furnace; two are used for the sample and reference materials, while the third is used to monitor the furnace temperature.

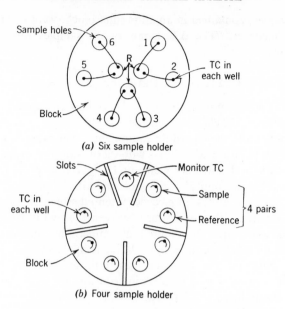

(a) Six sample holder

(b) Four sample holder

Figure VI.3. Multiple sample holder arrangements.

The DTA curves of samples in sealed tubes can be determined by use of the sample holder in (q).   Sealed tube holders are described in a later section of this chapter.

To increase the number of samples that can be studied at any one given time, various multiple sample holders have been described, capable of studying three, four, five, or six samples simultaneously.   Several arrangements of these holders are illustrated in Figure VI.3.

The arrangement in (a), which is due to Kulp and Kerr (6, 7), contained six sample wells and three reference wells.   Each sample and reference well was ¼ in. in diameter and ⅜ in. in deep, drilled in a nickel metal block.   A multipoint recorder was used to record the temperature differences for each sample-reference combination.

A more recent multiple sample holder, as shown in (b), is used in the Delta-therm instrument.   Four separate sample-reference pairs are employed, as well as a monitor thermocouple.   Slits are cut at various radial intervals in the block to prevent thermal gradients from one sample interfering with an adjacent reference well.   Four separate dc amplifiers are used in conjunction with a four-channel recorder.

A five-sample multiple sample block has also been described by Cox and McGlynn (8).   A somewhat different approach was used by Burr (9) in which five samples, contained in an aluminum block, were recorded using a multichannel recorder. The ΔT signals of the samples were recorded at 36-sec intervals.

3. ΔT and T Detection Systems

The choice of a temperature detection device depends upon the maximum temperature desired, the chemical reactivity of the sample, and the sensitivity of the dc amplifier and the recording equipment.   The most common means of differential temperature detection is with thermocouples, although thermopiles, thermistors, and resistance elements have been employed.   For high-temperature studies, an optical pyrometer may also be practical.

A thermocouple generates an electric potential which is roughly proportional to the difference in temperature between the two junctions (Seebeck effect), and is well suited for differential temperature measurements (10). It may also be used for absolute and relative temperature measurements by keeping one junction, the reference junction, at constant temperature. Thermocouples normally used in DTA instruments are shown in Table VI.1. Temperature limits listed are for relatively accurate measurements with 20-gauge wire in air.   The copper-constantan thermocouple is the most popular for use from -150 to $250°C$ and is highly stable and reproducible over this temperature range.   Noble metal thermocouples, especially the platinum-platinum, 10% rhodium, are preferred for applications requiring high accuracy in the range 500–1200°C.   The Platinel-type thermocouples are also useful in this range and have the advantage of a larger thermoelectric power ($\sim$40 $\mu V/°C$).   For temperatures up to 3000°C, tantalum carbide versus graphite has been suggested (11).

To increase the output signal from the differential thermocouples without the use of an amplifier, thermopiles have been employed (8, 12, 13).   The

TABLE VI.1
Characteristics of Some Typical Thermocouples at $25°C$ (10)

ISA type	Metal No. 1 (positive)	Metal No. 2 (negative)	Max. temp., °C	Thermoelectric power, ($\mu V/°C$)
S	Platinum	Platinum–10% Rhodium	1600	5.5
T	Copper	Constantan	250	40
Y, J	Iron	Constantan	450	51
E	Chromel	Constantan	1000	59
K	Chromel	Alumel	1000	41
—	Tungsten	Tungsten–26% Rhenium	2200	3.3

advantage of such a system is the greater output signal with a lower noise level, due to lack of electronic amplification. A five-thermocouple ΔT thermopile used in the Mettler DTA 2000 system has a ΔT sensitivity of 115 $\mu V/°C$.

The usual mechanical configuration of the thermocouple is that of two wires welded together to form a thermojunction. Other types of thermo-couple configurations that have been proposed are the thin-film type (14–16) and the disc (17, 18) type. The thin-film thermocouples eliminate the problems of attempting to match the thermojunctions formed from wire elements. The former can be made light in weight, and can be exactly matched by the evaporation of thin films of dissimilar metals which overlap to form the thermojunction. Preparation of nickel-gold thin-layer thermo-couples is illustrated in Figure IV.3 (15, 16). Gold was vacuum evaporated on the quartz plates, followed by electroplating of the nickel. Aluminum sample pans were then used to contain the samples in the temperature range from −125 to 500°C. The thermoelectric power of the thermocouples increased from 10 $\mu V/°C$ at 25°C to about 25 $\mu V/°C$ at 200°C.

The disk-type thermocouple and sample holder (18) are shown in Figure VI.4. The disk is made of constantan and serves as the major path of heat

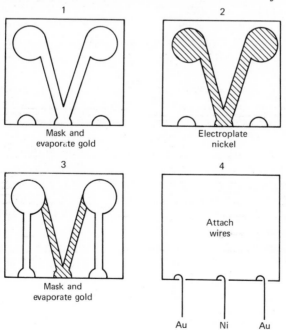

Figure VI.3. Preparation of thin film thermocouples (13, 14).

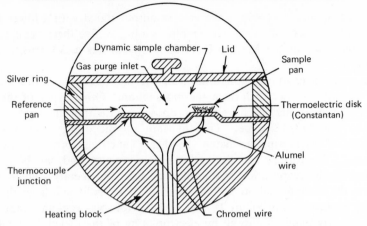

Figure VI.4. Disk-type thermocouple described by Baxter (16).

transfer to and from the sample and also as one-half of the ΔT-measuring thermocouple. A Chromel wire is connected to each raised indentation, thus forming a Chromel-constantan differential thermocouple system. This system is usable in the temperature range from -150 to $600°C$. Yamamoto et al. (17) used a dumbbell-shaped piece of Chromel which consisted of two circular disks connected by a narrow strip. Alumel wire, welded to the center of each disk, served as the other thermocouple junction. Other thermoelectric disks are shown in Section 2.

As with many other analytical techniques, the temperature axis used in differential thermal analysis (and DSC) must be calibrated with materials having known transition temperatures. The International Confederation of Thermal Analysis (ICTA) has been very active in developing a set of standard materials for this purpose (19) and has worked with the U.S. National Bureau of Standards to have these materials made commercially available (20). The temperature range from 125 to 900°C may be covered by use of the ten standard materials given in Table VI.2. The results of the ICTA "round-robin" study with 24 cooperating laboratories have been reported by Menis and Sterling (20).

A high-precision method for calibrating the temperature axis of a DTA apparatus has been discussed (20), using the NBS-ICTA Standard Reference Materials Nos. 758, 759, and 760. These standard materials comprise a total of eight inorganic compounds, which exhibit $solid_1 \rightleftarrows solid_2$ transitions, and two high-purity metals.

The calibration of the temperature axis of the Perkin-Elmer DSC-2 has been discussed by O'Neill and Fyans (24). Temperature readout in this

TABLE VI.2
NBS–ICTA Standard Reference Materials (20)[a]

No. 758 125–435°C		No. 759 295–675°C		No. 760 570–940°C	
Potassium nitrate	10 g	Potassium perchlorate	10 g	Quartz	3 g
Indium	3 g	Silver sulfate	3 g	Potassium sulfate	10 g
Tin	3 g	Quartz	3 g	Potassium chromate	10 g
Potassium perchlorate	10 g	Potassium sulfate	10 g	Barium carbonate	10 g
Silver sulfate	3 g	Potassium chromate	10 g	Strontium carbonate	10 g

[a] Available at $45 set from the Office of Standard Reference Materials, Room B314, Chemistry Building, National Bureau of Standards, Washington, D.C. 20234. A certificate is furnished with each set giving the mean extrapolated onset and peak temperatures for each sample. Definitions of these points on the differential thermal curve are included in the certificate.

particular instrument can be in °C, °K, or °F from 00.0 to 999.9°, and is switch-selected. A calibration curve obtained at a heating rate of 40°F/min is shown in Figure VI.5. Since many industrial processes operate in terms of °F, it is often convenient to use this scale in DSC studies.

The accuracy of temperature calibration, using the same metals shown in Figure VI.5, is illustrated in Table VI.3. Temperatures given are the indicated onset temperatures obtained by extrapolating the curve peak leading

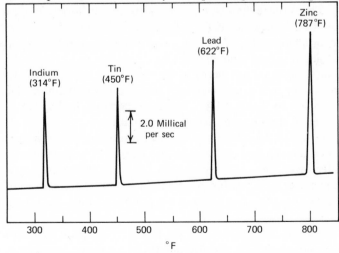

Figure VI.5. Typical temperature calibration curve for DSC (122).

TABLE VI.3
Accuracy of Temperature Calibration of Perkin-Elmer DSC-2 (24)

Standard	Literature M.P. °C	Found °C	Error °C
Indium	156.63	156.79	+0.16
Tin	231.97	231.95	−0.02
Lead	327.50	327.36	−0.14
Zinc	419.58	419.70	+0.12
Aluminum	660.37	659.53	−0.84

edge to the extrapolated base line.   To eliminate subjective bias, this was calculated by a computer analysis.

Fairly high-resistance thermistors, 100,000 Ω at ambient temperature, connected in a bridge circuit have been used to detect the differential temperature (22, 23).   This method does not normally require the use of a dc amplifier.   Because their resistance decreases rapidly with increase in temperature, thermistors are generally only useful up to about 300°C (23).

4. FURNACES AND FURNACE TEMPERATURE PROGRAMMERS

Again, as in the preceding section, the choice of furnace heating element and type of furnace depends upon the temperature range under investigation. DTA furnaces have been described which operate in the range from −190 to 2800°C.   The furnace may be mounted vertically or horizontally; it may be heated by a resistance element, by infrared radiation (25, 26), by high-frequency rf oscillation (11, 27), or by a coil of tubing through which a heated or cooled liquid or gas is circulated (28).

Resistance elements are perhaps the most widely used in furnace construction.   Some resistance elements and their approximate temperature limits are given in Table VI.4.   These temperature limits, are of course, dependent upon the furnace design and insulation.

The wide variety of DTA furnace configurations is shown in Figure VI.6. In (a), a design described by Vassallo and Harden (29), the furnace is heated by a heater cartridge.   It has provision for rapid cooling or for use below room temperature by passing a coolant through the cooling coils which surround the furnace.   Sample and reference materials are placed in glass capillary tubes.

More sophisticated furnaces, for use in the temperature range from −150 to 2400°C, are shown by the Mettler thermoanalyzer furnaces in (b)–(e). The furnace in (b) is for use from −150 to 400°C and uses a Kanthal resistance

TABLE VI.4
Maximum Temperature Limits for Furnace Resistance Elements

Element	Approximate temperature, °C
Nichrome	1000
Kanthal	1350
Platinum	1400
Platinum-10% rhodium	1500
Rhodium	1800
Tantalum	1330
Globar	1500
Kanthal Super	1600
Molybdenum	2200
Platinum-20% rhodium	1500
Chromel A	1100
Tungsten	2800

wire heater element. For use from 25 to 1000°C, the furnace in (c) is employed. This furnace also uses Kanthal heater elements and features high-vacuum operation. The high-temperature furnace (d) is for use from 25 to 1600°C, also under high-vacuum conditions. It contains a furnace winding composed of super-Kanthal. Recently introduced was the super-high-temperature furnace, (e), which can be used in the temperature range from 400 to 2400°C. The furnace heater elements are constructed of tungsten. The sample holders described in Section 2 can be used with the Mettler furnaces described here.

The requirements for a good DTA furnace include symmetry in heating and the ability of the heater elements to heat uniformly. The furnace temperature distribution must be uniform in the area of the sample holder for good results. Wiedemann (30) has reported the temperature distribution curves of the Mettler furnace, as illustrated in Figure VI.7. The gray area indicates the zone of homogeneous temperature in relation to the position of the sample holder. A temperature distribution study has been given by Yamamoto et al. (17) for a furnace used in the DTA apparatus they designed.

For operation at low temperature, the furnace may be surrounded by a Dewar flask and precooled with liquid nitrogen. The furnace is then heated by the furnace element using a heating-mode program. Another method is to use a gas as a heat-exchange medium, such as is illustrated in Figure VI.8 (31). Most temperature programmers do not function efficiently unless a thermal reservoir at least 30°C below the program temperature is available. The ultimate in low-temperature control should perhaps be thermoelectric cooling; however, such a system is not yet available.

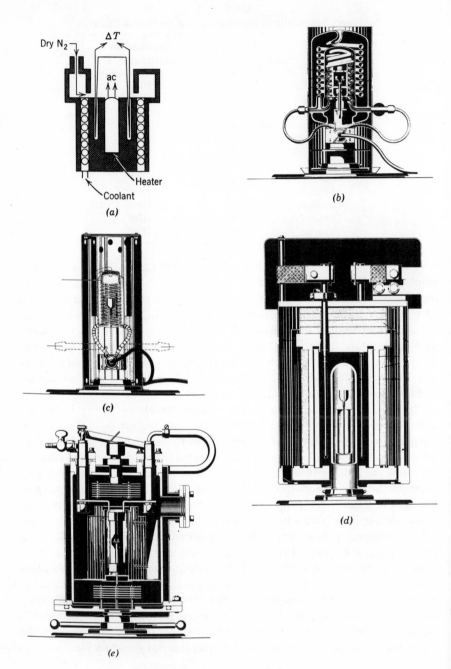

Figure VI.6. DTA furnace configurations.

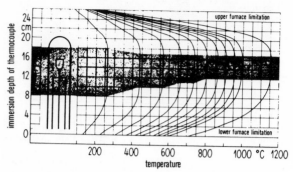

Figure VI.7. Temperature distribution in a DTA furnace (39).

In an attempt to control the atmosphere within the furnace and sample holder, various techniques have been employed. They include (a) flooding the furnace with a gaseous atmosphere (32–35); (b) vacuum furnaces (32, 33, 36–41); and (c) a dynamic-gas-flow atmosphere (32, 33, 42, 43). An elaborate high-vacuum system for DTA furnaces has been described by Wiedemann (30).

Although most DTA instruments have only one furnace, to increase the number of samples that can be run each day several furnaces may be used in conjunction with the sample holder, amplifier, and recording system. In fact, an instrument that contains four different furnaces (44) has been described.

The rate of temperature increase of the furnace is controlled by a temperature programmer. This programmer should be capable of linear temperature programming over a number of different temperature ranges, and hence must be compatible with several different thermocouple types. The heating rates should be linear and reproducible, as a nonlinear heating rate will influence the DTA curves. As shown by Theall (45), programmer output power cycling will cause variations in the DTA curve peaks as well as create spurious peaks. Another characteristic of the programmer is that it should be stable with respect to line voltage and ambient temperature variations. The programmer control thermocouple should be compensated either electrically or by an ice bath.

The type of temperature programmers varies from the simple variable-voltage transformer coupled to a synchronous motor to the more sophisticated feedback, proportional-type programmer. On–off-type programmers cannot be used because of the fluctuating power outputs which give rise to severe thermal gradients in the furnace and sample holder system. The solid-state, feedback-type, proportional programmer used in the DuPont

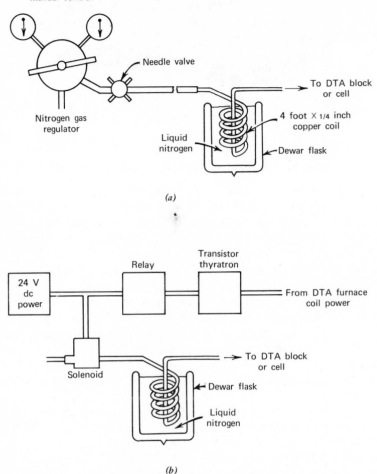

Figure VI.8. Simple cooling systems for DTA apparatus (38).

thermal analysis instruments is shown in Figure VI.9. In a proportional-type programmer, when the error signal between the command voltage and the output of the control signal differ by an amount more than the "dead band" of the control amplifier, the error signal is amplified, and this power is applied to the heater. The power applied to the heater is proportional to the error signal at the input to the control amplifier. The heating-rate accuracy of this programmer is said to be ±5% or 0.1°C/min, whichever is greater. The accuracy is governed mainly by the output of the control

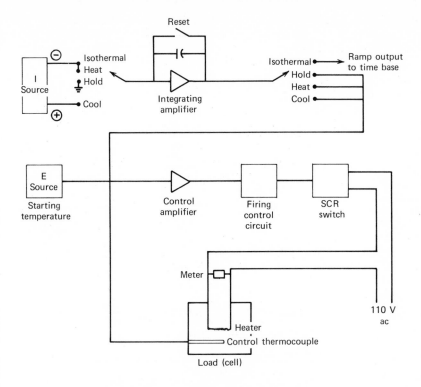

Figure VI.9. DuPont temperature programmer.

thermocouple. In the programmer itself, the limiting factor is the adjustment and drift of the power supply which controls the current which is integrated by the integrating amplifiers. The reproducibility is 0.1°C/min, while the heating-rate linearity is ±1% or 0.01°C/min. The former is dependent on the drift of the power supply and amplifier bias, while the latter depends on the output linearity of the control thermocouple.

In the Stone instruments, a stepper motor is used to control the rate at which nonlinear ramp functions of the output of the four most common types of thermocouples are generated. This nonlinear generated curve is compared with the nonlinear thermocouple output, and the signal difference led to an error amplifier which triggers an SCR circuit which controls the furnace voltage. The linearity is said to be better than 0.25% for any 100°C interval or 0.5% full scale.

The accurate determination of the linearity of the furnace heating rate is not an easy matter. From a plot of temperature versus time, as shown in

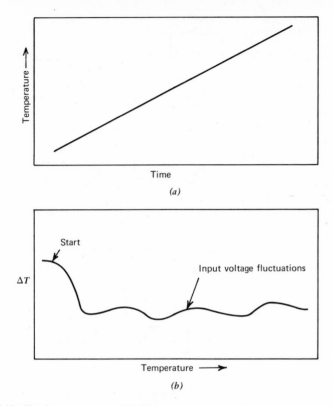

Figure VI.10. Heating rate curves; (*a*) Temperature *versus* time curve; (*b*) DTA curve of unbalanced system.

Figure VI.10*a*, the heating rate can be estimated to $\pm 5\%$ or better. To determine minute fluctuations of the furnace heating-rate curve, the curve in Figure VI.10(*b*) is used. A DTA curve is recorded with the reference chamber filled with an inert material (α-alumina, for example) while the sample chamber is empty. In this unbalanced condition, the fluctuations in the power input voltage are easily seen. This type of behavior is also seen if the ΔT thermocouples are unsymmetrically located in the furnace chamber (46). It should be noted that it is even more difficult to accurately measure heating rates of 1°C/min or less.

The heating rates of most commercial DTA furnaces can be varied from 0.5 to 50°C/min. Most DTA curves, however, are recorded at heating rates of 10 to 20°C/min. The higher heating rates are convenient for preliminary examination of the thermal behavior of a sample.

5. LOW-LEVEL VOLTAGE AMPLIFIER AND RECORDING SYSTEMS

The output voltage from the differential thermocouple is of the order of 0.1to100 μV, depending upon the type of thermocouples used (see Table VI.I) and the temperature difference between them.   Hence, unless a very sensitive recording system is used (<100 μV full scale), the ΔT signal must be amplified by a low-level microvolt dc amplifier.   The amplifier must have low noise, low drift, and high stability to be useful for DTA instrumentation. Instability of the amplifier will result in an unstable base line (45), while drift either by input voltage or ambient temperature changes will cause output fluctuations.   Pickup of 60 Hertz ac by the input wiring can cause output noise as well as an unstable base line.

Many times, in an effort to reduce amplifier noise, capacitors are added across the output of the amplifier, and occasionally at the input (45).   These capacitors frequently reduce the response time of the amplifier, which causes a shift in the curve peaks and also a loss of peak resolution.   A proper value of capacitor must be used, if noise is a problem, to form a compromise between noise reduction and loss of peak resolution.   Amplifier impedance mismatch can also cause nonlinear output voltages, which can distort the curve peaks.

Various recorders have been described, from photographic light-beam galvanometer types to modern electronic potentiometric recorders.   Burgess (47) gives an excellent discussion of some of the earlier recording devices. Probably the first to use modern potentiometric recorders, especially the multipoint type, were Kerr and Kulp (7) and Kauffman and Dilling (48). This type of recorder, plus the use of multiple sample holders, increased the usefulness of DTA for the qualitative identification of geological materials. Another technique, using a two-channel recorder, is to record both the differential temperature and the reference-material temperature as a function of time on the same chart paper.   More useful perhaps is the modern X-Y recorder in which the differential temperature is plotted directly as a function of reference-material temperature.   Thus, a direct recording of the differential temperature versus temperature is obtained.   A wide choice of recorders and recording ranges are available.

6. DTA INSTRUMENTS

a. Introduction

The modern DTA instrument is derived from the two-thermocouple design suggested by Roberts-Austen (49) in 1899.   Many instruments have been designed and constructed since that time, each slightly different in the design of the furnace, temperature programmer, recording equipment, sample-holder design, and so on. Smothers and Chiang (50) in 1958 described in detail some 255 DTA instruments located throughout the world. This list was deleted in the second edition of their book (51), which, however, included a bibliography of some 4248 references to the DTA literature, many of them describing the instrumentation employed by the investigators. Modern equipment is adequately summarized in various textbooks (52–54), while specifications on commercially available instruments are described elsewhere (56–58).

A number of DTA instruments are described here; an attempt is made to include only those instruments which possess some novelty in design or that have made important contributions to the development of DTA instrumentation.   In general, few commerical instruments will be included due to the lack of space to adequately describe them.   However, these are comprehensively described elsewhere in detail (56–58).

b. Sealed-Tube DTA Techniques

The enclosure of the sample in a sealed and sometimes evacuated chamber or tube is an old technique.   The technique is useful for phase-diagram investigations (59), especially those involving corrosive materials (chalcogenide reactions, for example), and in other areas such as organic reactions (60, 61), metal salt hydrates (62), metallurgical problems (63), molten salt equilibria (64), and numerous other problems.

A simple sealed-tube sample holder is shown in Figure VI.2(q).   Other sealed-tube sample and reference holders are shown in Figure VI.11 (59). In one case, (a), the thermocouples were sealed directly into the tubes, and this presented problems with metal-to-glass seals.   Heat transfer from the sample to the sensing thermocouple is fairly low in most of the examples given.   The system in (e) is much better than most of the others because the ring thermocouple provides a high-thermal-conductivity contact over an appreciable area.   None of the sample holders illustrated exhibit very high-resolution DTA curves, due to the thermal inertia of the system.   The preparation of the evacuated type of sealed-tube sample holders has been described in detail (67).

More recent work using sealed-tube sample holders has emphasized the use of milligram quantities of sample and, of course, semimicro tubes such as glass capillary melting-point tubes.   The furnace and sample holder used by Wendlandt (62) are illustrated in Figure VI.12.   Samples were enclosed in 0.9–1.1-mm-I.D. capillary tubes which were placed in thin aluminum heat-transfer sleeves.   An identical empty tube was used for the reference thermocouple.   After placing the sample (from 5 to 7 mg) in the tube, it was sealed off using a small oxygen-gas flame to a length of about 20 mm.

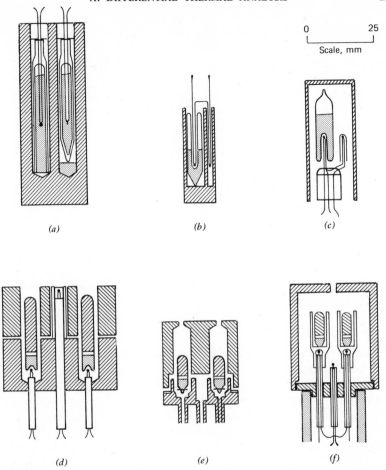

Figure VI.11. Sealed-tube sample holders in which the containers are evacuated; (a) Roberts (61); (b) Jensen (65); (c) Kracek (62); (d) Bollin and Kerr (63); (e) Bollin (55); Faktor and Hanks (64).

The sample loading procedure used by Barrett et al. (60, 61) was similar to that given above except that special precautions had to be taken due to the volatility of the organic samples. The capillary tube containing the sample was cooled by inserting it into a cooled aluminum block (dry ice–acetone bath coolant) and then sealing off the tube with a micro gas torch. Some precaution must be observed concerning the size of the sample; large samples may generate excessive pressure if gaseous decompositions are involved, and this may lead to minor explosions. The sample size should be adjusted so that the gas pressure will not exceed about 45 atm.

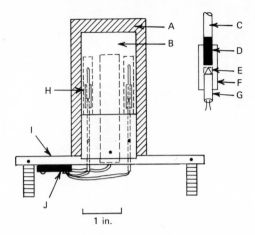

Figure VI.12. Sealed-tube DTA furnace and sample holder (62). A, insulated cover; B, aluminum block; C, glass capillary tube; D, sample; E, sample thermocouple; F, aluminum heat transfer sleeve; G, ceramic insulator tube; H, reference chamber; I, transite platform; J, terminal strip.

A sealed-tube DTA apparatus has been described by Gilpatrick et al. (64) for phase studies of the molten salt system, NaF-KF-BF$_3$. The sample holder, as shown in Figure VI.13, contains about 4 g of sample which is continuously agitated to avoid segregation and compositional changes during the heating and cooling cycles. Nickel was used as the material of construction for the capsules, which were evacuated and sealed under vacuum. The enclosed thermocouples gave reproducible results for the resulting phase transitions at ΔT sensitivities of 25–100 μV/cm of chart displacement. A similar metal sample-holder capsule was described by Etter et al. (63); the capsules were constructed of tantalum.

Barrall and Rogers (70) described a constant-pressure device (not a sealed-tube type sample holder) which is illustrated in Figure VI.14. Total cell volume was about 0.8 ml; a glass capillary tube provides a reservoir for the gases released during a reaction and prevents significant dilution of the sample atmosphere with air or the loss of gas. The small mercury seal provides for sample atmosphere expansion and maintains an essentially constant pressure in the cell.

Mention has already been made of the high-pressure metal sample holders employed by Bohon (3) for his DTA studies on explosive materials. A similar type of sample holder constructed from a Swagelok stainless-steel fitting has been described by David (71).

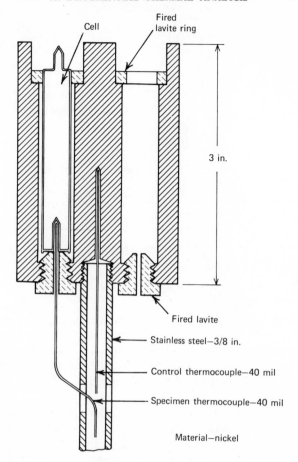

Figure VI.13. Sample holder for molten salt phase studies (64).

c. High-Pressure DTA Instruments

High-pressure DTA has been reviewed briefly by Locke (72). The previous studies have been concerned mainly with the effect of pressure on $solid \rightarrow liquid$ and $solid_1 \rightarrow solid_2$ equilibria. The equipment used in high-pressure studies employs piston-cylinder pressure-generating systems in the 60–80-kbar range, while lower-pressure equipment involves external pressurization by gases.

Harker (73) described an apparatus in which the sample and reference materials were sealed in platinum capsules. Wires welded to the capsules were led from the pressurized microreactor through a pressure packing.

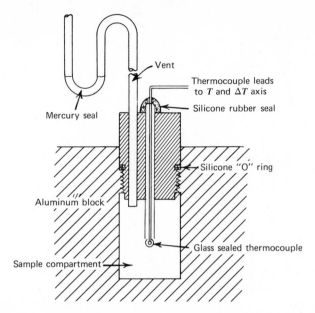

Figure VI.14. Sample holders for constant pressure in a self-generated atmosphere (70).

Nitrogen was used as the pressurization gas, although argon is to be preferred because the former causes severe embrittlement of platinum alloys. Several melting points in the CaO-Ca(OH)$_2$-Ca$_2$SiO$_4$ system at 15,000 p.s.i.g. were obtained.

Cohen et al. (74) described a piston-cylinder apparatus which could be used up to 50 kbar at temperatures to 1200°C. They discussed in detail the components of the high-pressure furnace, sample holder, and pressure transmission. A typical thermocouple assembly is shown in Figure VI.15. All wires, with the exception of the tip which is in contact with the sample container, are insulated with alundum cement. Sample capsules used were made of tantalum, niobium, platinum, graphite, and so on; they were in the shape of a cup $\frac{1}{16}$ in. in diameter by $\frac{1}{8}$ in. in length. The cap is made of the same material as the capsule body. Pressure is transmitted to the sample by the furnace assembly, a combination of parts made of talc, graphite, pyrophyllite, thermocouple tubing, thermocouples, sample container, boron nitride, and so on. The determination of the pressure on the sample is not a straightforward problem.

A DTA apparatus that could be used up to a pressure of 4000 bar and temperatures to 500°C was described by Kuballa and Schneider (75). This apparatus, which is illustrated in Figure VI.16, contained a DTA cell that was constructed of stainless steel; a maximum pressure of 4000 bar and a

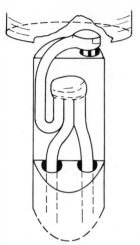

Figure VI.15. Three wire thermocouple assembly for high-
pressure DTA apparatus (74).

maximum temperature of 500°C were possible.   The cell was enclosed by two
Bridgman pistons which contained the thermocouples.   Cell wells for the
sample and reference materials were made from a 97% platinum–3% iridium
alloy; they were suspended on the sheathed thermocouple junctions.   Upper
and lower sides of the sample holder area were thermally heated by ZrO_2
blocks.   A furnace placed around the pressure vessel permitted linear
heating rates from 0.2 to 10°C/min.

DTA instruments for high-pressure hydrogenation reactions have been
described by various Japanese investigators (76, 77).   Bousquet et al. (78)
described a high-pressure DTA furnace and sample holder which could be
used up to 500 bar and a maximum temperature of 500°C.   This apparatus is
available from Netzsch-Geratebau Gmbh.   The Stone high-pressure DTA
sample holder has been described by Locke (72).   It can be used at pressures
up to 3000 p.s.i.g. and at a maximum temperature of 500°C.

d. High-Temperature DTA Instruments

The problems of a DTA apparatus for operation above 1200°C are quite
different from those that operate below this temperature.   The electrical
leakage from the increased electrical conductivities of the refractory components and of the air in the furnace becomes important.   Adequate shielding
of the low-level thermocouple circuits is much more critical.   The high
temperatures often cause melting of the sample, which destroys the thermocouple assembly as well as the sample holders.   Even with these problems,
DTA instruments which operate up to 3000°C or higher have been described.

A DTA apparatus capable of operation up to 1575°C is shown in Figure
VI.17 (79).

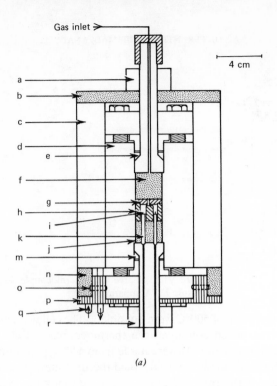

(a)

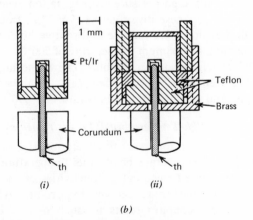

(b)

Figure VI.16. High-pressure DTA apparatus of Kuballa and Schneider (75). (*a*) a, r = coolers; b, n = pyrophillite insulating disk; c = heating block; d = high pressure vessel; e, m = copper seals; f, l = ZrO_2 blocks; g = copper shield; h = calorimeter block; i = thermocouple; k = corundum capillary; o = fastening screw; p = support; q = in- and outlet of refrigerant. (*b*)(i) Open Pt/Ir well; (ii) closed teflon well with brass holder (th = steel-sheathed thermocouple).

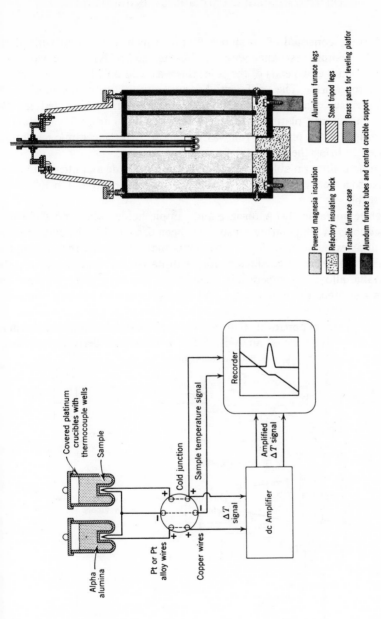

Figure VI.17. High-temperature DTA apparatus of Newkirk (79).

Covered platinum crucibles with thermocouple wells

Sample

Alpha alumina

Cold junction

Sample temperature signal

Pt or Pt alloy wires

Copper wires

Recorder

ΔT signal

Amplified ΔT signal

dc Amplifier

Aluminum furnace legs

Steel tripod legs

Brass parts for leveling platfor

Alundum furnace tubes and central crucible support

Powered magnesia insulation

Refractory insulating brick

Transite furnace case

The furnace consisted of an alundum core, which was wound with plati
num–20% rhodium resistance wire.   A booster coil of Nichrome wire wa
also wound on the two ends of the core; this was used only for the very-high
temperature work.   The differential temperatures were detected with
platinum versus platinum–10% rhodium thermocouples, inserted in the
indentations of the platinum sample and reference cups.   To shield the
thermocouple wires, platinum foil was wound around the ceramic insulating
tubes used to bring them into the furnace hot zone.

A commercially available temperature programmer, dc amplifier, and two
point recorder were employed.   Furnace heating rates of from 0 to 30°C/min
were possible.

A high-temperature DTA furnace and sample holder which does not use
thermocouples as temperature sensors has been described by Nedumov (80)
A critical assessment of various instruments that do employ thermocouples
found that they were unsatisfactory for quantitative DTA studies of metals
and metallic alloys.   Temperature detection in this apparatus is by use of
tungsten resistance thermometers.   The apparatus can be used to temperatures over 3000°C.

A multifunctional apparatus that permits the determination of the thermal
analysis, derivative thermal analysis, DTA, and thermal derivative thermal

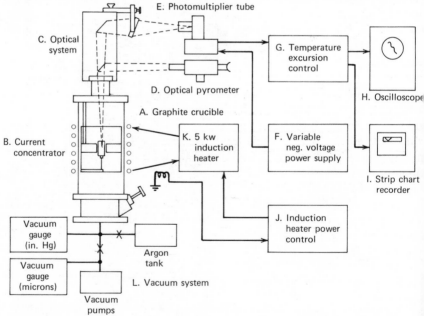

Figure VI.18. High-temperature multifunctional apparatus described by Rupert (81).

nalysis (temperature versus derivative of temperature) curves has been
escribed by Rupert (81–83). This apparatus is shown schematically in
Figure VI.18 (81). The sample is contained in crucible A located within the
ddy current concentrator B. The current concentrator receives power from
he induction heater K, whose output is controlled by the induction power
eater control J. Light from the sample emerges through a 0.070-in.-
iameter hole in the top of the crucible, and travels upwards through a Pyrex
r quartz window into the lower end of the beam splitter. Part of the light is
eflected at approximately a right angle to the axis of the beam splitter, by
he partially aluminized bottom mirror, to the optical pyrometer D used to
neasure the temperature of the sample. The light that passes through the
ottom mirror is reflected outward by the top mirror, and is focused by the
7.5-cm-focal-length achromatic lens into a 0.067-in.-diameter aperture in
ront of a photomultiplier tube, which was used at temperatures above
400°C, while for lower temperatures a 1P21 tube was used. Further
nodification of the instrument has been described (82).

e. Micro-DTA Instruments

The determination of DTA curves from microgram quantities of sample
as previously been described by Mazieres (4) (Section 2). A more compre-
ensive review of micro-DTA instrumentation is that by Sommer and Jochens
84). In this review, the entire area of high-temperature microscopy, coupled
vith DTA measurements, is discussed in detail. In most of the instruments
escribed, the thermocouple junction acts as a heater and temperature
letector, as well as the sample holder.

The micro-DTA apparatus developed by Miller and Sommer (85) is shown
n Figure VI.19. The circuit used a motor-driven variable voltage regulator
nd was capable of heating rates from 5°C/min to 1000°C/sec. Recordings
f the sample temperatures and the ΔT were made on a high speed recorder.
n an improvement of this apparatus, the sample and reference thermo-
ouples were placed in individual cells and only the cell housing the sample
vas retained in the optical system of the microscope (86).

A similar DTA apparatus has been described by Proks and Zlatovsky (87)
n which the sample is contained in the thermojunction. The thermocouple
 heated by a high-frequency current, which has been amplitude-modulated
y a low-frequency signal.

f. Automation of DTA Instrumentation

Present-day DTA instruments are capable of automatic operation in that
fter the sample has been manually inserted the temperature rise is controlled

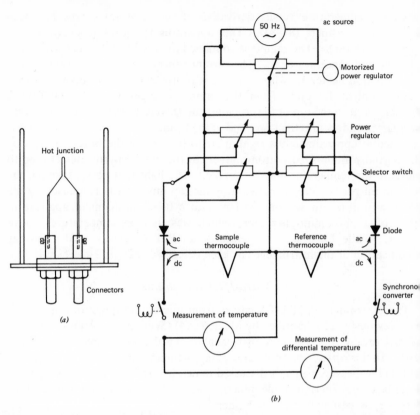

Figure VI.19. Micro DTA apparatus (2); (a) sample holder; (b) electrical circuit (85, 86).

by a temperature programmer which will turn off the instrument after a preselected temperature limit is attained. After cooling the furnace back to room temperature, the pyrolyzed sample is removed from the sample holder, a new sample is introduced, and the heating cycle is repeated.

Wendlandt and Bradley (88, 89) have described an automated instrument which is capable of studying eight samples in a sequential manner. The samples are automatically introduced into the furnace, pyrolyzed to a preselected temperature limit, and then removed. After the furnace has cooled back to room temperature, the cycle is repeated. Operation of the sample-changing mechanism, furnace-temperature rise and cooling, recording, and so on is completely automatic.

A line drawing of the sample-changing mechanism and the furnace platform is shown in Figure VI.20.

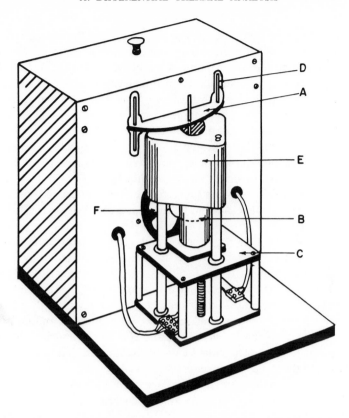

Figure VI.20. Automated DTA apparatus of Wendlandt and Bradley (88). A, Sample holder plate; B, furnace; C, furnace platform assembly; D, sample capillary tube; E, furnace insulation; F, cooling fan.

The powdered samples are contained in glass capillary tubes, D, of 1.6–1.8-mm I.D., which are placed in the circular sample holder plate, A. The aluminum sample holder plate is 8.0 in. in diameter by $\frac{1}{8}$ in. thick and has provision for retaining eight glass capillary tubes. The glass tubes are held in their respective positions by means of small spring clips. The plate is rotated by a small synchronous electric motor equipped with an electromagnetic clutch. The rotation of the plate by the motor is controlled by a lamp-slit-photocell arrangement. Adjacent to each sample-holder position is a 0.50 × 0.06-in. slit cut in the aluminum plate. Alignment of the plate slit between the lamp and photocell by the drive motor permits exact positioning of each capillary tube with the furnace cavity.

After the sample capillary tube is in position, the furnace platform, C is raised so that the tube is positioned into the aluminum heat-transfer sleeve, located on the sample thermojunction.   Movement of the furnace platform is controlled by a reversible electric motor connected to the platform by a screw drive.   Upper and lower limits of travel are controlled by two micro-switches.   The furnace is insulated from the platform by a 0.25-in. layer of transite and, while in the heating position, by a Marinite sleeve, E.   The rotation interval for sample changing is 15 sec, while it takes 50 sec to raise the furnace platform to the full upper limit.

After the sample has been heated to the upper temperature limit, the furnace is lowered, the sample holder plate rotates to a new position, and a cooling fan is activated to direct air on the hot furnace.   Cooling time for the furnace, from 450°C to room temperature, takes about 20 min.   After the furnace has been cooled to room temperature, the above cycle is repeated with a new sample.

A schematic diagram of the furnace and sample chamber is shown in Figure VI.21.   The cylindrical furnace, E, is 1.5 in. in diameter by 3.3 in. in length, and is heated by a 210-watt stainless-steel heater cartridge, K.   The upper temperature limit of the furnace is about 500°C.   The sample and reference cavities are about 0.25 in. in diameter by 1.5 in. in length.   Thermal contact between the sample and reference capillary tubes, A and G, is made by the aluminum heat-transfer sleeves, C and H.   The cylindrical sleeves are about 0.7 in. in length.   The ends of the sleeves are drilled out so that the sample tube and the $\frac{1}{16}$-in.-diameter ceramic insulator tube, D or J, fit closely within the sleeve.   To minimize heat leakage from the furnace to the sample-holder plate, B, a transite cover, F, is used to enclose the top of the furnace.

Since the automated DTA apparatus has an upper temperature limit of

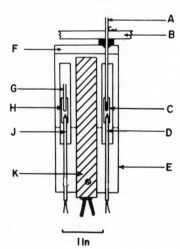

Figure VI.21. Furnace and sample chamber. A, Glass capillary tube for sample; B, sample holder plate; C, sample heat transfer sleeve; D, sample thermocouple; E, furnace block; G, reference capillary tube; H, reference heat transfer sleeve; J, reference thermocouple; K, heater cartridge.

about 500°C, its use has been restricted to intermediate-temperature applications such as the deaquation of metal salt hydrate systems. It should find wide use in the routine DTA examination of both organic and inorganic samples. The automated features should permit convenient computer interfacing so that reaction temperatures, peak areas, purity calculations, ΔH calculations, and so on can be easily carried out.

g. Miscellaneous DTA Instruments

One of the first precise vacuum or inert-atmosphere instruments was designed and constructed by Whitehead and Breger (37). The furnace was constructed from an alundum core, 9 in. in length by 2 in. I.D., wound with Chromel A resistance wire. The core was shielded by four sheet-nickel cylinders, mounted on three posts, and the entire assembly was placed inside a 12 × 24-in. Pyrex bell jar. All electrical connections were made through the bottom of the bell jar mounting base. The sample block was made in the dimensions shown from Type 446 or 309 stainless steel. The furnace heating rate was controlled by a Leeds and Northrup Micromax controller; the differential temperatures were recorded on a Beckman Photocell recorder.

Wendlandt (39) has also described a simple vacuum or controlled-atmosphere DTA furnace and sample holder. The furnace tube was 19 cm in length, 2.5 cm in diameter, and partially constructed of fused silica. The lower end of the tube contained two Pyrex glass 25-mm-I.D. O-ring joints, which were attached to the fused silica tube by a Nylon seal, machined to the dimensions of the glass O-ring joint, and attached to it by a compression clamp. The Chromel-versus-Alumel thermocouple wires were brought into the furnace zone by use of 3.0-mm-diameter, two-holed ceramic tubing. The sample and reference cups made of Inconel were 7 mm in diameter and 10 mm in length, and had a volume of 0.8 ml. The cups fit snugly on the insulator tube and were in intimate contact with the thermojunctions. The furnace was wound, either on the insulated fused silica tube or on an external ceramic tube, which fit closely about the silica tube, with Nichrome resistance wire. The furnace temperature programmer consisted of a variable-voltage transformer driven by a synchronous motor. The differential temperature signal was amplified by a dc microvolt amplifier and recorded against temperature on an X–Y recorder.

A low-temperature DTA apparatus, capable of operation in the temperature range −190 to 400°C, has been described by Reisman (40); the apparatus is illustrated in Figure VI.22. With the Dewar container filled with liquid nitrogen, the heating rate of the sample block was controlled by increasing the voltage into the heater coils, while cooling was accomplished

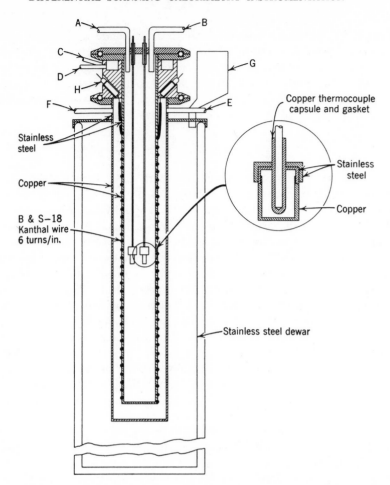

Figure VI.22. DTA apparatus for the temperature range from −190 to 400°C (40).

by varying the pressure of the gas present in the outer chamber. Commercially available dc amplifiers and recorders were employed in the apparatus to record the DTA curve.

An extremely rugged DTA furnace and sample holder has been described by Bohon (35) and is illustrated in Figure VI.23. In this apparatus, the differential thermocouples are isolated from the sample to avoid destruction from explosions or chemical reaction with the sample. This was accomplished by employing Chromel-versus-Alumel thermocouples encased in an Inconel sheath. The furnace tube and auxiliary pressure manifold were made from Monel metal. Pressures up to 1000 p.s.i.g. have been sustained in

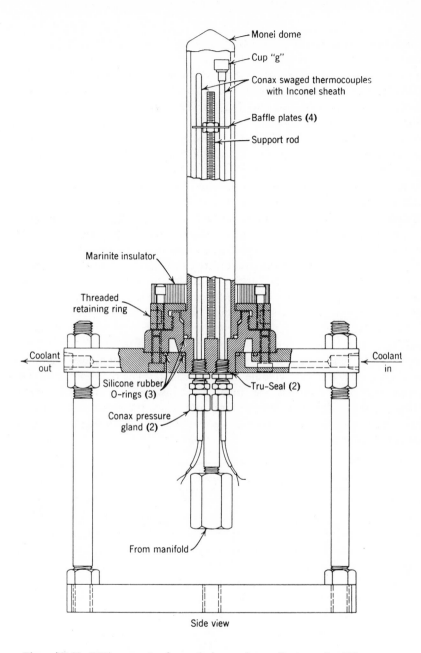

Figure VI.23. DTA apparatus for explosives and propellant samples (35).

235

the furnace tube assembly at 350°C, and 400 p.s.i.g. at temperatures up to 500°C.

A highly sensitive DTA apparatus which permitted the determination of phase-transition temperatures of ±0.5°C over a wide range of heating rates has been described by Vassallo and Harden (29). The sample holder and heating block are illustrated in Figure VI.5(A). The apparatus is conventional in the differential-temperature-measuring circuit, but in the furnace temperature measuring circuit a dc amplifier and zero-suppressing circuit are used when temperature accuracies of better than ±0.5°C are required. The heating block permits the use of 1.5–2.0-mm × 30-mm melting point capillary tubes as sample and reference material containers, with the thermocouples inserted into the sample through the top of the tube. The block was heated by a 30-W cartridge heater placed in the center of the block. The double coil of copper tubing served as a cooling line. One coil was immersed in a coolant contained in the lower part of the flask, while the second coil surrounded the block. The block was cooled by a flow of air or nitrogen through the lower coil, thence to the coil surrounding the block. A temperature range from −150 to 450°C could be covered by the apparatus.

For studying the kinetics of homogeneous reactions in solution, Borchardt and Daniels (91) used the glass apparatus illustrated schematically in Figure VI.24. The cells consisted of two Pyrex tubes, 1.25 in. in diameter and 5 in. in length, each having a volume of about 60 ml. The thermocouples, contained in Kel-F covered copper tubes, were inserted into the tube contents from the top. The bath temperature, and hence the sample and reference materials, was gradually increased by use of a heater connected to a variable-voltage transformer.

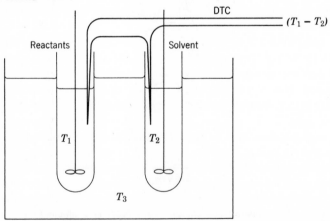

Figure VI.24. Solution DTA apparatus used by Borchardt and Daniels (91).

A DTA apparatus containing thermistors as the differential temperature detection devices has been described by Pakulak and Leonard (22) and is illustrated in Figure VI.25a, while the dc thermistor bridge circuit is shown in Figure VI.25b. The matched thermistors, 100,000Ω at 25°C, were contained in glass tubes and centered in each of the sample and reference tubes. A third thermistor was used to detect the temperature of the furnace. All thermistors and tubes were placed into a furnace constructed of aluminium. The heating rate of the furnace was controlled by increasing the furnace-windings voltage by means of a variable-voltage transformer. A heating rate of 2°C/min was normally employed.

It should be noted that a dc amplifier was not required since the bridge unbalance voltage signal was large enough to be recorded by the recorder.

One of the first controlled-atmosphere–controlled-pressure DTA instruments was described by Stone in 1960 (42, 43). A schematic diagram of the furnace and sample holder is shown in Figure VI.26. The sample holder permits a gas flow through the sample and reference materials during the heating cycle. Gas enters through the sample-holder manifold and diffuses through the porous disks in order to provide a minimum of turbulence and a maximum of gas uniformity in and around the sample. The gas composition may be changed at any time during the heating process or may be cycled between two or more gases. Pressure within the furnace chamber may vary from 10^{-2} Torr to 100 p.s.i.g.

An oven (15, 16) in which the thin-film thermocouples, previously described in Sections 3, are used is shown in Figure VI.27. It permitted programmed temperature operation from $-120°$ to 500°C. For cooling, previously cooled nitrogen was flowed through the coil surrounding the oven chamber. By controlling the flow of nitrogen gas, linear temperature programming was possible.

Miller and Wood (92) described the calorimetric accuracy and performance of the DuPont high-temperature (1200°C) DTA cell. This furnace and sample holder are shown in Figure VI.28. Samples are contained in platinum cups inserted over platinum–platinum, 10% rhodium thermocouples and surrounded by an alumina tube which is heated with a Kanthal-wire-wound furnace.

It is frequently necessary to isolate the DTA furnace and sample holder from the control console because of explosion hazards, radioactivity, toxic environments, and so on. The remote DTA sample holder and furnace must be capable of normal operation and must be easily loaded with the sample. Such a remote system has been described by Graybush et al. (93) in which a DuPont standard DTA cell was modified for remote operation for the study of primary explosives. The need to protect samples from the slightest oxidizing environment necessitated the cell being evacuated to 10^{-6} Torr. This required removal of the porous disk and redesign of the support and also resealing of the electrical and gas connectors. Thermocouple connections were protected from the development of thermal gradients by shielding with glass tubing.

Several DTA instruments have been described by Barrall et al. (94, 95). A DTA calorimeter cell in which the ΔT-sensing thermocouples are attached to the sample container is shown in Figure VI.29. In this apparatus, copper-constantan thermocouples are soldered to a 4-mm-O.D. copper cup fitted

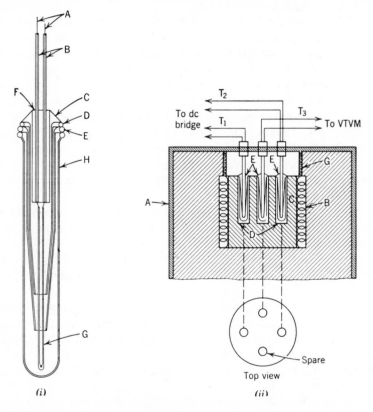

Figure VI.25. Thermistor DTA apparatus of Pakulak and Leonard (22); (a) sample holder and furnace; (b) thermistor bridge circuit. (a) (i) A, output leads; B, teflon sleeving; C. electric resistor cement; D, 1-mm centrifuge tube; E, 1.5-mm centrifuge tube; F, two-hole ceramic tube; G, thermistor; H, test tube. (ii) A, furnace; B, furnace coil and well; C, block, aluminum; D, sample tubes; E, thermistor containers; T_1, T_2, T_3 thermistors; G, glass ring 1 in. × $3\frac{1}{4}$ in. (b) R_1, R_2, 100 Ω potentiometer; R_3, R_4, 1000 Ω ± 1%; R_5, R_6, 2000 Ω ± 1%; T_1, T_2, thermistors-100,000 Ω at 25°C; B, $1\frac{1}{2}$ V dry cell.

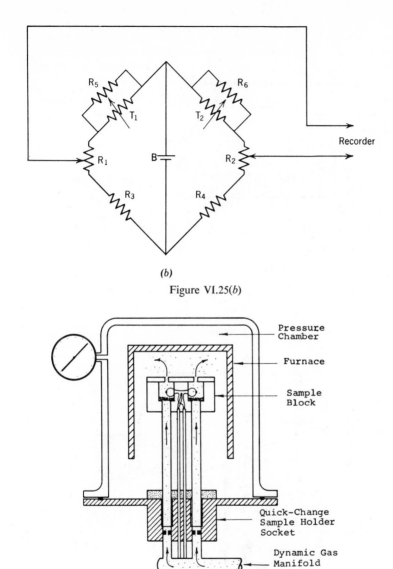

(b)

Figure VI.25(*b*)

Figure VI.26. Stone's DTA apparatus, schematic (42, 43).

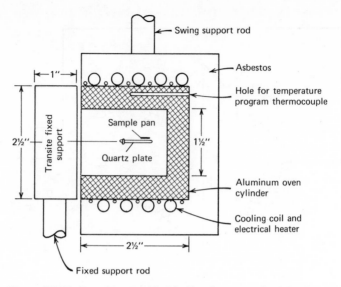

Figure VI.27. Oven for use with thin-film thermocouples (15, 16).

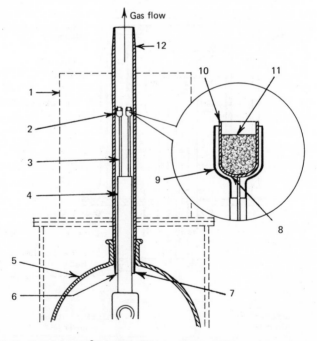

Figure VI.28. DuPont 1200°C DTA cell according to Miller and Wood (92); 1, furnace 2, shoulder; 3, ceramic insulator; 4, ceramic support; 5, bell jar; 6, gas flow; 7, tapered joint; 8, thermocouple junction; 9, platinum cup; 10, liner; 11, sample; 12, alumina furnace tube.

240

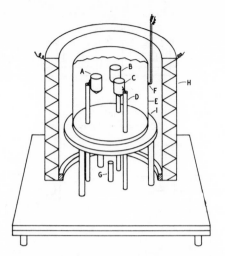

Figure VI.29. DTA calorimeter cell described by Barrall et al. (94). A, B, Copper sample cups, 4-mm. O.D. by 6 mm; C, Copper reference cup; D, Two-conductor ceramic supports, 3-mm diameter by 50 mm; E, Copper radiator shield, 35-mm diameter by 53 mm; F, Program-sensing thermocouple; G, Liquid CO_2 cooling gas jet; H, Electric furnace, 45-mm *i.d.* by 100 cm; I, Copper base plate, 38-mm diameter.

with a copper lid. The thermocouples and sample cups are supported on ceramic insulator tubes which are attached to a metal base. All of the cups were heated by thermal radiation received from the blackened copper radiation shield; this prevented radiation hot spots due to furnace windings. The entire DTA cell was enclosed by a glass bell jar which provided a controlled atmosphere from reduced pressures to about 2 atm.

A controlled-pressure system for a DTA apparatus has been described by Kemme and Kreps (96). This apparatus permitted the determination of sample vapor pressures by DTA.

Instrumentation used to evaluate the catalytic properties of a material by DTA has been described (97, 98). An earlier apparatus has been described by Stone and Rase (42).

Radio-frequency heating of large samples of rubber has been described by Wald and Winding (99). The sample is placed between two parallel plates which are connected to a rf generator. Sample temperature is detected by a thermocouple located in the center of the sample.

B. Differential Scanning Calorimetry

Perhaps the most widely used DSC instrument has been the Perkin-Elmer differential scanning calorimeter, Model DSC-1, introduced in 1963 (100,

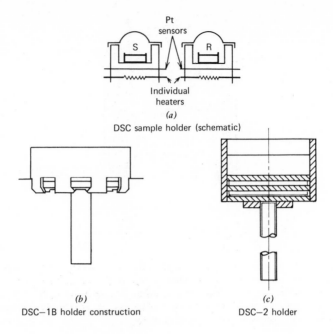

Figure VI.30. Sample holder construction used in the Perkin-Elmer DSC instruments (24).

101). Recently introduced (102, 24, 103) was the Model DSC-2, which has a temperature range extended to 725°C, improved base line repeatability and linearity, and higher calorimetric sensitivity. A comparison (24) of the sample holders used in the DSC-1, DSC 1B, and DSC-2 instruments is shown in Figure VI.30. In the DSC 1 cell, the sample and reference holder consisted of a stainless-steel cup and support, a platinum-wire sensor, an etched Nichrome heater, and other thermal parts. All of these components were mechanically crimped together in a very tight sandwich. This sample holder operated well over the temperature range −125° to 500°C. In the DSC-2 sample holder, the materials of construction used are a platinum-iridium alloy for the body and structured members of the holder, a platinum wire for both the heater and sensor, and α-alumina for electrical insulation. All parts of the holder are spot-welded together.

A schematic diagram of the calorimeter is shown in Figure VI.31. The apparatus, unlike DTA, maintains a sample temperature isothermal to a reference substance (or furnace block) by supplying heat to the sample or reference material. The amount of heat required to maintain these isothermal conditions is then recorded as a function of time (or temperature). In addition to recording the enthalpy curve, if the sample evolved a volatile

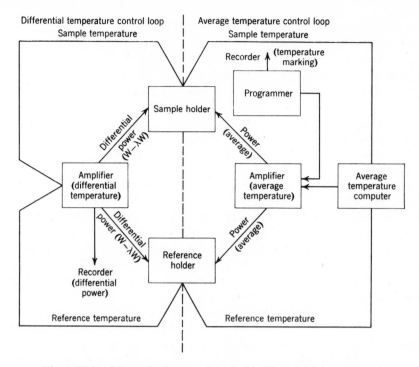

Figure VI.31. Schematic diagram of the Perkin-Elmer DSC instrument.

material during the heating process, the gas evolved by the sample is recorded.

The instrument contains two "control loops," one for the average-temperature control and the other for the differential-temperature control. In the former, a programmer provides an electrical output signal that is proportional to the desired temperature of the sample and reference holders. The programmer signal which reaches the average-temperature amplifier is compared with signals received from platinum-resistance thermometers permanently embedded in the sample and reference holders via an average-temperature computer.

In the differential temperature loop, signals representing the sample and reference temperatures, as measured by the platinum-resistance thermometers, are fed to the differential-temperature amplifier via a comparator circuit, which determines whether the reference or the sample temperature is greater. The differential-temperature-amplifier output then adjusts the differential power increment put into the reference and sample heaters in the direction and magnitude necessary to correct any temperature difference between them. A signal proportional to the differential power is also transmitted to the pen of a galvanometer recorder, giving a curve of differential power versus time (temperature). The area under a peak, then, is directly proportional to the heat energy absorbed or liberated in the transition.

The DSC-2 features a temperature readout accuracy of $\pm 1.0°C$ with a precision of $\pm 0.1°C$. Calorimetric sensitivity ranges from 0.1 to 20.0 mcal sec^{-1} full scale deflection using a 10-mV recorder. The furnace atmosphere may be N_2 or Ar, static or dynamic, at a pressure of 0.5 to 3 atm. Helium gas is required for low-temperature operation.

Various sample holders have been described for the Perkin-Elmer DSC instrument. A sealed metal cell with a removable screw-on cap has been described by Freeberg and Alleman (104). Metals used were brass, stainless steel, and aluminum. Wendlandt (105) described a capillary-tube sample holder which used 1.6–1.8-mm-diameter glass capillary tubes. The tubes were contained in an aluminum holder which were set in the sample and reference cells of the calorimeter. Sample holders for measuring the vapor pressure of a liquid (106) as well as for heats of mixing (107) have been described. Enclosure of the sample-holder chamber in a vacuum chamber has been described by Morie et al. (108).

A discussion of the theory and operational characteristics of the DuPont DSC cell has been given by Baxter (18). The DSC cell, which has been illustrated in Figure VI.4, is based on a thermoelectric disk made of constantan which serves as the major path of heat transfer to and from the sample and also as one-half of the ΔT-measuring thermocouples. A Chromel wire is connected to each platform, thus forming the Chromel-constantan differential thermocouple. The temperature range of the instrument is -150 to $600°C$. The utility of this DSC cell has been extended by its enclosure in a pressure chamber capable of operation to 67 atm. This high-pressure system has been described by Levy et al. (109).

A new DSC cell, based on the DTA principle (as is the DuPont DSC cell previously described) has been described by David (110). The calorimeter cell, as shown in Figure VI.32, contains a differential thermocouple of a new thin-form design that is isolated from the cell wall and bottom to provide greater sensitivity. This thermocouple consists of a sheet of negative Platinel II type thermocouple alloy coupled to a positive Platinel II alloy. Flat shallow containers are employed for the sample and reference materials. Two additional thermocouples are used for measuring the temperature of the cell and are used for the furnace programmer, limit switch, and temperature readout. The maximum temperature of the cell is $1000°C$.

Schematic diagrams of the Deltatherm dynamic adiabatic calorimeter, as discussed by Dosch and Wendlandt (111), are shown in Figure VI.33. As illustrated in (c), the temperature difference between the sample and adiabatic

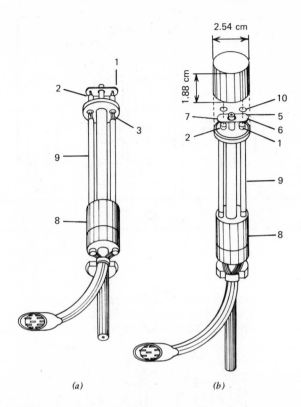

Figure VI.32. DSC cell by David (110).   (1) Thermocouple for x axis or system temperature readout; (2) limit switch thermocouple (3) programming or furnace thermocouple; (4)
dynamic gas port entry; (5) dynamic gas port exit; (6) sample side of differential thermocouple; (7) reference side of differential thermocouple; (8) ceramic thermal insulator; (9)
ceramic support rods; (10) sample pans.

enclosure is detected by the differential thermocouple, TC1.   This temperature difference controls the output of the power driver which applies power
to heater H1.   Power input to the heater is measured by a Hall effect
multiplier watt meter.   The calorimeter assembly (b) is mounted on a
ceramic pillar and consists of a massive copper block, A, enclosed by copper
covers B, C, and D.   The sample chamber, F, and its cover are made of
silver and thermally isolated from the block.   Sheathed thermocouples, J
and L, are used for temperature detection.   Samples may be in the form of
solid blocks, powders, or liquids.

Other calorimeters described include a microcalorimeter (112) similar to
the Calvet instrument (113), high-temperature differential calorimeters (115–
118), and others (119–122).

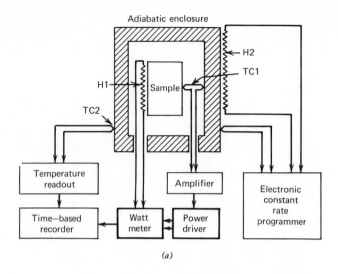

(a)

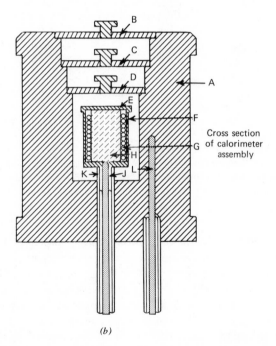

Cross section
of calorimeter
assembly

(b)

Figure VI.33. Deltatherm calorimeter (111). (a) Block diagram of entire system; (b) Cross section of calorimeter: A, Copper block; B, C, and D copper covers; E, sample chamber cover; F, sample chamber; G, heater; H, sample; J, thermocouple; K, ceramic support; L, thermocouple.

References

1. Wendlandt, W. W., "Differential Thermal Analysis," in *Technique of Inorganic Chemistry*, H. B. Jonassen and A. Weissberger, eds., Interscience, New York, 1963, Vol. 1, Chap. 6.
2. Garn, P. D., *Anal. Chem.*, **37**, 77 (1965).
3. Bohon, R. L., *Anal. Chem.*, **35**, 1845 (1963).
4. Mazieres, C., *Anal. Chem.*, **36**, 602 (1964).
5. Miller, B., *Thermochim. Acta*, **2**, 225 (1971).
6. Kulp, J. L., and P. F. Kerr, *Science*, **105**, 413 (1947).
7. Kerr, P. F., and J. L. Kulp, *Am. Minerologist*, **33**, 387 (1948).
8. Cox, D. B., and J. F. McGlynn, *Anal. Chem.*, **29**, 960 (1957).
9. Burr, J. T., in *Thermal Analysis*, R. F. Schwenker and P. D. Garn, eds., Academic, New York, 1969, Vol. 1, p. 301.
10. Wilhoit, R. C., *J. Chem. Educ.*, **44**, A571 (1967).
11. Brewer, L., and P. Zavitsanos, *J. Phys. Chem. Solids*, **2**, 284 (1957).
12. Lodding, W., and E. Sturm, *Am. Minerologist*, **42**, 78 (1957).
13. Joncich, M. J., and D. R. Bailey, *Anal. Chem.*, **32**, 1578 (1960).
14. Van Tets, A., and H. G. Wiedemann, in *Thermal Analysis*, R. F. Schwenker and P. D. Garn, eds., Academic, New York, 1969, Vol. 1, p. 121.
15. King, W. H., C. T. Camilli, and A. F. Findeis, *Anal. Chem.*, **40**, 1330 (1968).
16. King, W. H., C. T. Camilli, and A. F. Findeis, in *Analytical Calorimetry*, R. S. Porter and J. F. Johnson, eds., Plenum, New York, 1968, p. 261.
17. Yamamoto, A., K. Yamada, M. Maruta, and J. Akiyama, in *Thermal Analysis*, R. F. Schwenker and P. D. Garn, eds., Academic, New York, 1969, Vol. 1, p. 105.
18. Baxter, R. A., in *Thermal Analysis*, R. F. Schwenker and P. D. Garn, eds., Academic, New York, 1969, Vol. 1, p. 65.
19. McAdie, H. G., in *Thermal Analysis*, R. F. Schwenker and P. D. Garn, eds., Academic, New York, 1969, Vol. 1, p. 693.
20. Menis, O., and J. T. Sterling, *Status of Thermal Analysis*, O. Menis, ed., National Bureau of Standards Special Publication 338, Washington, D.C., Oct., 1970.
22. Pakulak, J. M., and G. W. Leonard, *Anal. Chem.*, **31**, 1037 (1959).
23. Weaver, E. E., and W. Keim, *Proc. Indiana Acad. Sci.*, **70**, 123 (1960).
24. O'Neill, M. J., and R. L. Fyans, *Eastern Anal. Symposium*, New York, Nov., 1971.
25. Hill, J. A., and C. B. Murphy, *Anal. Chem.*, **31**, 1443 (1959).
26. Hogan, V. D., and S. Gordon, *Anal. Chem.*, **32**, 573 (1960).
27. Campbell, C., and G. Weingarten, *Trans. Faraday Soc.*, **55**, 2221 (1959).
28. Clampitt, B. H., *Anal. Chem.*, **35**, 577 (1963).
29. Vassallo, D. A., and J. C. Harden, *Anal. Chem.*, **34**, 132 (1962).
30. Wiedemann, H. G., *Chem. Ing. Tech.*, **36**, 1105 (1964).
31. Barrall, E. M., in *Guide to Modern Methods of Instrumental Analysis*, T. H. Gouw, ed., Wiley-Interscience, New York, 1972, Chap. XII.
32. Lodding, W., and L. Hammell, *Rev. Sci. Instr.*, **30**, 885 (1959).
33. Lodding, W., and L. Hammell, *Anal. Chem.*, **32**, 657 (1960).
34. Rudin, A., H. P. Schreiber, and M. H. Waldman, *Ind. Eng. Chem.*, **53**, 137 (1961).
35. Bohon, R. L., *Anal. Chem.*, **33**, 1451 (1961).
36. Martin, A. J., and K. L. Edwards, *J. Sci. Instr.*, **36**, 170 (1959).
37. Whitehead, W. L., and I. A. Breger, *Science*, **111**, 279 (1950).

38. Stone, R. L., *J. Am. Ceram. Soc.*, **35**, 76 (1952).
39. Wendlandt, W. W., *J. Chem. Educ.*, **40**, 428 (1963).
40. Reisman, A., *Anal. Chem.*, **32**, 1566 (1960).
41. Chihara, H., and S. Seki, *Bull. Chem. Soc. Japan*, **26**, 88 (1953).
42. Stone, R. L., and H. F. Rase, *Anal. Chem.*, **29**, 1273 (1957).
43. Stone, R. L., *Anal. Chem.*, **32**, 1582 (1960).
44. Levandowsby, J., and N. Sacovy, *Inst. franc. petrole XI*, 818 (1956).
45. Theall, G. G., in *Thermal Analysis*, R. F. Schwenker and P. D. Garn, eds., Academic, New York, 1969, Vol. 1, p. 97.
46. Barrall, E. M., and L. B. Rogers, *Anal. Chem.*, **34**, 1101 (1962).
47. Burgess, G. K., *U.S. Bur. Std., Bull.*, **5**, 199 (1908–1909).
48. Kauffman, A. J., and E. D. Dilling, *Econ. Geol.*, **45**, 22 (1950).
49. Roberts-Austen, W. C., *Metallographist*, **2**, 186 (1899).
50. Smothers, W. J., and Y. Chiang, *Differential Thermal Analysis: Theory and Practice*, Chemical Publishing Co., New York, 1958, pp. 294–399.
51. Smothers, W. J., and Y. Chiang, *Differential Thermal Analysis: Theory and Practice*, Chemical Publishing Co., New York, Second Ed., 1966.
52. Garn, P. D., *Thermoanalytical Methods of Analysis*, Academic, New York, 1965, Chap. IX.
53. Barrall, E. M., and J. F. Johnson, in *Techniques and Methods of Polymer Evaluation*, P. E. Slade and L. T. Jenkins, eds., Marcel-Dekker, New York, 1966, Chap. I.
54. Mackenzie, R. C., and B. D. Mitchell, in *Differential Thermal Analysis*, R. C. Mackenzie, ed., Academic, Gordon, 1970, Chap. III.
55. Anon., *Industrial Research*, Nov., 25, (1969).
56. Wendlandt, W. W., *Thermal Analysis Techniques*, in *Handbook of Commercial Scientific Equipment*, C. Veillon and W. W. Wendlandt, eds., Marcel-Dekker, New York, Vol. 2, in press.
57. Wendlandt, W. W., *Lab. Management*, Oct., 26 (1965).
58. Wendlandt, W. W., *J. Chem. Educ.*, **49**, A571, A623 (1972).
59. Bollin, E. M., in *Differential Thermal Analysis*, R. C. Mackenzie, ed., Academic, London, 1970, Chap. VII.
60. Barrett, E. J., H. W. Hoyer, and A. V. Santoro, *Mikrochim. Acta*, **1970**, 1121.
61. Santoro, A. V., E. J. Barrett, and H. W. Hoyer, *J. Thermal Anal.* **2**, 461 (1970).
62. Wendlandt, W. W., *Thermochim. Acta*, **1**, 419 (1970).
63. Etter, D. E., P. A. Tucker, and L. J. Wittenberg, in *Thermal Analysis*, R. F. Schwenker and P. D. Garn, eds., Academic, New York, 1969, Vol. 2, p. 829.
64. Gilpatrick, L. O., S. Cantor, and C. J. Barton, *Thermal Analysis*, R. F. Schwenker and P. D. Garn, eds., Academic, New York, Vol. 1, p. 85.
65. Roberts, H. S., *J. Am. Chem. Soc.*, **57**, 1034 (1935).
66. Kracek, F. C., *Trans. Am. Geophys. Un.*, **27**, 364 (1946).
67. Bollin, E. M., and P. F. Kerr, *Am. Mineralogist*, **46**, 823 (1961).
68. Faktor, M. M., and R. Hanks, *Trans. Faraday Soc.*, **63**, 1122 (1967).
69. Jensen, E., *Am. J. Sci.*, **240**, 695 (1942).
70. Barrall, E. M., and L. B. Rogers, *J. Inorg. Nucl. Chem.*, **28**, 41 (1966).
71. David, D. J., *Anal. Chem.*, **37**, 82 (1965).
72. Locke, C. E., *Proceedings of the Third Toronto Symposium on Thermal Analysis*, H. G. McAdie, ed., Chemical Institute of Canada, Toronto, 1969, p. 251.
73. Harker, R. I., *Am. Mineralogist*, **49**, 1741 (1964).
74. Cohen, L. H., W. Klement, and G. C. Kennedy, *J. Phys. Chem. Solids*, **27**, 179 (1966).

75. Kuballa, M., and G. M. Schneider, *Ber. Bunsen Gesellschaft fur physik. Chem.*, **75**, 513 (1971).
76. Takeya, G., T. Ishii, K. Makino, and S. Ueda, *Kogyo Kagaku Zasshi*, **69**, 1654 (1966).
77. Ueda, S., S. Yokoyama, T. Ishii, and G. Takeya, *Kogyo Kagaku Zasshi*, **74**, 1377 (1971).
78. Bousquet, J., J. M. Blanchard, B. Bonnetot, and P. Claudy, *Bull. Soc. Chim. France*, **1969**, 1841.
79. Newkirk, T. F., *J. Am. Ceram. Soc.*, **41**, 409 (1958).
80. Nedumov, N. A., in *Differential Thermal Analysis*, R. C. Mackenzie, ed., Academic, London, 1970, p. 168.
81. Rupert, G. N., *Rev. Sci. Instr.*, **34**, 1183 (1963).
82. Rupert, G. N., *Rev. Sci. Inst.*, **36**, 1629 (1965).
83. Rupert, G. N., *Proceedings of the First International Conference on Thermal Analysis*, J. P. Redfern, ed., Macmillan, London, 1965, p. 19.
84. Sommer, G., and P. R. Jochens, *Minerals Sci. Eng.*, **3**, 3 (1971).
85. Miller, R. P., and G. Sommer, *J. Sci. Instr.*, **43**, 293 (1971).
86. Sommer, G., P. R. Jochens, and D. D. Howat, *J. Sci. Instr.*, Ser. 2, **1**, 1116 (1968).
87. Proks, I., and I. Zlatovsky, *Chem. Zvesti*, **23**, 620 (1969).
88. Wendlandt, W. W., and W. S. Bradley, *Anal. Chim. Acta*, **52**, 397 (1970).
89. Wendlandt, W. W., *Chimia*, **26**, 2 (1972).
90. Wendlandt, W. W., *J. Chem. Educ.*, **38**, 571 (1961).
91. Borchardt, H. J., and F. Daniels, *J. Am. Chem. Soc.*, **79**, 41 (1957).
92. Miller, G. W., and J. L. Wood, *J. Thermal Anal.*, **2**, 71 (1970).
93. Graybush, R. J., F. G. May, and A. C. Forsyth, *Thermochim. Acta*, **2**, 153 (1971).
94. Barrall, E. M., J. F. Gernert, R. S. Porter, and J. F. Johnston, *Anal. Chem.*, **35**, 1837 (1963).
95. Barrall, E. M., R. S. Porter, and J. F. Johnson, *Anal. Chem.*, **36**, 2172 (1964).
96. Kemme, H. R., and S. I. Kreps, *Anal. Chem.*, **41**, 1869 (1969).
97. Macak, J., and J. Malecha, *Anal. Chem.*, **41**, 442 (1969).
98. Dimitrov, R., *Compt. rend. acad. bulgare sci.*, **23**, 1215 (1970).
99. Wald, S. A., and C. C. Winding, *Anal. Chem.*, **37**, 1622 (1965).
100. O'Neill, M. J., *Anal. Chem.*, **36**, 1238 (1964).
101. Watson, E. S., M. J. O'Neill, J. Justin, and N. Brenner, *Anal. Chem.*, **36**, 1233 (1964).
102. *Perkin-Elmer Thermal Analysis Newsletter*, No. 10, Feb., (1972).
103. O'Neill, M. J., and A. P. Gray, ICTA III, Davos, Switzerland, Aug. 23–28, 1971, paper I-24.
104. Freeberg, F. E., and T. G. Alleman, *Anal. Chem.*, **38**, 1806 (1966).
105. Wendlandt, W. W., *Anal. Chim. Acta*, **49**, 187 (1970).
106. Farritor, R. E., and L. C. Tao, *Thermochim. Acta*, **1**, 297 (1970).
107. Mita, I., I. Imai and H. Kambe, *Thermochim. Acta*, **2**, 337 (1971).
108. Morie, G. P., T. A. Powers, and C. A. Glover, *Thermochim. Acta*, **3**, 259 (1972).
109. Levy, P. F., G. Nieuweboer, and L. C. Semanski, *Thermochim. Acta*, **1**, 429 (1970).
110. David, D. J., *J. Thermal Anal.*, **3**, 247 (1971).
111. Dosch, E. L., and W. W. Wendlandt, *Thermochim. Acta*, **1**, 181 (1970).
112. Evans, W. J., E. J. McCourtney, and W. B. Carney, *Anal. Chem.*, **40**, 262 (1968).
113. Calvet, R., *Compt. rend.*, **226**, 1702 (1948).
114. Sale, F. R., *J. Phys. E: Sci. Instr.*, **3**, 646 (1970).
115. Thomasson, C. N., and D. A. Cunningham, *J. Sci. Instr.*, **41**, 308 (1964).
116. Berger, C., M. Richard and L. Eyraud, *Bull. Soc. Chim. France*, **1965**, 1491.

117. Roux, A., M. Recbaid, L. Eyraud, and J. Elston, *J. Phys. et Phys. Appl.*, **25**, 51A (1964).
118. Nicholson, P. S., *Lawrence Radiation Lab. Report*, UCRL-17820, Sept. 1967.
119. Speros, D. M., and R. L. Woodhouse, *J. Phys. Chem.*, **67**, 2164 (1964).
120. Speros, D. M., and R. L. Woodhouse, *J. Phys. Chem.*, **72**, 2846 (1968).
121. Garski, H., *Z. angew. Chemie*, **24**, 206 (1968).
122. Muller, W., and D. Schuller, *Ber. Bunsen Gesellschaft fur Physik. Chem.*, **75**, 79 (1971).

CHAPTER VII

APPLICATIONS OF DIFFERENTIAL THERMAL ANALYSIS AND DIFFERENTIAL SCANNING CALORIMETRY

A. Introduction

As previously discussed in Chap. V, the DTA or DSC curve consists of a series of peaks in an upward or downward direction on the ΔT or heat-flow axis. The positions (on the temperature or X axis), shape, and number of peaks are used for purposes of qualitative identification of a substance, while the areas of the peaks, since they are related to the enthalpy of the reaction, are used for quantitative estimation of the reactive substance present or for thermochemical determinations. Because of the various factors which affect the DTA or DSC curve of a sample, the peak temperatures and the shape of the peak are rather empirical. Generally, however, the curves are reproducible for any given instrument, so that they can be useful in the laboratory. By use of various calibration substances, the areas enclosed by the curve peaks can be related to heats of reaction, transition, polymerization, fusion, and so on. Or, if the heat of the reaction is known, the amount of reacting substance can be determined.

Some origins of the endothermic or exothermic curve peaks are summarized in Table VII.1. Any phenomenon that produces an enthalpic change or a change in heat capacity (second-order transitions) can be detected by DTA or DSC techniques provided that the instrument has the required sensitivity. These phenomena are caused by fundamental changes in state, chemical composition, molecular reactivity of the substances, and so on. The shape of the peaks, and also the peak maximum ($\Delta T_{\max}$) and peak minimum ($\Delta T_{\min}$) temperatures, are controlled basically by the reaction kinetics, although they are also influenced by the sample packing and geometrical parameters, the heating rate, the furnace atmosphere, and the reference temperature source. Even more subtly, the changes of the base line can be related to the change in specific heat of the sample; this is an important parameter in the detection of glass transition temperatures, T_g, in polymers. The area of the peak is determined by the enthalpic change and also by the instrumental factors such as the size, thermal conductivity, and the specific heat of the sample.

TABLE VII.1
Physicochemical Origin of DTA and DSC Curve Peaks (57)

| | Enthalpic change | |
Phenomena	Endothermal	Exothermal
Physical		
Crystalline transition	x	x
Fusion	x	
Vaporization	x	
Sublimation	x	
Adsorption		x
Desorption	x	
Absorption	x	
Curie point transition	x	
Glass transition	Change of base line, no peaks	
Liquid crystal transition	x	
Heat capacity transition	Change of base line, no peaks	
Chemical		
Chemisorption		x
Desolvation	x	
Dehydration	x	
Decomposition	x	x
Oxidative degradation		x
Oxidation in gaseous atmosphere		x
Reduction in gaseous atmosphere	x	
Redox reactions	x	x
Solid-state reaction	x	x
Combustion		x
Polymerization		x
Pre-curing (resins)		x
Catalytic reactions		x

The technique of DTA has been employed by geologists, ceramicists, and metallurgists for many years. DTA proved to be a rapid analytical tool for the determination and identification of clays and other minerals, phase transitions and phase diagrams, high-temperature kiln reactions, and so on. Normally, the technique is supplemented by X-ray diffraction, dilatometry, thermogravimetry, electrical conductivity, and other techniques. Only in fairly recent times has the chemist become interested in this technique, although many classic chemical studies were carried out in the 1930s. Mention should be made of the early studies by Kracek (58, 59). The list of applications to chemical problems has grown very rapidly, with applications being made to all of the sciences. Specific areas of investigation are summarized

TABLE VII.2
Specific DTA and DSC Applications in Chemistry

Materials	Types of studies
Catalysts	Decomposition reactions
Polymeric materials	Phase diagrams
Lubricating greases	Reaction kinetics
Fats and oils	Solid-state reactions
Coordination compounds	Dehydration reactions
Carbohydrates	Radiation damage
Amino acids and proteins	Catalysis
Metal salt hydrates	Heats of adsorption
Metal and nonmetal oxides	Heats of reaction
Coal and lignite	Heats of polymerization
Wood and related substances	Heats of sublimation
Natural products	Heats of transition
Organic compounds	Desolvation reactions
Clays and minerals	Desolvation reactions
Metals and alloys	Solid-gas reactions
Soil	Curie point determinations
Biological materials	Purity determinations
	Thermal stability
	Oxidation stability
	Glass transition determinations
	Comparison

in Tables VII.2 and VII.3. Indeed, nearly every chemical field has been touched by this technique, although much of the recent emphasis has been in the area of polymer chemistry. Perhaps the largest use of DSC has been in some phase of polymer characterization.

Because of the large number of applications of the techniques of DTA and DSC, the applications described here will be concerned mainly with analytical chemistry problems. In this area, DTA and DSC can be used as a control or a routine tool for comparing similar but not identical materials. As a control technique, it may be used to distinguish between raw materials quickly and easily in those cases in which the treatment of the material must be modified if slight changes in the material are encountered. As a comparison technique, DTA and DSC may be used in some cases to detect materials that yield anomalous results by other tests. Lastly, by suitable calibration of the instruments, these techniques may be used for the quantitative estimation of a substance or mixture of substances, or for purity determinations (see Chap. X). An attempt is made here to include only illustrative types of applications, and it is not intended to be comprehensive.

TABLE VII.3 Some Industrial Uses of DTA and DSC

To determine or evaluate	Ceramics	Cermets	Chemicals	Elastomers	Explosives	Forensic chemistry	Fuels	Glass	Inks	Metals	Minerals	Paints	Pharmaceuticals	Phosphors	Plastics	Propellents	Soaps	Soils	Textiles
Identification	√√		√√√	√	√	√	√		√	√	√√	√	√√	√√	√√	√√	√√	√	√
Composition (quantitative)	√√	√	√√√	√	√	√			√	√√	√√	√	√√	√√	√√	√√	√√	√	√
Phase diagrams			√√√	√	√	√		√				√	√√	√√√	√√	√√			√√√
Solvent retention	√		√√	√	√				√			√	√	√	√	√		√	√√√
Hydration–dehydration			√√	√√√	√√		√√√		√√√		√√	√√√	√√√	√√	√√	√√	√√		√√√
Thermal stability			√√	√√√	√√		√√√		√√	√		√√√	√√	√√	√√	√√	√√		
Oxidative stability			√	√√	√		√√				√	√√	√	√	√√	√√			
Polymerization				√	√	√									√	√			
Curing		√√	√√			√√		√√		√√	√√	√	√√	√	√	√	√√		
Purity		√√	√√																
Reactivity	√						√									√			
Catalytic activity		√√		√	√			√√		√√	√√	√			√	√√			√√√
Glass transitions								√√							√√	√√			√√√
Radiation effects	√√	√		√	√	√√	√√	√√	√	√√	√√	√√	√	√√	√√√	√√√			√√√
Thermochemical constants	√√	√	√	√	√								√			√√	√	√	

254

B. DTA Card Indexes

Several card indexes containing DTA reference curves are available commercially. The Sadtler Research Laboratories collection consists of over 2000 DTA curves, of which about 450 are commercial compounds, 150 pharmaceuticals and steroids, 1000 pure organic compounds, and 360 pure inorganic compounds. An alphabetical index and a molecular formula index permit rapid and convenient location of curves in the collection. The curves, which were recorded with a DuPont Model 900 DTA apparatus, are

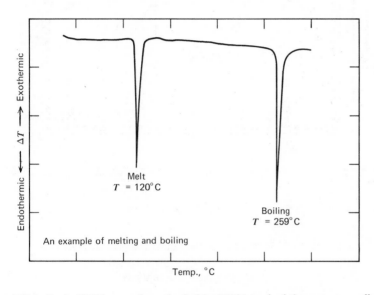

Figure VII.1. Typical DTA curve from the Sadtler DTA standard thermograms collection.

illustrated by the example shown in Figure VII.1. Each curve was recorded at the same heating rate, starting temperature, and furnace atmosphere.

A somewhat different approach was used in the SCIFAX Differential Thermal Analysis Data Index, which was edited by Mackenzie. Punched cards are used, such as the one illustrated in Figure VII.2, which are arranged in three sections: mineral, inorganic, and organic. There are a total of 1662 cards, of which 1012 are mineral, 287 are inorganic, and 311 are organic, plus various code cards.

DTA curves are also recorded in the collection of Derivatograph curves edited by Liptay (1).

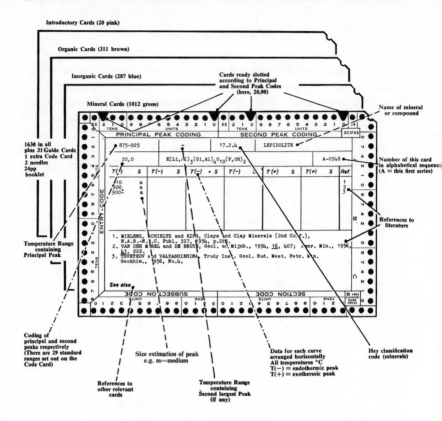

Figure VII.2. SCIFAX DTA Data Index compiled by Mackenzie.

C. Organic Compounds

The applications of DTA and DSC techniques to organic compounds are quite diverse, as is seen in Figure VII.3. It is difficult to point to one of the applications as the most important, or, for that matter, the most widely used. In the pharmaceutical and organic compound manufacturing industries, purity determination is perhaps the most important application. In other areas, identification only may be of vital interest. Only recently, due to the use of sealed-tube DTA and DSC sample holders, the study of organic reactions has assumed some importance. Information can be obtained from a single run which would normally take hours or days to complete by standard methods.

The enthalpic changes which occur in organic compounds are considerably less complex than those for organic polymers. However, they may exhibit various polymorphic changes which can be detected by DTA and DSC. The

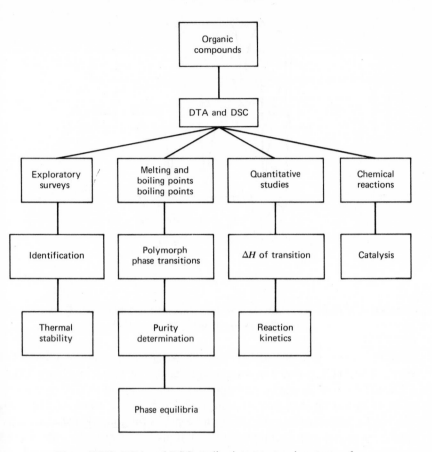

Figure VII.3. DTA and DSC applications to organic compounds.

main sources of endothermic and exothermic enthalpic changes in organic compounds are fusion, vaporization, solid-solid transitions, sublimation, dehydration, decomposition, and combustion.

The applications of these techniques to organic compounds have been extensively reviewed (2–8). Specific DTA applications are reviewed by Mitchell and Birnie (2), while DSC techniques are discussed by Gray (4). .

The thermal decomposition of a number of organic acids has been studied by DTA by Wendlandt and Hoiberg (60, 61). Since the acids were decomposed in an argon atmosphere, only endothermic peaks were observed in the DTA curves. These peaks were caused by such reactions as dehydration, decarboxylation, sublimation, decomposition, and phase transitions from the solid to the liquid state. The maximum peak temperatures for the phase

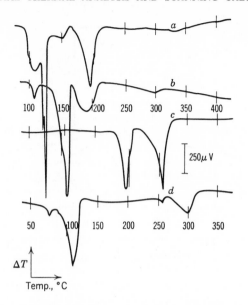

Figure VII.4.   DTA curves of some organic acids: (*a*) oxalic acid dihydrate; (*b*) malonic acid; (*c*) succinic acid; and (*d*) glutaric acid (60).

transitions were 10 to 30° higher than the reported melting-point temperatures.   The DTA curves for some of the acids are given in Figure VII.4.
The DTA curve for oxalic acid dihydrate, the only acid studied containing
water of hydration, had dehydration peaks with ΔT_{min} values of 110, 120,
and 125°C, respectively.   All other curve peaks for the organic acids were
caused by fusion and decomposition reactions.   For example, the second
endothermic peak in the succinic acid curve was probably caused by
dehydration reaction, resulting in anhydride formation, of the type.

$$
\begin{array}{c}
\mathrm{O} \\
\parallel \\
\mathrm{C-OH} \\
\mid \\
\mathrm{CH_2} \\
\mid \\
\mathrm{CH_2} \\
\mid \\
\mathrm{C-OH} \\
\diagdown \\
\mathrm{O}
\end{array}
\quad \xrightarrow{\Delta} \quad
\begin{array}{c}
\mathrm{O} \\
\parallel \\
\mathrm{CH_2-C} \\
\mid \qquad \diagdown \\
\qquad \mathrm{O} + \mathrm{H_2O} \\
\mid \qquad \diagup \\
\mathrm{CH_2-C} \\
\diagdown \\
\mathrm{O}
\end{array}
$$

It has long been known that meconic acid (**I**) (3-hydroxy-4-oxo-4H-
pyrane-2,6-dicarboxylic acid) can be heated in air to 120–220°C to form
comenic acid (**II**) (37).   DTA was used to establish the best preparative

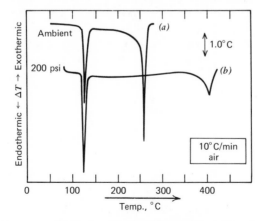

conditions for this reaction. It was found that comenic acid is formed by decarboxylation of meconic acid in a reaction giving an exothermic peak at ΔT_{max} of 240°C. At and above this temperature the product sublimes.

The boiling point of benzoic acid is shifted to higher temperatures by use of high-pressure DTA. Levy et al. (38) obtained the DTA curves of pure benzoic acid at ambient pressure and at a pressure of 2000 p.s.i.g., as shown in Figure VII.5. Curve (a) indicates that benzoic acid melts at about 122°C. Under 200 p.s.i.g. nitrogen pressure [curve (b)], the melting-point endothermic peak remains unchanged while the boiling point is elevated to 378°C. In order to avoid sublimation and evaporation and to ensure equilibrium conditions, the samples were run in a small hermetically sealed aluminum pan which contained a small hole (~0.002 in.) punched in the top to equalize the pressure.

The DTA curves of a number of pharmaceutical compounds have been described by Brancone and Ferrari (9) in which qualitative information concerning purity, solvation, structural configuration, and polymorphism were obtained. The DTA curve of triamcinolone diacetate (9), as shown in Figure VII.6, aided in the establishment of its proper drying temperature. The solvent peaks at ΔT_{min} of 142 and 166°C, respectively, disappear on drying.

Figure VII.5. DTA curves of benzoic acid at ambient pressure (a) and at a pressure of 200 p.s.i.g. of nitrogen (b) (38).

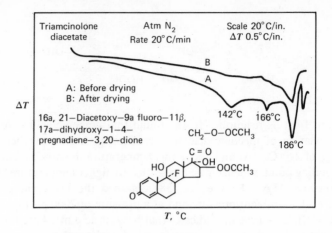

Figure VII.6. DTA curves of triamcinolone diacetate (9).

The precise determination of melting and boiling points by DTA was first discussed in detail by Vassallo and Harden (10). They obtained a precision of $\pm 0.3°C$ over a wide range of heating rates and were able to make determinations in the temperature range from -150 to $450°C$. The temperatures estimated for melting point, T_m, or boiling point, T_B, were selected from the most often recommended portions of the DTA peak, as shown in Figure VII.7. Point A is the intersection of the extrapolated straight-line

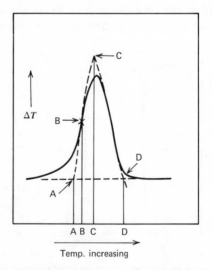

Figure VII.7. Transition temperature estimation methods (10).

TABLE VII.4

Comparison of Various Methods for Transition Temperature Measurement (10)

Temperature measured	Benzoic acid,[a] melting point, °C			
	A	B	C	D
Reference	121.1	125.0	128.0	131.0
Sample	120.3	121.0	121.7	131.9
	Toluene[b] boiling point, °C			
Method	A	B	C	D
Reference	112.5	114.2	115.9	116.2
Sample	114.6	113.2	111.2	118.8

[a] NBS, m.p. = 121.8°C.
[b] Merck, b.p. = 1.0° boiling range including 110.6°C.

portion of the low temperature side of the peak with the base line, and point B is the inflection point of the low-temperature side. Point C is the extrapolated peak temperature, while D is the extrapolated return to the base line. The sample and reference temperatures at which the various points occurred during the melting of benzoic acid and the boiling of toluene are shown in Table VII.4. With the exception of D, T_m estimates using sample temperature are generally closer to the true value. The closest estimates of T_m were achieved by use of sample temperature at point C or reference temperature at point A. The T_b estimated for toluene at points A, B, and D were higher than at point C, and this is probably due to superheating effects.

A low-temperature DTA curve of n-butane (10) is shown in Figure VII.8. In the temperature range $-150°$ to $10°C$, the endothermic peaks for boiling ($-0.5°C$) and melting ($-135°C$) are narrow and well defined. Melting and boiling points of a number of organic compounds are given in Table VII.5 (10).

Barrall (39) has discussed in great detail the precise determination of melting and boiling points by DTA and DSC. Many different methods are available for these determinations; however, Barrall's method gives a consistent technique which is readily adaptable to commercial instruments and will cover a wide range of organic and inorganic materials. The data obtained are more consistent than those measured with the hot-stage, oil-bath, or capillary-tube techniques. Boiling-point data obtained by DTA and DSC are far more reproducible and usually more closely comparable to equilibrium-still results than data from microebulliometry, and they are certainly more rapid.

The experimental conditions of DTA necessitate that the ΔT parameter be plotted as a function of sample temperature, while in DSC the differential

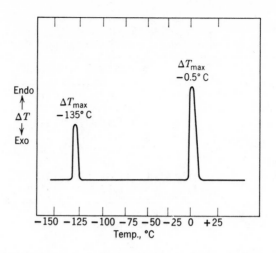

Figure VII.8. Low temperature DTA curve of n-butane (10).

TABLE VII.5
Transition Temperatures Determined by DTA (10)

Compound	Melting point, $T°C$		Boiling point, $T°C$	
	Found	Reported	Found	Reported
n-Butane	−135.0	−135.5	−0.5	−0.55
n-Pentane	−129.5	−129.7	36.2	36.0
n-Hexane	−94.5	−95.3	69.0	68.8
n-Heptane	−90.3	−90.6	98.2	98.4
n-Octane	−57.0	−56.8	125.6	125.6
n-Nonane	—	—	150.2	150.7
n-Decane	—	—	173.0	174.0
n-Dodecane	—	—	215.5	216.0
Benzoic acid	121.8	121.8	—	—
Water	0.0	0.0	100.0	100.0
Toluene	—	—	111.1	110.6
Benzene	5.2	5.5	80.5	80.1
Acetic acid	16.5	16.6	118.4	118.1
Alathon 10 polyethylene resin	110.5	110.5[a]	—	—
Marlex 50 polyethylene resin	134.2	134.5[a]	—	—
Teflon TFE fluorocarbon resin	327.5	327.0[a]	—	—
Teflon FEP fluorocarbon resin	272.0	272.5[a]	—	—
Delrin acetal resin	170.5	171.0[a]	—	—

[a] Melting points taken with Kofler hot stage microscope, or by X-ray techniques.

power curve is recorded as a function of time. For melting-point determinations, the sample may be encapsulated; in the case where the thermocouple must be inserted into the sample, a sample diluent mixture must be employed. The encapsulations are usually in a tightly sealed metal container of high thermal conductivity. In the case of boiling-point determinations, provision must be made for (a) equilibration of liquid and vapor and (b) control of atmospheric pressure. Barrall (39) describes in detail how these two criteria can be met.

If an encapsulation procedure is employed for determining the melting point, an extrapolation procedure is used to correct for the thermal lag in the system, as shown in Figure VII.9a. The true melting point of the compound is obtained by the extrapolation of the leading edge of the peak curve to the isothermal base line. If the sample is in direct contact with the thermocouple, the true melting point is T_b in Figure VII.9b. Other terms to be defined in the latter are the minimum in the peak, T_m, and the temperature at the end of the peak, T_e. The range between T_b and T_m is a function of the purity of the sample. Due to base line drift, T_b is occasionally difficult to locate in impure samples.

In boiling-point measurements, T_m is of little significance. It is a function only of various instrument parameters, the rate of vapor diffusion, and the amount of sample present at the onset of boiling. The temperature at the beginning of the curve, T_b, corresponds to the boiling point of the compound (after suitable corrections).

A microboiling and melting point procedure using DTA was described by Kerr and Landis (11). The 2–5-μl samples were trapped at the exit port of a GC column and transferred with a 10-μl syringe to a capillary tube for DTA study. Examples of boiling points of various organic compounds are illustrated in Figure VII.10. Micromelting point determinations permitted the estimation of the relative purity (>90%) of m- or p-xylenes.

Transition-temperature determinations for liquid crystals have been the subject of numerous investigations (12–20). Not only can the melting points be determined, but also the liquid-crystal transitions. One of the first such determinations was that by Barrall et al. (12) on anisaldazine, which is illustrated in Figure VII.11. A stable, linear heating rate is necessary because a 1°C sudden departure for linearity can appear as a "glass transition point" on the DTA curve. The two curves in Figure VII.11 illustrate the use of temperature-axis amplification in order to separate closely spaced peaks. In (A), the temperature span is 0 to 280°C, while in (B) it is from 135.80 to 183.34°C. The temperature scale in (B) was amplified about eight times over that in (A). Amplification of up to 1000 times in any 300° interval was possible in the temperature range from −100 to 500°C.

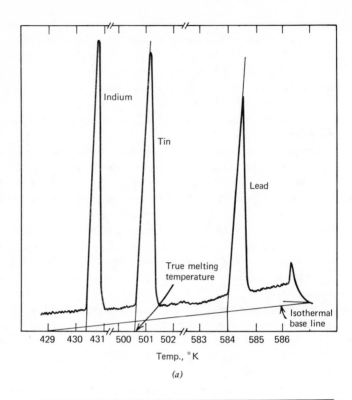

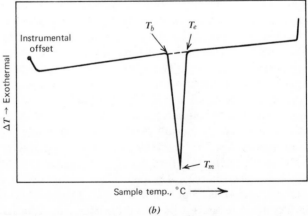

Figure VII.9. Melting point temperature determination; (a) encapsulated samples of 99.99999 mole-% purity; (b) thermocouple embedded directly in the sample (39).

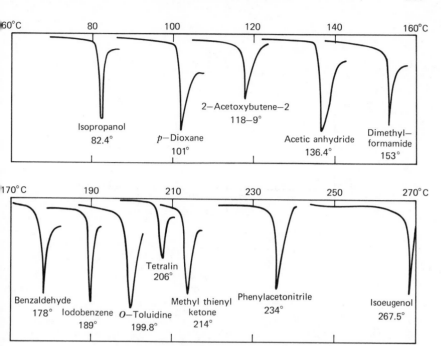

Figure VII.10. DTA curves of 2–5-μl samples of various organic compounds (11).

Transition temperatures for a series of cholesteryl esters (13) are shown in Table VII.6. These temperatures are compared with those obtained by Gray (21). Temperatures observed by DTA endothermic peak minima agree to within ±2° with those observed visually. The absolute accuracies of the temperatures determined was ±0.1° in all cases, while the reproducibility was ±0.05°C for an individual sample on successive remeltings.

Heats of transition for liquid crystals have been determined by a number of investigators (14, 15, 16). These heats of transition are very small, ranging from 0.5 to 1.8 cal/g in most cases.

The phase transitions and dissociation reactions of organic explosive materials have been extensively investigated by DTA and DSC. Hall (40) studied the processes of solid phase transition, fusion, and decomposition in several nitramines by DSC. Compounds investigated include N-picryl-N-methylnitramine, 1,3,5-trinitroso-1,3,5,7-tetraazacyclooctane, and others. Rogers and Smith (41) studied a similar group of compounds and estimated the preexponential factor from the DSC curve. More recently, Rogers and Dinegar (42) determined the heat of fusion and other parameters of pentaerythritol tetranitrate (PETN) by DSC.

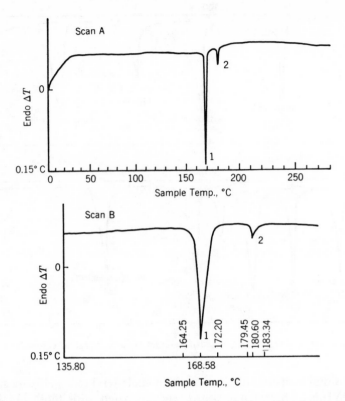

Figure VII.11. DTA curves of anisaldazine (12); 1, Fusion point; 2, Liquid crystal transition point. Curves A and B at $\Delta T = 0.05°C = 1$ in.; T in curve $A = 1$ mV per in.; T in curve $B = 1$ mV per 8 in.

Differential scanning calorimetry was used by Murrill and co-workers (43–45) to elucidate *solid → solid* phase transitions in a large number of organic compounds.   First-order transitions were reported for tetrahedral compounds of the type $CR^1R^2R^3R^4$, where R is methyl, methylol, amino, nitro, and carboxy, as well as for octahedral-type compounds.   This technique was also used to detect phase transitions in alkali metal stearates (46), some dibenzazepines, carbazoles, and phenothiazines (16), and the half esters of 0-phthalic acid (31).   The solid-state decomposition kinetics and activation parameters of N-aryl-N'-tosyl-oxydi-imide N-oxides were determined using DSC by Dorko et al. (49).
    Calorimetric heats of transition for other physical processes (melting, boiling, solid-solid, and so on) have been reported by several investigators

TABLE VII.6
Transition Temperatures for Cholesteryl Esters (13)

Cholesteryl ester	DTA[a]			Gray[a]		
	T_1	T_2	T_3	S	C	I
Formate	—	—	97.3		(60.5)	97.5
Acetate	44	81–87	118.4		(94.5)	116.5
n-Propionate	99 ± 1	110?	115.3		102	116
n-Heptylate			114.1	(<92.5)	(95.5)	114
n-Nonanoate	74.0	80.8	93.0	(77.5)	80.5	92
n-Decanoate		85.7	91.2	(81.5)	85.5	92.5
Myristate	73.6	79.7	85.5	71	81	86.5
Palmitate		79.7		(78.5)	779	83
Stearate			85.1	(75.5)	(79.5)	83

[a] S = smectic; C = cholesteric; I = isotropic liquid; T_1 = lowest transition; T_2 = intermediate transition; T_3 = transition to isotropic liquid; parentheses indicate monotropic metastable transition.

(22–25, 31). Very careful calibration is necessary to obtain accuracies of the order of ±5%.

Chiu (48) investigated the formation of an organic derivative by DTA. He replaced the traditional method of preparing the derivative from the sample and reagent with a one-step process. The sample was heated with a specific reagent at a programmed heating rate in a selected atmosphere. The DTA curve showed the derivative forming reaction, the physical transitions of the sample or the reagent in excess, and the physical transitions of the intermediates and products. Glass capillary tubes were employed as the sample holder.

The formation of the acetone hydrazone derivative with p-nitrophenylhydrazine is illustrated in Figure VII.12. Curve (a) shows the endothermic peak for the boiling of acetone, with a $\Delta T_{\max}$ of 58°C. For p-nitrophenylhydrazine, the endothermic peak at a $\Delta T_{\max}$ of 160°C was caused by the fusion of the compound. A mixture of acetone and p-nitrophenylhydrazine, however, in the 54–80°C temperature range, gave a complex endothermic peak which was attributed to the net result of evaporation of excess acetone, solution of p-nitrophenylhydrazine in acetone, and hydrazone formation. The fusion of the hydrazone was indicated by the endothermic peak with a $\Delta T_{\max}$ of 153°C. A rerun of the residue gave only a single endothermic peak, with a $\Delta T_{\max}$ of 153°C. The reported melting point of the hydrazone derivative is 152°C. Similar examples, such as the reactions of triethylamine with picric acid and dextrose with propylamine, were illustrated.

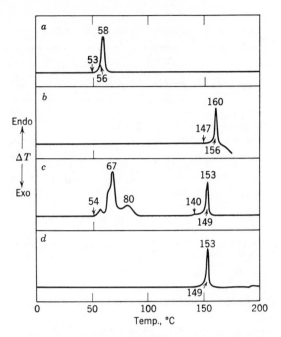

Figure VII.12. DTA curves showing formation of *p*-nitrophenylhydrazone of acetone. (*a*) acetone; (*b*) *p*-nitrophenylhydrazine; (*c*) reaction mixture of acetone and *p*-nitrophenylhydrazine; (*d*) rerun of residue from (*c*) (48).

The method described is rapid and dynamic in nature, and requires that (*a*) a specific reagent should form a derivative with the sample rapidly; (*b*) the derivative so produced should show a discernible physical transition or a characteristic DTA curve; (*c*) one reactant more volatile than the other should be used in excess; (*d*) one reactant should serve as the solvent for the other; and (*e*) a catalyst may be used.

A Diels-Alder diene synthesis, using maleic anhydride and anthracene, was carried out using DTA by Harmelin et al. (26). This technique permits the determination of the temperature at which reaction occurs, the melting point of the adduct formed, and the decomposition of the adduct.

The thermal properties of explosives and propellant compositions are widely studied by DTA and DSC. Fauth (47) recorded the DTA curves of some hydrazine, guanidine and guanidinium picrates, styphnates, and sulfates. The decomposition temperatures found were generally considerably lower than those reported in the literature. Other picrates, those with thallium, ammonium, tetramethylammonium, and tetraethylammonium, were studied by Stammler (27). David (28) and Bohon (29) examined the thermal behavior of explosives and propellants under various external pressures up to 3000 p.s.i.g. using DTA. Heats of explosion and/or decomposition were determined. Decomposition of primary explosives using a remotely operated DTA cell was described by Graybush et al. (30).

Using sealed-tube sample holders, Santoro and co-workers (32–35) investigated a wide variety of organic reactions. Examples are the cis → trans isomerization of stilbene and oleic acid, polymerization of styrene, Diels-Alder reactions, and others. Unstable intermediates in an organic reaction have been detected using DTA techniques by Koch (36). If a solution of an unstable compound is heated, temperature changes characteristic of reactions of the intermediate can be detected. Conversely, the absence of thermal effects indicates that no unstable product is present.

The determination of the relative purity of an organic compound by DTA and DSC methods will be discussed in Chapter X. Most of the analytical methods are based on the DSC technique, although, as can be seen in the DTA curves in Figure VII.13, DTA may also be used.

Differential thermal analysis has been applied to the degradation of hydrocarbon liquids (50–52) and has also been proposed for the quality control of grease (53, 54). Noel (55) showed that DSC could be used to characterize petroleum products and that the results could be correlated with some ASTM tests. The advantages of DSC over previous tests to characterize petroleum products are as follows:

(*a*) Information is obtained quickly and reproducibly.

(*b*) Very small samples are required (milligram quantities versus quart samples).

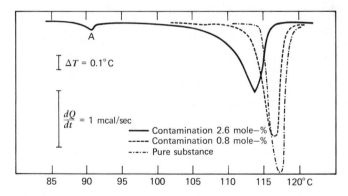

Figure VII.13. DTA curves of acetanilide containing various amounts of an impurity. The eutectic melting peak, A, was only recorded for the sample with 2.6 mole-% contamination.

(*c*) The equipment has multiple uses rather than being able to do only one test.

(*d*) The information is fundamental and low on empiricism; thus it has research value as well as use for control and trouble shooting.

Noel and Corbett (56) applied DSC to the study of glass and melting transitions in asphalts. The T_g values obtained agreed well with those previously determined by dilatometry. Heats of fusion of the asphalt waxes ranged from 26 to 32 cal/g.

The oxidation of an automotive brake fluid and a motor oil was studied by high-pressure DSC (38). An air pressure of 600 p.s.i.g. indicated the relative oxidative stability of each of the types of materials.

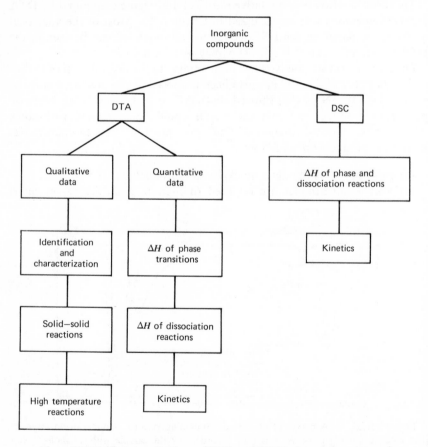

Figure VII.14. Some applications of DTA and DSC to inorganic compounds.

D. Inorganic Compounds

The applications of DTA and DSC to inorganic compounds are similar to those previously discussed for organic compounds. Endothermic and exothermic peaks are caused by phase transitions (melting, boiling, polymorphic changes), dehydration, dissociation, isomerization, oxidation-reduction reactions, and so on. These applications are summarized in Figure VII.14, while the general classes of compounds studied are given in Figure VII.15. A large number of DTA applications to inorganic compounds are reviewed

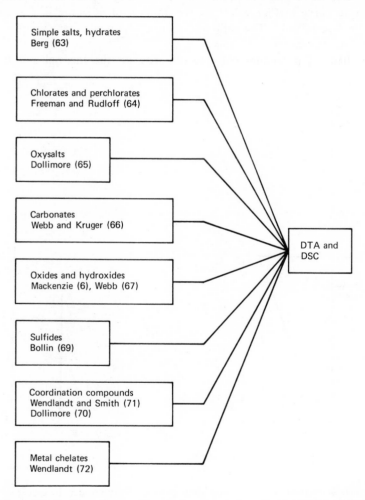

Figure VII.15. Some classes of inorganic compounds studied by DTA and DSC.

by Mackenzie et al. (62). As in the case of organic compounds, only analytical chemistry applications will be included here.

A rapid method for the determination of the moisture content of "nearly dry" powdered substances had been described by Stone (97). Using a dynamic-gas-flow apparatus, Stone placed samples of the substance in the DTA sample holder at room temperature and atmospheric pressure. When the system was evacuated at a fixed rate, the DTA curve peak began at the point where the external water vapor pressure was less than the partial pressure of the water vapor pressure of the sample. From the peak height, using the calibration curve in Figure VII.16, the amount of moisture in the sample could be determined. The calibration curve was made from samples containing known amounts of water.

The effect of potassium chlorate impurity on the thermal stability of

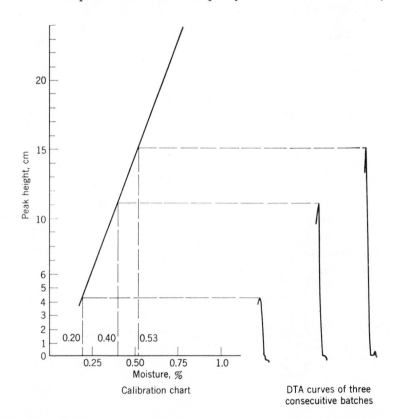

Calibration chart   DTA curves of three
                                     consecuitive batches

Figure VII.16. Method of determining water content of dry powders by evacuation and DTA (room temperature) (97).

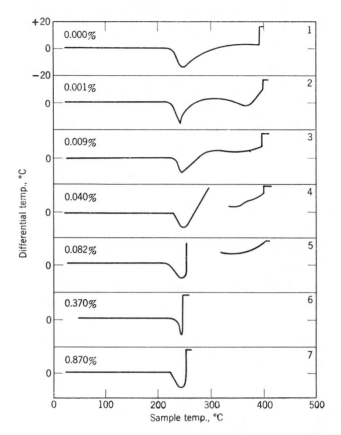

Figure VII.17. DTA curves of NH_4ClO_4 containing various amounts of $KClO_3$ (98).

ammonium perchlorate has been studied by a DTA method by Petricciani et al. (98). The effect of this impurity on the DTA curve of ammonium perchlorate is illustrated in Figure VII.17.

The curves clearly illustrate the presence of an increasingly large exothermic reaction after the 244°C lattice transition of the ammonium perchlorate. In the 0.1% $KClO_3$ region of impurity, the heat evolved after the lattice transition was great enough to initiate complete thermal decomposition of the sample. This represents an effective 150° lowering of the thermal decomposition temperature of the pure material, which normally decomposes at about 400°C. The DTA technique could be used to detect the approximate amount of impurity in the ammonium perchlorate.

Wendlandt et al. (99), in studying the thermal decomposition of the thorium, uranium, and rare-earth metal oxalate hydrates by DTA, also

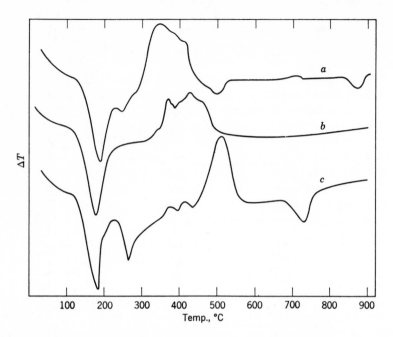

Figure VII.18. DTA curves of rare-earth oxalate hydrate mixtures. (*a*) 1:1 La–Ce oxalate mixture (physical); (*b*) 1:1 La–Ce oxalate mixture (precipitated); 1:1 Pr–Nd oxalate mixture (precipitated) (99).

studied mixtures of the rare-earth oxalate hydrates.   The DTA curves of mixtures of lanthanum-cerium and praseodymium-neodymium oxalate mixtures are given in Figure VII.18.

In curve *a*, a 1:1 physical mixture of La-Ce oxalates is presented, while curve *b* represents a 1:1 mixture of La-Ce oxalates that was precipitated by the homogeneous precipitation method.   The physical mixture exhibited dehydration endothermic peaks at the appropriate temperatures, but the broad exothermic peak from 300 to 450°C was different from the individual lanthanum and cerium oxalate curves.   However, the 875°C endothermic peak was still present.   The 1:1 precipitated mixture of La-Ce gave only a single endothermic dehydration peak, while the broad exothermic decomposition peak began at a slightly higher temperature.   The curve was devoid of any peaks beyond 500°C, in contrast to curve *a*.   Curve *c* is a 1:1 precipitated mixture of Pr-Nd oxalates.   The double dehydration peaks were present, although shifted to slightly higher temperatures, while the praseodymium oxalate exothermic peak at 460°C was shifted to a higher temperature, 510°C, in the mixture.   In contrast, the 768°C neodymium endothermic peak shifted to a lower temperature, 670°C, while the 675°C neodymium oxalate exothermic peak disappeared entirely. It is apparent from the precipitated mixture curves that the resultant curves are not the same as the sum of the curves for the individual oxaltes.

Erdey and Paulik (100), in a simultaneous DTA-TG study, investigated the thermal decomposition of barium, strontium, manganese(II), calcium, magnesium, and zinc oxalates in air and nitrogen atmospheres. It was found that the evolved carbon dioxide formed in the reaction played an important part in that it may inhibit the progress of the reaction and shift the peak temperatures to higher values.

The changes in enthalpy which occur when $CaHPO_4 \cdot 2H_2O$ is heated up to 1300°C were determined by a DTA method by Mesmer and Irani (73). An internal standard of $CaCO_3$ was used to calculate the enthalpy changes which occurred during the reaction,

$$CaHPO_4 \cdot 2H_2O \rightarrow CaHPO_4 + 2H_2O\,(g) \tag{1}$$
$$CaHPO_4 \rightarrow \tfrac{1}{2}Ca_2P_2O_7\,(\gamma) + \tfrac{1}{2}H_2O\,(g) \tag{2}$$
$$\tfrac{1}{2}Ca_2P_2O_7\,(\gamma) \rightarrow \tfrac{1}{2}Ca_2P_2O_7\,(\beta) \tag{3}$$
$$\tfrac{1}{2}Ca_2P_2O_7\,(\beta) \rightarrow \tfrac{1}{2}Ca_2P_2O_7\,(\gamma) \tag{4}$$

Results of these enthalpy changes are given in Table VII.7.

Heats of transition of a number of inorganic compounds were determined using several new methods of quantitative DTA (74, 75). Results for one of the methods are summarized in Table VII.8 (75). It is interesting to note that a value of 25.2 cal/g was obtained for the melting transition in KNO_3, which agrees fairly well with the 27.7 cal/g reported previously (76). The value obtained with the DuPont DSC cell was 22.7 cal/g, which is in agreement with another literature value, 22.7 cal/g (77).

An investigation was made by Barrall and Rogers (78) to determine whether the precision of DTA measurements could be increased by using a reference compound as a thermochemical standard. The reference standard, in this case, was silver iodide. The DTA curve obtained using this procedure is shown in Figure VII.19. The ΔH of transition found was 216 ± 20 cal,

TABLE VII.7
Enthalpy Changes for $CaHPO_4 \cdot 2H_2O$ (73)

Reaction	Heating rate, °C/min	T, °C	ΔH_T, kcal/mole
1	4	135	21.3, 20.9
2	4	430	6.9, 7.5
3	10	850	−0.17, −0.23
4	10	1220	0.85, 0.68

TABLE VII.8
Heats of Transition Obtained by Ozawa et al. (75)

Sample	Present work		Lit.	
	Temp. (°C)	ΔH (cal/g)	Temp. (°C)	ΔH (cal/g)
Benzoic acid	122	38.0	—	35.2
			—	34.8
KNO_3	129	13.2	128	13.8
			127.5	11.78
			—	12.83
			127.9	12.05
			127	13.0
KNO_3	336	25.3	338	27.7
			334.3	22.75
			—	27.20
			331	28.1
$AgNO_3$	161	3.42	160	3.9
			160	3.44
			158.9–160.6	3.49
			164	3.03
$AgNO_3$	206	17.7	211	16.2
			210	17.57
			207	17.0
Al	658	102.5	659	95.2
			660	96.3

which is compared with a literature value of 205 ± 30 cal. Using an inert substance as the reference, a ΔH of 200 ± 90 cal was obtained. Thus, there was some improvement in the calorimetric accuracy obtained using this method.

Wendlandt (79) found that by using the sealed-tube technique, heats of dehydration of metal salt hydrates could be obtained which would be impossible using conventional open tubes or crucibles. This approach is illustrated by the DTA curves for $CuSO_4 \cdot 5H_2O$ shown in Figure VII.20. In the open-tube curve, the peaks are due to the following reactions:

$$CuSO_4 \cdot 5H_2O \rightarrow CuSO_4 \cdot 3H_2O + 2H_2O \text{ (l)} \tag{a}$$

$$2H_2O \text{ (l)} \rightarrow 2H_2O \text{ (g)} \tag{b}$$

$$CuSO_4 \cdot 3H_2O \rightarrow CuSO_4 \cdot H_2O + 2H_2O \text{ (g)} \tag{c}$$

For the sealed-tube curve, the first peak is due to reaction (a), while the origin of the second peak is not known but may be due to the reaction

$$CuSO_4 \cdot 3H_2O \rightarrow CuSO_4 \cdot H_2O + 2H_2O \text{ (l)} \tag{d}$$

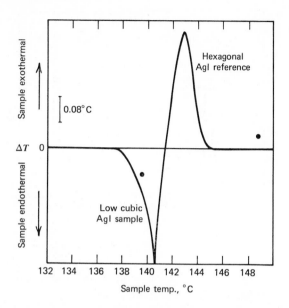

Figure VII.19. DTA curve of cubic AgI *versus* hexagonal AgI (78). A 0.0800-g specimen of 15.1% low-temperature cubic silver iodide on carborundum as sample and a 0.0603-g specimen of 20% hexagonal silver iodide on carborundum as reference heated from 0 to 200°C at 8.0°C/min.

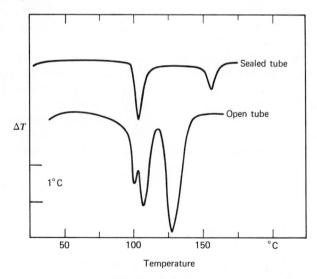

Figure VII.20. Open and sealed tube DTA curves of $CuSO_4 \cdot 5H_2O$ (79).

277

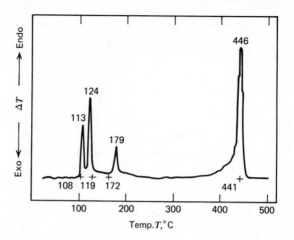

Figure VII.21. DTA curve of sulfur showing various phase transitions (82).

Thus, the sealed-tube sample holder permitted the determination of the ΔH of reaction (a), which was 12.9 ± 0.6 kcal/mole.

Similar studies were carried out on the deaquation of $[Cr(NH_3)_5H_2O]X_3$ (80) and $[Co(NH_3)_5H_2O]X_3$ (81) complexes.

The DTA curve of sulfur, as recorded by Chiu (82), is shown in Figure VII.21. The enantiotropic change from the *rhombic* to the *monoclinic* form is indicated by the 113°C peak, while melting was observed during the 124°C peak. Further transformations in liquid sulfur were observed at 179°C, and finally the boiling peak at 446°C.

Detection of organic contamination in ammonium nitrate is shown by the two DTA curves in Figure VII.22 (83). The exothermic peak begins at a lower temperature in the sample with organic-material contamination.

The amount of metallic nickel in catalysts was determined by a DTA method by Macak and Malecha (84). Nickel produced by the reduction of nickel oxide was reoxidized by oxygen and the ΔT of the oxidation reaction determined by the apparatus. The maximum value of ΔT between the reactor for catalytic reaction and that with inert SiO_2 packing was proportional to the amount of nickel in the catalyst sample. Accuracy of the method was about $\pm 4\%$.

Reaction of hydrogen, under pressure in a DSC cell, was used as a method for the determination of platinum or palladium in various catalysts (85). About 5 mg of catalyst are introduced into a DSC pan and the cell is pressurized to 150 p.s.i.g. with helium. The temperature is then increased to 75°C and the helium gas replaced with hydrogen at 200 p.s.i.g. From the area of the DTA curve peak (versus time), the amount of platinum or

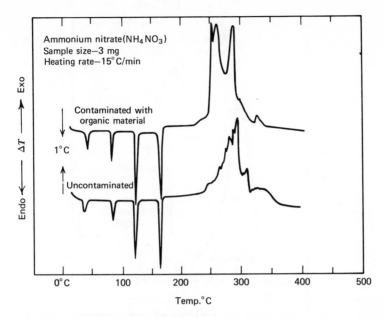

Figure VII.22. DTA curves of NH_4NO_3 in the pure state and contaminated with organic material (83).

palladium can be calculated. Such a DTA curve for the reaction with a carbon catalyst containing 5% palladium is shown in Figure VII.23.

The DTA curves of a number of inorganic compounds proposed as "standards" have been recorded by Garn (86). Each of these compounds exhibits a $solid_1 \rightarrow solid_2$ transition which has been well characterized. The heating (or cooling) behavior of the compounds are summarized in Table VII.9.

The amount of tricalcium silicate in Portland cement can be obtained from a large reversible transition which occurs at 915° (87). The degree of hydration of this compound can also be estimated. In another investigation, Ramachandran (88) described the DTA determination of chloride content in concrete compositions.

The thermochemistry of a large number of transition-metal complexes of the type ML_nX_2 have been investigated using DSC by Beech et al. (89–92). These results as well as others are discussed in a book by Mortimer and Ashcroft (93). The overall decomposition reactions of the complex ML_4X_2 are as follows:

$$ML_4X_2 (c) \rightarrow ML_2X_2 (c) + 2L (g)$$
$$ML_2X_2 (c) \rightarrow MLX_2 (l) + L (g)$$
$$MLX_2 (l) \rightarrow ML_{2/3}X_2 (l) + \tfrac{1}{3}L (g)$$
$$ML_{2/3}X_2 (l) \rightarrow MX_2 (c) + \tfrac{2}{3}L (g)$$

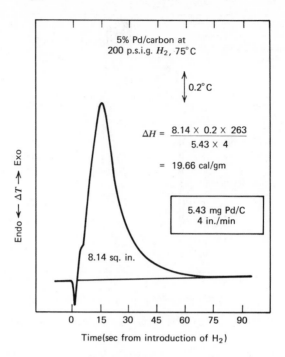

Figure VII.23. Determination of palladium content in a carbon catalysts (85).

The decomposition reactions recorded by DSC for the reaction (89)

$$ML_2X_2 \text{ (c)} \rightarrow MLX_2 \text{ (l)} + L \text{ (g)}$$

are shown in Figure VII.24.   The area, $(a + b)$, is a measure of this reaction, at a mean temperature, T_m.   In the absence of a sample, no additional heat is required to raise the temperature of the empty sample pan, compared with the reference pan, as indicated by the horizontal baseline marked "empty pan."   With a sample of mass m_{react} and specific heat $C_{p,react}$ present in the pan, the baseline is displaced by an amount corresponding to $m_{react} \cdot C_{p,prod}$, the product of the mass and specific heat of MLX_2.   The ligand, L, of mass m_L and specific heat $C_{p,L}$, has been lost as a gas during the reaction.   The mean temperature, T_m, is equal to $\frac{1}{2}(T_f - T_i)$.   Total heat absorbed is represented by the area $(a + b + c + d)$, so that the heat ΔH_{T_i} at temperature T_i of the reaction is given by

$$\Delta H_{T_i} = (a + b + c + d) - (T_f - T_i)[m_{prod} \cdot C_{p,prod} + \tfrac{1}{2}m_L \cdot C_{p,L}] \tag{VII.1}$$

TABLE VII.9

Average Temperatures of the Departure, Intersection, and Peak for Tentative Standards
(86)

	Heating rate °C/min	Heating			Cooling		
Material		Departure	Inter- section	Peak	Departure	Inter- section	Peak
Barium carbonate	2.8	796.0	802.0	810.0	753.0	753.0	750.0
	8.0	797.0	804.0	808.0	770.0	769.0	775.0
Potassium chromate	2.8	664.5	666.3	669.3	669.4	665.4	664.3
	8.0	661.8	664.2	667.7	670.5	665.7	665.1
Potassium nitrate	2.8	128.0	128.9	131.3	123.1	122.8	122.2
	8.0	128.8	129.3	132.6	123.8	123.4	123.1
Potassium perchlorate	2.8	298.2	299.3	301.0	293.7	293.5	293.8
	8.0	297.4	298.3	300.2	292.7	292.7	292.4
Potassium sulfate	2.8	577.9	580.0	582.3	580.4	577.9	578.3
	8.0	577.9	583.9	587.4	585.9	585.1	584.3
Silicon dioxide	2.8	566.2	568.1	569.9	572.0	569.9	569.0
	8.0	561.4	567.7	570.9	577.8	570.9	569.2
Silver sulfate	2.8	421.6	424.4	428.4	405.8	495.6	413.6
	8.0	422.2	427.8	433.7	409.3	408.6	414.3

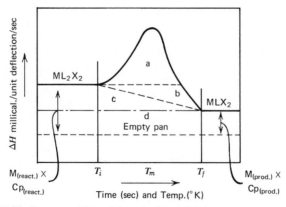

Figure VII.24. Decomposition of the complex, ML_2X_2, as recorded by DSC (89).

Only one-half of the term $m_L \cdot C_{p,L}$ is included because the ligand is liberated from the system, between T_i and T_f, at a rate which is approximately constant. The heat, ΔH_{T_m}, at some other temperature, T_m, is given by

$$\Delta H_{T_m} = \Delta H_{T_i} + (T_m - T_i)[m_{\text{prod}} \cdot C_{p,\text{prod}} + m_L C_{p,L} - m_{\text{react}} \cdot C_{p,\text{react}}]$$
(VII.2)

since

$$(T_m - T_i)[m_{\text{prod}} \cdot C_{p.\text{prod}}] = \tfrac{1}{2}d$$
(VII.3)

$$(T_m - T_i)[m_1 C_{p,L}] = \tfrac{1}{2}(T_f - T_i)[m_1 C_{p,L}]$$
(VII.4)

$$(T_m - T_i)[m_{\text{react}} C_{p,\text{react}}] = \tfrac{1}{2}(b + c + d)$$
(VII.5)

and $b = c$ (approximately), the heat, ΔH_{T_m}, can be written as:

$$\Delta H_{T_m} = a + b$$
(VII.6)

Thermochemical data for the compounds $Co(py)_2 X_2$ are given in Table VII.10.

Differential scanning calorimetry has been used to measure the heats of transition of some sodium, potassium, and silver compounds (94). The agreement of the experimental values with the literature for KSCN and KNO_2 was very poor.

Simchen (95) pointed out that the dissociation temperature of $NaHCO_3$ of 270°C commonly reported in many handbooks, is grossly in error. By use of DSC, the decomposition temperature was found to be about 100°C.

Block (96) reported an analytical method for chloride-bromide mixtures utilizing DSC. The fact that the heat of fusion of an ideal solid solution of the type $A_m X_n - B_m X_n$ or $A_m X_n - A_m Y_n$ is directly proportional to the concentration of solute ion was used to determine chloride-bromide mixtures

TABLE VII.10
Thermochemical Data for $Co(py)_2 X_2$ Complexes (89)

Parameter	Cl	Br	I
M.P., °K	—	440	450
ΔH, kcal/mole	28.5 ± 0.5[a]	27.3 ± 0.9	12.3 ± 0.3[b]
Heating rate, °K/min	8	16	16
T_i, °K	420	overlaps m.p.	590
T_p, °K	510	520	610
T_f, °K	600	540	630

[a] Refers to blue form; transition of violet → blue is $\Delta H = 3.02 \pm 0.07$ kcal/mole.
[b] Heat of fusion of $Co(py)_2 I_2$ (c) is 6.3 ± 0.3 kcal/mole.

in the concentration range 0–100%. Solutions containing both chloride and bromide are precipitated with silver nitrate, forming solid solutions of silver chloride-bromide. The heat of fusion of the mixed crystal is then determined, and the percent chloride or bromide obtained from a previously prepared standard curve.

E. Clays and Minerals

One of the early fields of application of DTA was in the area of clays and minerals. These compounds, which gave birth to the theory and instrumentation of the technique, have been widely investigated. DTA was used to identify clays from various locations throughout the world and was widely used to determine the free quartz content of minerals. Numerous other applications were made of DTA; DSC was little used due to the low-temperature capability of the latter. Most of the interesting thermal behavior of clays and minerals occur above 500°C, and frequently above 1000°C.

The applications of DTA to these materials, as discussed by Mackenzie et al. (62), are illustrated in Figure VII.25.

Using the DTA technique for quality control purposes, Garn and Flaschen (113) studied the thermal decomposition of different samples of magnesium carbonate and talc. The DTA curves for the different magnesium carbonate

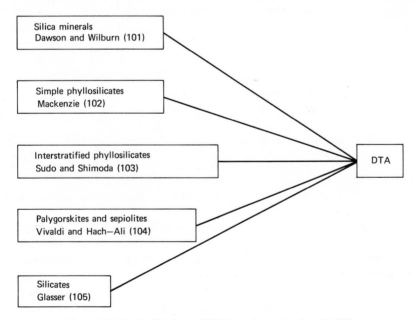

Figure VII.25. Applications of DTA to clays and minerals (62).

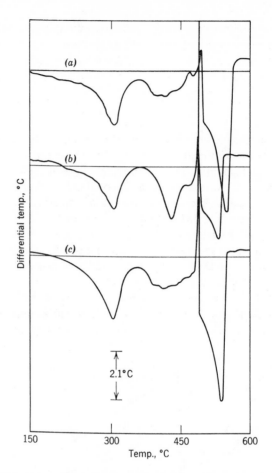

Figure VII.26. DTA curves of commercial magnesium carbonate samples (113). (a) Lot No. 6, Baker; (b) Lot No. 3, Merck; (c) Lot No. 2, Merck.

samples are given in Figure VII.26, while the talc curves are shown in Figure VII.27.

The curves for the magnesium carbonate samples showed distinct differences due to their different thermal histories. Each of the talc curves exhibited a strong exothermic peak, starting at about 850°C. The magnitude of the reaction was about the same in each case, but differences in impurities caused pronounced differences in the curves. The Montana and Sierramic talcs gave a small endothermic peak at about 570°C, while the latter talc gave a pronounced endothermic peak at about 700°C.

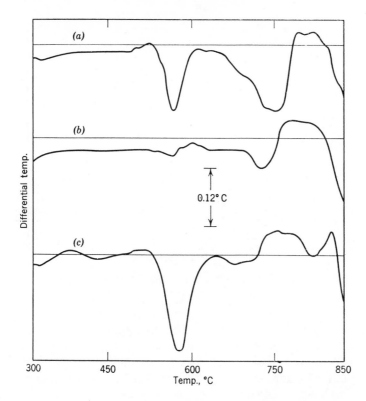

Figure VII.27. DTA curves of: (*a*) Sierramic; (*b*) Yellowstone; and (*c*) Montana talcs using platinum cups (113).

The determination of goethite (α-FeO·OH) and gibbsite [Al(OH)$_3$] by themselves, and in mixtures, has been carried out by a DTA method by Lodding and Hammell (114). If goethite is heated in the controlled-atmosphere DTA apparatus in a reducing atmosphere (hydrogen), it dehydrates below 300°C and the iron(III) ion present is immediately reduced to amorphous Fe$_3$O$_4$ which recrystallizes to magnetite between 300 and 360°C. If the hydrogen atmosphere is now replaced by nitrogen after reaching 400°C, and then by air, an exothermic peak is formed due to the oxidation of magnetite to maghemite, γ-Fe$_2$O$_3$. A second exothermic peak, due to the conversion of γ-Fe$_3$O$_3$ to α-Fe$_2$O$_3$, occurs at 775–836°C. This peak is usually a doublet and the integrated area under it is proportional to the amount of newly formed hematite, which is, therefore, equal to the amount of hydrated iron oxide present in the original sample. The amount of gibbsite can then be determined by the difference from the area under the

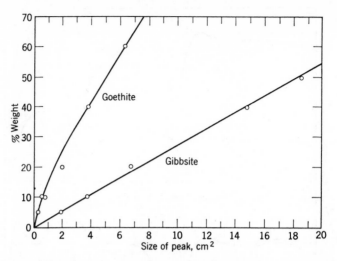

Figure VII.28. Calibration curves of peak areas for dehydration of gibbsite and geothite (114).

dehydration peak.   The hematite or magnetite present in the sample was said to have a negligible influence on the area of the conversion peaks.

The calibration curves of peak area to amounts of goethite and gibbsite present in the mixtures are given in Figure VII.28.   A similar curve for the $\gamma \rightarrow \alpha$-$Fe_2O_3$ conversion was also presented by Lodding and Hammell (114).

Berg and Rassonskaya (115) proposed the use of a high-heating-rate DTA apparatus for rapid analysis of minerals and clays.   The high heating rate, 80–100°C/min, was obtained by placing the sample holder into a previously heated furnace, preferably 300°C higher than the final sample temperature desired.   It was claimed that the peak temperatures at this high heating rate for melting and boiling transitions were the same as those obtained at the 3–6°C/min. heating rate.   Judging from previous studies, it is difficult to see how this could be true.   Similar results were found for the peaks resulting from the dissociation of metal carbonates and for solid-state crystalling phase transition.   The advantages claimed for this technique are (a) speed of investigation, 3–10 min; (b) small quantities of sample required, 20–100 mg; (c) simple regulation of heating rate; and (d) cheapness of analysis.

Using sealed glass or fused silica tubes for sample holders, Bollin et al. (116) followed the reaction of two or more solid substances by the technique of DTA.   The procedure they used, which was later elaborated on by Bollin and Kerr (117), was called pyrosynthesis.   The pyrosynthesis curves of the CuS-Cu_2S system are given in Figure VII.29.

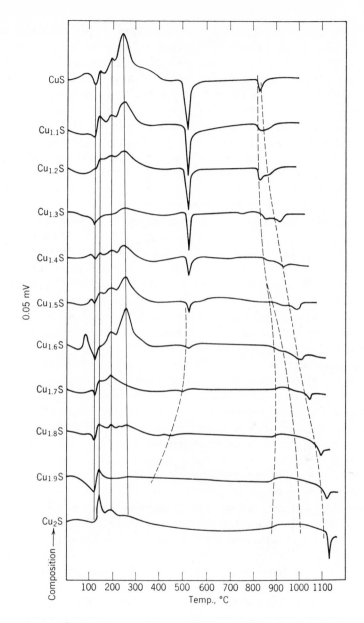

Figure VII.29. Pyrosynthesis curves for the system CuS–Cu₂S (117).

287

In the eleven curves illustrated, the amounts of copper and sulfur were varied to give the compositions indicated. The series was synthesized, starting with the CuS end where the exothermic peak was accompanied by a endothermic peak at 115°C, caused by the melting of sulfur. The endothermic peak at 505°C was the incongruent melting point of covellite (CuS) to digenite + liquid + vapor. The magnitude of this peak decreased to zero at a composition of $Cu_{1.8}S$, which is the composition of digenite.

Other systems studied were $FeS–FeS_2$, $CuFeS_2$, $FeS_{1.5}$, and other sulfides, selenides, tellurides, arsenides, and antimonides. The many variable parameters, such as design of the sample holder, sample tubes, heating rate, and particle sizes of the sample were determined. It is expected that pyrosynthesis would be useful in inorganic chemistry as well.

The technique of DTA can be used to determine the amount of uncalcined gypsum ($CaSO_4 \cdot 2H_2O$) in plaster of Paris ($CaSO_4 \cdot 0.5H_2O$). From the curve for the dehydration of gypsum in Figure VII.30a, it is seen that the peak at ΔT_{min} of 142°C is due to the dehydration of the first 1.5 moles of water per mole of salt. The second endothermic peak, at ΔT_{min} of 198°C, is due to the evolution of the remainder of the water. Thus, the presence of gypsum in plaster of Paris could be determined from the DTA curve in Figure VII.30b if a peak at about 142°C appeared. The peak area would be proportional to the amount of gypsum present in the sample (106).

Crystalline quartz, when heated, undergoes a $solid_1 \rightarrow solid_2$ phase transition ($\alpha \rightarrow \beta$ quartz) at about 573°C. Keith and Tuttle (107) found that in a study of 250 quartz samples, the inversion temperature range was 38° in natural quartz, although most of the samples were within 2.5° of 573°C. This inversion temperature was attributed to solid solution of varying amounts of other ions in the quartz. Since the amount of solid solution is influenced by the temperature during formation, the inversion temperature can be used as a criterion of formation temperature for samples crystallized under similar chemical environments.

Quartz samples from various deposits are shown in the DSC curves in Figure VII.31 (108). Sample sizes ranged from 32 to 37 mg and a heating rate of 1.5°C/min was used in order to minimize thermal gradients within the sample.

The quantitative analysis of the clay minerals, kaolinite, gibbsite, and goethite was described by Davis and Holbridge (109). The heats of dehydroxylation (dehydration) of two kinds of alunites were determined by DTA by Cohen Arazi and Krenkal (110). Reddick (111) determined the ΔH of decomposition for calcite, magnesite, rhodochrosite, and siderite as well as for ankerite. The relationship between the magnesite peak temperature and the magnesite content for various mixtures was reported by Warne

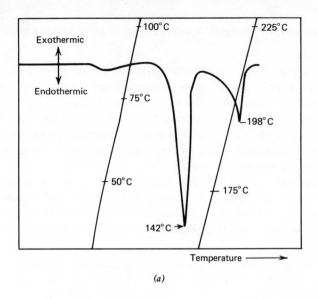

(a)

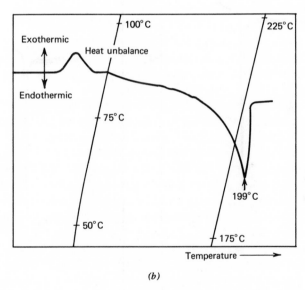

(b)

Figure VII.30. (*a*) DTA curve of gypsum; (*b*) DTA curve of plaster of Paris (106).

and Mackenzie (112). The lowering of the peak temperature on dilution is virtually identical whether the diluent of the magnesite is alumina or other carbonates.

F. Biological Materials

Most of the applications of DTA to biological materials have been for identification and characterization. However, perhaps this is to be expected because of the complexity and hetereogeneity of these materials; compare a sample of peat with benzoic acid or other simple organic compounds. The DTA curves are frequently quite broad and are devoid, in many cases, of narrow endothermic and exothermic peaks. Many of the older investigations in this area were done under rather uncontrolled conditions of furnace atmosphere and heating rates, so they cannot be compared with data obtained with modern instrumentation.

The application of DTA (and other thermal analysis techniques) to biological materials has been reviewed recently by Mitchell and Birnie (118) and Pfeil (119). The former is mainly concerned with the DTA studies of fresh plant material, bacteria, partially decomposed plant material, peat, and

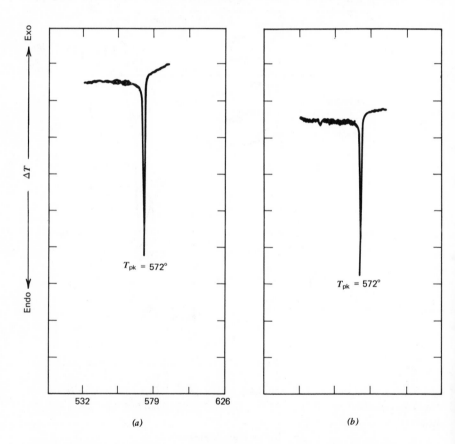

Figure VII.31. $\alpha \rightarrow \beta$ inversion in quartz samples (108). Source of each sample indicated. (a) Brazil; (b) Cornog, Pa.; (c) Chester Co., Pa.; (d) Hot Springs, Ark.

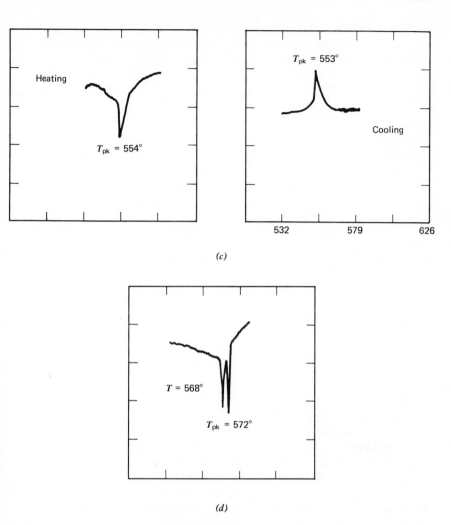

(c)

(d)

Figure VII.31. (c and d).

soil organic matter.   Pfeil (119) discussed the application to human materials such as the liver, endema in burns, bones, and so on.

A summary of some of the applications of DTA and DSC to biological materials is given in Table VII.11.

Labowitz (138) reported a phase transition in anhydrous cholesterol at 37°C which had $\Delta H = 0.66$ kcal/mole and $\Delta S = 2.1$ cal °K^{-1}.   This transition has been discussed by others.

TABLE VII.11
Applications of DTA and DSC to Biological Materials

Material	Techniques	References
Fats, Oils, and waxes	DTA, DSC	120–122, 132, 133
Human bone and hemoglobin	DTA	123
Tobacco	DTA	124
Polynucleotides	DTA	125
Protein denaturation and others	DSC	126, 137
Skin, skin constituents	DTA	127
Biopolymers	DTA	128, 129
Yeast and blood cryobiology	DTA	130, 131
Human liver-hepatoma	DTA, EGA, TG	119
Edema in burns (human skin tissue)	DTA, TG, MS	119
Grain (corn, wheat, oats, etc.)	DTA, TG	119
Fresh plant material	DTA	118
Bacteria and actinomycetes	DTA	118
Peat and partially decomposed plant material	DTA	118
Soil organic matter	DTA	118
Wood	DTA	134
Cellulose	DTA	135
Lichens	DTA	136
Cholesterol	DTA, DSC	138
Lactose	DSC	139

The heat capacity of anhydrous ovalbumin and β-lactoglobulin was determined using DSC by Berlin et al. (137). A linear relationship was found between the specific heat and moisture content with hydrated samples which contained 0.03–0.21 g sorbed water per gram of protein. Berlin et al. (139) also determined the heat of desorption of water vapor from amorphous and crystalline lactose by DSC.

Olafsson and Bryan (141) determined the thermal stability of 19 amino acids using DSC. They used the ΔT_{min} temperatures as the decomposition temperatures of the acids. In some cases, the unique shape of the curves and the number of peaks into which it could be resolved were used to characterize the compound.

The DTA of cellulose, cellulose nitrate, pentaerythritol, pentaerythrityl trinitrate, and other compounds of this type has been studied by Pakulak and Leonard (135). Using a thermistorized DTA apparatus, the upper temperature limit of the instrument was only about 200°C; hence, cellulose and cellulose acetate did not give any peaks, while cellulose nitrate gave an exothermic peak with a ΔT_{max} of 180°C. Similar results were noted for the pentaerythritol series.

TABLE VII.12
Summary of DTA Curve Peaks for Air-Dried Wood (134)

Peak, °C	Compound
(1) Endotherm at 145	Alcohol-water extract
(2) Endotherm at 163	Alcohol-water extract
(3) Exotherm at 210	Unaccounted
(4) Exotherm at 265	Possibly acid lignin
(5) Exotherm at 285	Benzene-alcohol extract
(6) Exotherm at 300	Sum of benzene-alcohol extract and acid lignin
(7) Exotherm at 330	Cellulose
(8) Exotherm at 360	Same as (6)

The DTA curves of several bacterial dextrans have been determined by Morita (142) in order to study certain relationships between the DTA curve peaks and their molecular constitution.

The technique of DTA has been used to study the thermal degradation of balsam fir wood by Arseneau (134). Using air-dried wood and also various samples of wood that had been extracted with several reagents, Arseneau attributed the various DTA curve peaks to the reactions summarized in Table VII.12.

The characterization of starch and related polysaccharides by DTA has been carried out by Morita (143). The DTA curves obtained for several samples of potato and corn starch are given in Figure VII.32. The samples were prepared into a compressed "sandwich"-type packing prepared by placing 150 mg of sample between two 200-mg layers of calcined alumina and compressing at 200 psi.

The DTA curves of the starches were characterized by endothermic peaks in the 135–310°C region, followed by two distinct exothermic peaks in the 375–520°C range. The curves illustrate very nicely the effect of pretreatment on the starches.

Since starch is a polymeric glucoside composed of α-1,4- and α-1,6-linked glucopyranosidic units, it was of interest to examine the thermal properties of the linear polymeric fraction of starch, namely, that of amylose. The DTA curves for various amylose fractions, prepared from the same starches have been reported. Examination of the three fractions reveals three distinct features: the endothermic peaks with ΔT_{min} of about 150 and 225°C, and a shoulder peak with a ΔT_{min} of 315°C. There were pronounced exothermic peaks in the 490–510°C temperature range.

The mechanism of the thermal degradation reactions is not known and is probably quite complicated. The DTA curves serve not merely to characterize or identify these carbohydrates, but will eventually lead to information

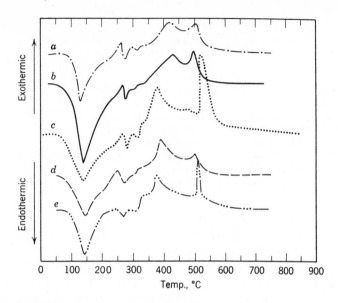

Figure VII.32. DTA curves of potato and corn starch. (*a*) Potato starch; (*b*) potato starch, duplicate run; (*c*) corn starch; (*d*) methanol-extracted corn starch; (*e*) ammonia-pregelatinized corn starch (143).

pertaining to the relationship between molecular composition and chemical properties.

Morita (144) also studied the DTA of several α- and β-linked polyglucosans, as well as rice starch. An interesting feature of this investigation was the study of the effect of moisture on the DTA curves obtained. This was illustrated by the study of rice starch stored in various types of atmospheres such as vacuum, 100% relative humidity water vapor, and so on. The presence of moisture altered the endothermic peak with a ΔT_{min} of 130°C, but not the 275 or 310°C peaks. The results suggest that the original 130°C peak is not entirely due to the loss of residual moisture, and that the dehydration process is not completely reversible.

G. Polymers

Perhaps the greatest number of applications of DTA and DSC in recent years has been in the area of polymeric materials. These two techniques are routinely used to measure glass transition temperatures, T_g; melting points, T_m; degree of crystallinity; heats of fusion and/or crystallization; decomposition temperatures; and numerous other parameters. Several commercial DTA and DSC instruments were developed mainly for use in polymer measurements.

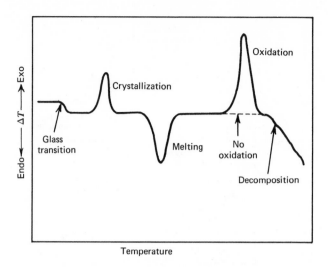

Figure VII.33. Schematic DTA curve of a typical polymer (145).

The DTA curve in Figure VII.33 illustrates how the various thermal processes appear on a DTA curve (145). In actual practice, however, all of these transitions are not so well defined on the same curve. It is necessary to make slight variations in procedure in order to show the transition of particular interest to best advantage. For example, oxidation is measured on a sample of smaller than normal size, and the run is carried out in the presence of either oxygen or air. For other measurements, nitrogen or low pressures are normally employed. The DTA or DSC equipment must be designed for programmed cooling so that the crystallization temperature on cooling can be measured.

The applications of these two techniques are schematically shown in Figure VII.34. Excellent reviews on these applications are given in books by Ke (146), Slade and Jenkins (147), Schwenker (148), Porter and Johnson (149, 150), Reich and Stivala (151), Schwenker and Garn (152), and others, review articles by Murphy (5–8, 153) and numerous others.

The identification of polymer blends is illustrated by the DTA curve in Figure VII.35. Chiu (154) studied a physical mixture of seven commercial polymers: high-pressure polyethylene (HPEE), low-pressure polyethylene (LPPE), polypropylene (PP), polyoxymethylene (POM), Nylon 6, Nylon 66, and polytetrafluoroethylene (PTFE). Each component shows its own characteristic melting endothermic peak, at 108°, 127°, 165°, 174°, 220°, 257°, and 340°C, respectively. Polytetrafluoroethylene also has a low-temperature crystalline transition at about 20°C. The unique ability of

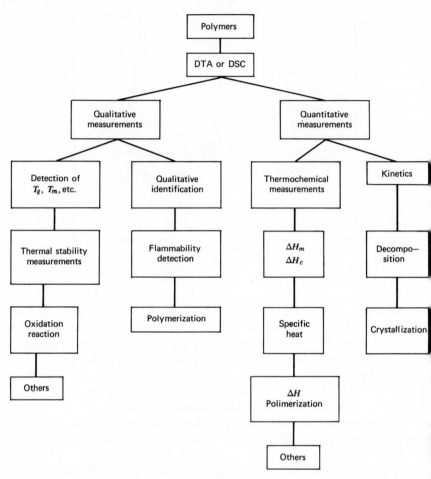

Figure VII.34. Applications of DTA and DSC to polymers.

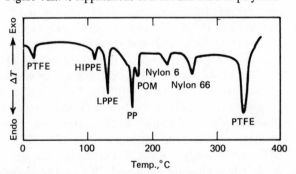

Figure VII.35. DTA curve of a seven component polymer mixture (154).

)TA to identify this polymer mixture is only exceeded by the fact that only
} mg of sample was employed in the determination.

Anderson (162) studied the DTA of six different epoxides, both reacted
ınd unreacted, with various amines and anhydride polymerizing agents.
The samples, varying in mass from 1 to 3 g, were intimately mixed with
:qual amounts of aluminum oxide. After placing the mixture in the sample
.ube, the tube was weighed before and after the heating cycle so that the loss
ın mass of the sample could be obtained.

The DTA curves of three catalyzed and uncatalyzed epoxides are given in
Figure VII.36. The epoxides studied were Epon 1310 [tetraglycidyl ether
ör tetrakis (hydroxphenyl)ethane], Diepoxide AG-13E (bis-epoxydicyclo-
ɔentyl ether or ethyleneglycol), and UC Endo isomer (dicyclopentadiene
łioxide). All the uncatalyzed epoxides, except the UC Endo isomers,

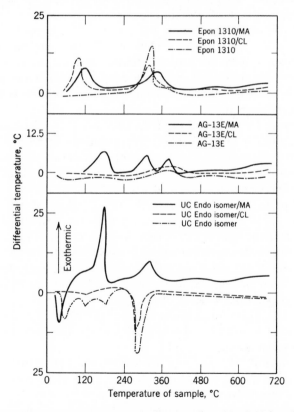

Figure VII.36. DTA curves of catalyzed and uncatalyzed epoxides; MA is maleic anhydride;
CL is m-phenylenediamine; heating rate of 2.5°C/min (162).

exhibited exothermic peaks in the 300–400°C region. These peaks were believed to be due to the isomerization of the epoxy group to carbonyl groups (aldehydes for primary epoxides and ketones for secondary epoxides). The appearance of vapors in the tubes indicated that volatilization and decomposition also occurred simultaneously with isomerization and polymerization. The Endo isomer showed an endothermic peak because of the heat absorbed by volatilization, and decomposition masked any heat resulting from the slower rate of isomerization and etherification polymerization of its epoxy groups. The peak at ΔT_{min} of 184°C corresponded to the melting point of the Endo isomer.

When the above three epoxides were mixed with the catalysts (maleic anhydride or m-phenylenediamine), except for the AG-13E/C1 and UC Endo isomer/CL, all of the mixtures exhibited two exothermic peaks and only one endothermic peak. This latter peak corresponded to the boiling points and/ or decomposition points of both the epoxide and the catalyst.

The DTA curves obtained on the above system may be used confidently as a characterization index. This technique offers unique advantages over other instrumental methods, especially those involving insoluble and amorphous crosslinked epoxy systems which exhibit diffuse X-ray patterns and with, because of their inherent intractable physical state, do not give reproducible infrared spectra.

Murphy et al. (163) studied the DTA of Vibrin 135 resins, the results of which are shown in Figure VII.37. Three samples of resin were studied two of them contained 2%. $tert$-butylperbenzoate catalyst, the other 0.5%. Each catalyst-resin mixture was then heated (cured) for a definite period of time. The DTA curves in (1) and (2) showed that two low-temperature exothermic peaks were observed, with ΔT_{max} values of 150 and 180°C respectively. This first peak was missing from the post-baked (180°C for 24 hr) sample, although the 320°C peak was found in all three curves. The presence of the low-temperature exothermic peak was attributed to the further polymerization of the undercured resin, especially the polyester portion of the resin. The high-temperature exothermic peak was caused by the curing of the triallyl cyanurate portion of the resin.

Murphy et al. (164) further demonstrated the effect of different catalyst on the curing of Vibrin 135 resin by a DTA method. The catalyst, benzoyl peroxide, effected the most complete cure for the resin.

The technique of DTA has been used by Clampitt (165) for the estimation of the linear content of polyethylene blends. The DTA curves for several polyethylene blends are given in Figure VII.38.

Careful examination of the unannealed polyethylene sample curve indicated a peak with a ΔT_{min} of 134°C, and also a shoulder peak. On annealing the

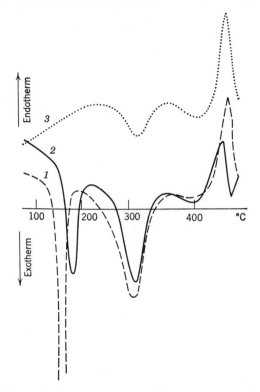

Figure VII.37. DTA curves of Vibrin 135 resins (163).

ample 30 min at 120°C, the shoulder peak was resolved into two peaks, with
T_{min} values of 115 and 124°C, respectively. Using the above annealing
rocedure, but varying the percentage of the linear content of the samples, the
)TA curves in Figure VII.38 were obtained. The curves contained endothermic peaks with ΔT_{min} values of 115, 124, and 134°C, respectively. For
he pure components, however, only one peak was obtained with high-pres-
ure polyethylene, with a ΔT_{min} of 110°C peak decreased and the area under
he 134°C peak increased as the amount of linear polymer increased. The
15°C peak was associated with the presence of crystals of high-pressure
olyethylene, while the 134°C peak presumably was due to crystals of linear
ontent.

Schwenker and Beck (166) studied by DTA the thermal degradation of
olymeric materials used in textile manufacturing in air and nitrogen atmos-
heres. The polymers studied were dacron, nylon 66, neoprene W, and
rlon. From the results obtained, the reactions, such as rearrangements,
ross-linking, and depolymerization taking place in the polymers on thermal

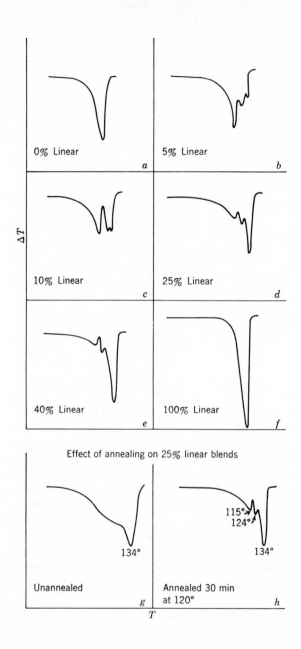

Figure VII.38. DTA curves of linear high-pressure polyethylene blends (165).

degradation can be detected and identified. DTA can detect relatively small changes in polymer composition or the presence of substituents on the polymer backbone, as well as prove quite valuable for thermal degradation mechanism studies.

The DTA curves of nylon 66 fabric and neoprene W, in air and in nitrogen, are given in Figure VII.39.

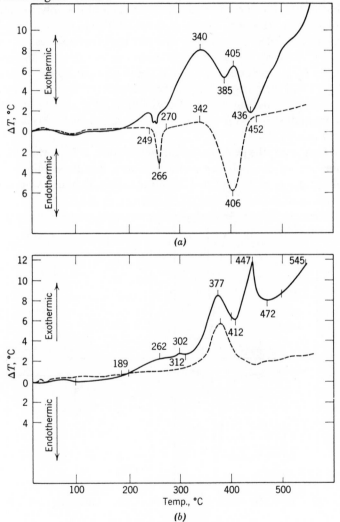

Figure VII.39. DTA curves of polymeric materials (166) ———, in air; – – – –, in N$_2$.
(a) Nylon 66; (b) Neoprene W.

At about 100°C, a weak endothermic peak due to the loss of sorbed water was observed in the nylon 66 curve.   In air there was an exothermic reaction initiating at about 185°C and forming a small endothermic peak at a ΔT_{min} of about 250°C, the latter being caused by the fusion of the polymer (m.p. about 255°C).   In nitrogen, the exothermic peaks were not present, suggesting that the air reactions were due to an oxidation reaction.   The two endothermic peaks in the nitrogen curve were due to the fusion of the polymer and to the depolymerization reaction.   It is obvious that the thermal degradation mechanisms are different for the air and nitrogen atmospheres.

In the DTA curves for neoprene W, both curves exhibited an exothermic peak with a ΔT_{min} of about 377°C.   This peak was attributed to the elimination of HCl and the cross-linking of the residue.

The melting points and degree of crystallinity of a number of polyolefins have been studied by DTA by Ke (167).   Five polyolefin DTA curves are given in Figure VII.40.

From the curves, the peak maximum temperature, ΔT_{min}, was used for the determination of the polymer melting point.   Results obtained by DTA were within ±1°C of the reported literature values, although several of them

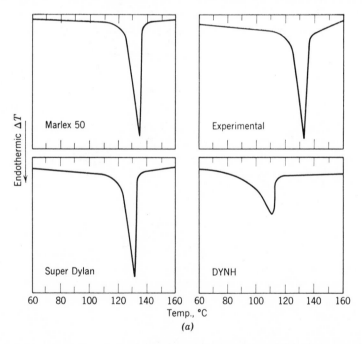

Figure VII.40. DTA curves of polyolefins (167).   (a) Polyethylenes; (b) polypropylene.

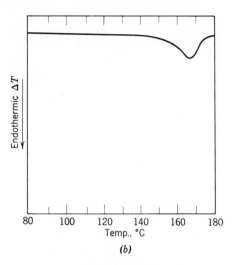

Figure VII.40. (b).

had a 15°C melting-point range, as indicated by the distance between the initial departure from the base line and the peak. Isotactic polypropylene gave a somewhat broader endothermic peak at ΔT_{min} of 169°C. The end point of the transition was somewhere beyond the peak at a point not known exactly.

Ke (167) also determined the DTA curve of mixtures of polyolefins and found that the components could be identified if the melting points were sufficiently far apart. The peak areas were proportional to the amount of each component present in the mixture.

The degree of crystallinity of polyethylenes was calculated by comparing the area of the respective endothermic peak with the double peak of dotria-contane. The curve contains two peaks, the first of which is due to a chain-rotational transition a few degrees below the melting point. The resulting degree-of-crystallinity values agreed well with the literature values, as shown in Table VII.13.

The effect of diluents on the melting behavior of polyethylenes has also been studied by Ke (168). A comparison between the melting transitions of solution and melt-crystallized polyethylene (169) has been made. The measurement of the melting and second-order transitions of polyethylene terephthalate by DTA has been studied by Scott (170). Rudin et al. (171) measured the oxidation resistance of various polymers and rubbers by a DTA method. A comparison of the melting and freezing curves before and after oxidation provided the indication of the extent to which the polymer had been damaged or oxidized.

TABLE VII.13
Degree of Crystallinity of Polyethylenes (167)

Polyethylene	Crystallinity, %	
	Found	Literature
Marlex 50	91	93
Super Dylan	81	65–85
Experimental (polyethylene)	86	87
DYNH	52	40–60

The melting and glass transitions in commercial Nylons and both homo- and copolyamides prepared by interfacial polycondensation have been studied by DTA by Ke and Sisko (172). The DTA curves for a number of the polyadipamides and polysebacamides are given in Figure VII.41.

The polyadipamides were made from diamines containing both even and odd numbers of carbon atoms, and the polysebacamides from diamines containing an even number of carbon atoms. All curves exhibited a peak caused by the melting of the polymer, the melting point of which decreased with an increase in the number of carbon atoms in the diamine chain.

The application of DTA to the detection of changes induced in biphenyl, polyvinyl chloride, Teflon, and Versalube F-50 has been reported by Murphy and Hill (173). The curves for biphenyl and irradiated biphenyl are shown in Figure VII.42.

The nonirradiated sample gave a curve with two endothermic peaks which were caused by the fusion (70°C peak) and volatilization (175°C peak) of the compound. The irradiated sample gave the first two peaks, as well as an exothermic peak at about 370°C. The melting peak occurred at a slightly lower temperature. It was assumed that the 370°C exothermic peak was caused by air oxidation of the nonvolatile, radiation-induced biphenyl polymer remaining in the sample holder after volatilization of low-molecular weight materials. The lowering of the melting point was also caused by the irradiation of the sample. Similar results were noted for polyvinyl chloride samples. It was noted that by proper selection of materials on the basis of the relationship of peak area to radiation dose, DTA might be applied to dosimetry over a wide range of energy levels.

A common application of DSC is the determination of the weight fraction of crystalline material in semicrystalline polymers. The method is based upon the measurement of the polymer sample's heat of fusion, ΔH_f, and the plausible assumption that this quantity is proportional to the crystalline content. If by some process of extrapolation the heat of fusion, ΔH_f^*, of a

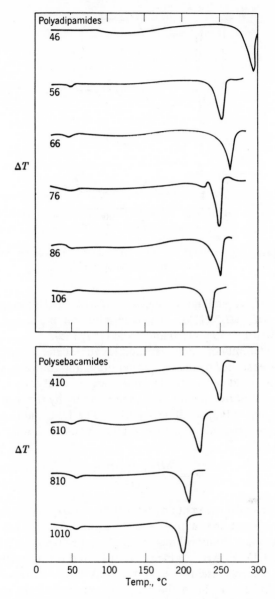

Figure VII.41. DTA curves of some polyadipamides and polysebacamides (172).

305

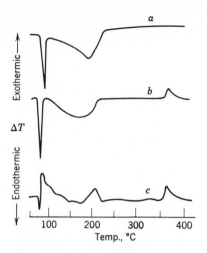

Figure VII.42.   DTA curves of biphenyl.   (a) nonirradiated; (b) irradiated; and (c) irradiated versus nonirradiated samples (173).

hypothetical 100% crystalline sample is known, then the weight fraction of crystallinity is $\Delta H_f/\Delta H_f^*$ (155).   The determination of polymer crystallinity has been reviewed by Gray (156) and Dole (157, 158).

Thus, the crystallinity of a polymer sample can be determined by measuring the total energy absorbed by the sample per gram and subtracting the amount of energy which would be absorbed by one gram of totally amorphous sample in the same temperature interval, and then dividing by the heat of fusion of one gram of a perfectly crystalline sample.   The DSC data obtained for a semicrystalline polymer are shown in Figure VII.43.   Use is made of the equation

$$x = \frac{\Delta H_{2,1} - \Delta H_{a(2,1)}}{\Delta H_F^\circ}$$

where x is the weight fraction at any temperature, ΔH_F° is the heat of fusion of the perfectly crystalline sample, and $\Delta H_{2,1}$ and $\Delta H_{a(2,1)}$ are the heats of fusion of the sample and amorphous material, respectively.   From the curve, $\Delta H_{2,1}$ is area ACDEF, $\Delta H_{a(2,1)}$ is area ABEF, and the difference is BCDG. Note that the correct baseline under the peak is the extrapolation of the recorded baseline from above the final melting point.   It should not be drawn tangent to the pre- and post-melting lines, as is the common practice. Other procedures have been discussed by Gray (155, 156) for determining the values of the curve areas, extrapolation procedures, and so on.   A computer program was also developed to aid in the crystallinity calculations.

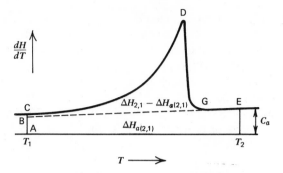

Figure VII.43. Typical DSC polymer melting curve and instrumental base line (156).

Another method for the determination of polymer crystallinity was discussed by Duswalt (159). It is based upon the ability of the instrument to cool a molten sample rapidly and reproducibly to a reselected temperature where isothermal crystallization is allowed to occur. A number of crystallization curves for polyethylene obtained isothermally at different, preset crystallization temperatures are shown in Figure VII.44. Differences in polymer crystallizability that may be caused by branching, nucleation, or molecular weight effects can be observed. The sensitivity and speed of the method allow pellet-to-pellet variations in a lot of polymer to be examined.

The quantitative measurement of the effect of annealing of poly(vinyl chloride) near the glass transition temperature was described by Foltz and McKinney (160). The method is based upon the use of a quenched sample of the polymer as the reference material; the DSC curve so obtained is then

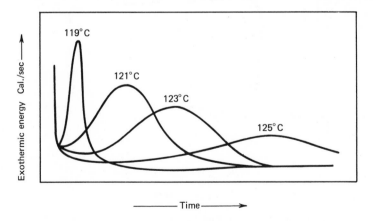

Figure VII.44. Isothermal crystallization curves for polyethylene (159).

the difference in heat energy between the sample and the annealed material. By this technique it is possible to measure small energy differences which in the usual procedure appear as minor inflections on the slope of the T_g curve or as irregularities prior to the start of the glass transition.

Activation energies for styrene polymerization were determined by a DTA method by Hoyer et al. (161). The DTA curve for the thermally induced polymerization of pure styrene consists of a single exothermic peak corresponding to the onset of polymerization at 140°C and a ΔT_{max} of 250°C. An E_a of 21.3 ± 0.6 kcal/mole was calculated for the reaction.

H. Miscellaneous Applications

The specific heat of a substance can be determined conveniently and rapidly using the techniques of DTA and DSC (174, 182). The method (174) is illustrated by the DuPont DSC curves for α-alumina, as given in Figure VII.45. A curve for the empty sample container is first run, as indicated by the upper curve. The sample is then placed in the sample container and its curve recorded, using the same instrument adjustments. The relationship between the "blank" (empty container) and the "sample" (empty container plus sample) then is

$$(C_p)_T \left(\frac{mcal}{mg°C}\right) = \frac{(\Delta T_x + \Delta T_{blank})E_T}{Ma} \qquad (VII.8)$$

where $(C_p)_T$ is the specific heat at temperature T; ΔT_x is the absolute differential temperature for sample in °C; ΔT_{blank} is the absolute differential

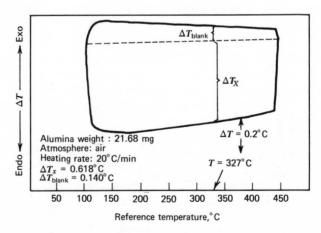

Figure VII.45. Specific-heat determination curves of α-alumina (174). $(C_p)_{327°} = 0.279$ mcal/mg.

temperature of empty container; E_T is the calibration coefficient at temperature T in mcal/°C min; M is the sample mass in mg; and a is the heating rate in °C/min.   The $(C_p)_{327}$ of α-alumina found was 0.279 mcal/mg.

DTA and DSC can be used to construct simple phase diagrams, as shown by the naphthalene-benzoic acid phase diagram in Figure VII.46.

An eutectic melting point is formed at 50–50 mixture of the two components. The phase diagram was constructed from the melting endothermic peaks of the various mixtures, also shown in Figure VII.46.   The melting temperatures in the phase diagram are the extrapolated temperatures for the onset of the melting (175, 176).

Vapor-pressure and heat-of-vaporization measurements are easily carried out using DTA (177) or DSC (178, 179) techniques.   The heat of vaporization and heat of mixing of various organic liquids were obtained using DSC (179) by a small modification the sample holder.   The metallic cover on the holder was replaced by a glass cover with a glass tube at its center.   This glass tube was designed to hold a microsyringe which contained the liquid sample.   By this device, liquid sample could be added to the sample holder without any disturbance of the system temperature.

The curves for the endothermic mixing of benzene-ethanol obtained by operating the DSC cell isothermally are shown in Figure VII.47.   Successive amounts of 2 to 4 μl of ethanol were added to 40 μl of benzene contained in the sample; the higher the ethanol concentration in the benzene, the smaller the heat of mixing for the addition of the same quantity of ethanol.   The molar heats of mixing,

$$\Delta H_M = \Delta H_{\exp}(n_1 + n_2) \qquad \text{(VII.9)}$$

where n_1 and n_2 are moles of ethanol and benzene, respectively, are plotted in Figure VII.48.   Agreement with previously reported values appears to be satisfactory.

A similar procedure was used (179) to determine the heat of vaporization of a liquid organic sample.   The sample was added to the sample cell by means of the microsyringe and the curve area is proportional to the heat of vaporization.   The relative standard deviation obtained for five determinations of the heat of vaporization of benzene was about $\pm 2\%$.   This method cannot be used to determine ΔH_v at the boiling point of the sample, however. Boiling point values can be obtained by extrapolation procedures.

Heats of sublimation can also be obtained by the DSC technique (180). Samples were placed in aluminum pans and the space between the bottom and the domed aluminum cover was filled with powdered aluminum.   The cover contained a small hole to permit evolved gases to escape.   For heat-of-fusion measurements, the cover did not have a hole in it.   Results of these measurements are illustrated in Table VII.14.

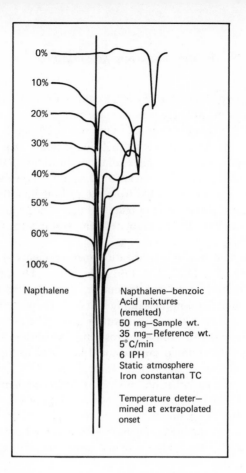

Napthalene

Napthalene—benzoic
Acid mixtures
(remelted)
50 mg—Sample wt.
35 mg—Reference wt.
5°C/min
6 IPH
Static atmosphere
Iron constantan TC

Temperature deter—
mined at extrapolated
onset

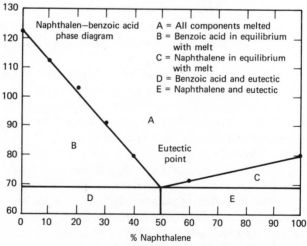

Naphthalen—benzoic acid
phase diagram

A = All components melted
B = Benzoic acid in equilibrium
    with melt
C = Naphthalene in equilibrium
    with melt
D = Benzoic acid and eutectic
E = Naphthalene and eutectic

Eutectic
point

% Naphthalene

Figure VII.46. Phase diagram of the naphthalene-benzoic acid mixture (175, 176). Points on phase diagram taken from eight DTA curves.

310

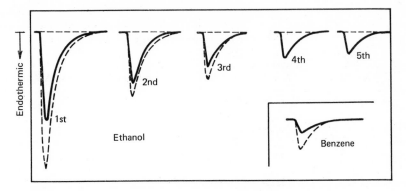

Figure VII.47. Successive additions of 4 μl of ethanol to 40 μl of benzene.   Lower right:
2 μl of benzene added (179).

Critical temperatures of organic liquids can be determined by a DTA method if sealed sample holders are used (181).   The determination is made by use of cooling curves; a discontinuity in the curve is observed at the critical temperature, T_c.   The sample, 20–50 μl, was sealed in 4-mm-diam. glass capillary tubes and heated to a preselected temperature and then allowed to cool while recording the ΔT signal as a function of sample temperature.

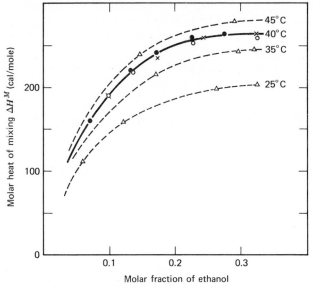

Figure VII.48. Molar heat of mixing of ethanol with benzene; ——— exp. values (179);
– – – – liter.

TABLE
Heats of Fusion, Vaporization, and

Compound	Heating rate °K/min	ΔH_{fus}	ΔH_{vap} (kJmol^{-1})	ΔH_{sub}
Benzoic acid	8, 16, 32			100 ± 5
Anthraquinone	32			127 ± 3
Phthalic anhydride	16			81 ± 1(0.5)
Thymol	16	20.5 ± 0.6	46.5 ± 3.0	67.0 ± 3
Ferrocene	16, 32	18.5 ± 0.1	65.5 ± 2	84 ± 2
Anthracene	16			126 ± 4
Naphthalene	16	18.9 ± 0.2	59 ± 2	78 ± 2
8-Hydroxy quinoline	16	22.1 ± 0.4	68 ± 3	90 ± 4

A sharp break in the curve corresponded to a point on the coexistence curve. The temperature corresponding to the coexistence point was determined as a function of increasing sample volume until a constant temperature (critical temperature) was achieved. The average deviation from literature values for compounds studied was ±0.16°C.

Curie-point temperatures can also be determined by DTA and DSC techniques. As illustrated in Figure VII.49, the specific heat of nickel

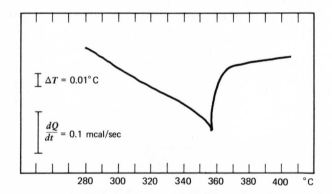

Figure VII.49. Curie point determination of nickel.

VII.14
Sublimation Determined by DSC (180)

Temp. range of vaporization (°K)	Literature value, ΔH_{sub} (kJmol^{-1})	Quoted temp. or temp. range of literature value measurement (°K)
420–480	91.5	343–387
	89.1	299–329
470–590	112.0	298
390–470	88.7	303–333
420–480	91.3	273–313
	69.0	299–312
385–455	73.4	298
	85.3	298
420–540	97.6	338–353
	98.6	342–359
355–490	62.0	298
	72.7	298
355–450	109	308–328

increases gradually up to the Curie point at 357°C, making a sudden change at this point. The sample size used was 75 mg.

Williams and Chamberland (140) discussed the application of DSC to the determination of Curie temperatures of ferromagnetic materials and Néel temperatures of antiferromagnetic and ferrimagnetic materials.

DTA studies on coal and related substances have been carried out by a large number of investigators, including Breger and Whitehead (183) and Gamel and Smothers (184).

Breger and Whitehead (183), using a vacuum DTA apparatus, studied the thermal properties of cellulose, wood, lignite, and various coals. It was found that the low-temperature peaks for lignin disappeared or were masked

TABLE VII.15
Heating Values of Selected Arkansas Coals as Measured DTA and Peroxide Bomb Methods in BTU/lb (184)

Sample	Area under curve, in.^{2}	Btu/lb coal
Paris mine	0.345	13,347
Utah mine	0.460	14,476
Jerome mine	0.415	13,994
Quality excelsior mine	0.470	14,531

in peat and then reappeared in the lignites. The decomposition peaks for lignin were suppressed in bituminous coals and were absent in the curves for anthracites.

Gamel and Smothers (184) related the areas under the decomposition peaks to the concentration of a Utah Mine coal in a coal-alumina mixture. A linear area versus concentration curve was obtained for 0–12.0% coal mixtures. They also found that the area under the curve peak was directly proportional to the BTU/lb values for the coal. This relationship is illustrated in Table VII.15.

References

1. Liptay, G., *Atlas of Thermoanalytical Curves*, Akademiai Kiado, Budapest, Vol. 1, 1971.
2. Mitchell, B. D., and A. C. Birnie, in *Differential Thermal Analysis*, R. C. Mackenzie, ed., Academic, London, 1970, Chap. 22.
3. Manning, M., *Industrial Res.*, Feb., 18 (1966).
4. Gray, A. P., *Am. Lab.*, Jan., 43 (1971).
5. Murphy, C. B., *Anal. Chem.*, **38**, 443R (1966).
6. Murphy, C. B., *Anal. Chem.*, **40**, 380R (1968).
7. Murphy, C. B., *Anal. Chem.*, **42**, 268R (1970).
8. Murphy, C. B., *Anal. Chem.*, **44**, 513R (1972).
9. Brancone, L. M., and H. J. Ferrari, *Microchem. J.*, **10**, 370 (1966).
10. Vassallo, D. A., and J. C. Harden, *Anal. Chem.*, **34**, 132 (1962).
11. Kerr, G., and P. S. Landis, *Anal. Chem.*, **44**, 1176 (1972).
12. Barrall, E. M., J. F. Gernert, R. S. Porter, and J. F. Johnson, *Anal. Chem.*, **35**, 1837 (1963).
13. Barrall, E. M., R. S. Porter, and J. F. Johnson, *J. Phys. Chem.*, **70**, 385 (1966).
14. *ibid.*, **68**, 2810 (1964).
15. Johnson, J. F. and G. W. Miller, *Thermochim. Acta*, **1**, 373 (1970).
16. Gipstein, E., E. M. Barrall, K. Bredfeldt, and O. U. Need, *Thermochim. Acta*, **3**, 253 (1972).
17. Ennulat, R. D., in *Analytical Calorimetry*, R. S. Porter, and J. F. Johnson, eds., Plenum, New York, 1968, p. 219.
18. Young, W. R., E. M. Barrall, and A. Aviram, in *Analytical Calorimetry*, R. S. Porter, and J. F. Johnson, eds., Plenum, New York, 1968, Vol. 2, p. 113.
19. Barrall, E. M., Ref. 18, p. 121.
20. Gipstein, E., E. M. Barrall, and K. E. Bredfeldt, Ref. 18, p. 127.
21. Gray, G. W., *J. Chem. Soc.*, **1956**, 3733.
22. Barrall, E. M., R. S. Porter, and J. F. Johnson, *Anal. Chem.*, **36**, 2172 (1964).
23. Ozawa, T., *Bull. Chem. Soc. Japan*, **39**, 2071 (1966).
24. Pacor, P., *Anal. Chim. Acta*, **37**, 200 (1967).
25. David, D. J., *Anal. Chim. Acta*, **36**, 2162 (1964).
26. Harmelin, M., C. Duval, and N. D. Xuong, *Proceedings of the Third Analytical Chemical Conference*, Budapest, 1970, Akademiai Kiado, 1970, p. 325.
27. Stammler, M., *Explosivstoffe*, **7**, 154 (1968).
28. David, D. J., *Anal. Chem.*, **37**, 82 (1965).
29. Bohon, R. L., *Anal. Chem.*, **35**, 1845 (1953).

30. Graybush, R. J., F. G. May, and A. C. Forsyth, *Thermochim. Acta*, **2**, 153 (1971).
31. Barrall, E. M., *Thermochim. Acta*, **3**, 55 (1971).
32. Barrett, E. J., H. W. Hoyer, and A. V. Santoro, *Mikrochim. Acta*, **1970**, 1121.
33. Santoro, A. V., E. J. Barrett, and H. W. Hoyer, *J. Thermal Anal.*, **2**, 461 (1970).
34. Barrett, E. J., H. W. Hoyer, and A. V. Santoro, *Tetrahedron Letters*, **5**, 603 (1968).
35. Santoro, A. V., E. J. Barrett, and H. W. Hoyer, *Tetrahedron Letters*, **19**, 2297 (1968).
36. Koch, E., *Angew. Chem. Inter. Ed.*, **9**, 288 (1970).
37. Atkinson, G. F., and I. J. Itzkovitch, *Anal. Chim. Acta*, **49**, 195 (1970).
38. Levy, P. F., G. Nieuweboer, and L. C. Semanski, *Thermochim. Acta*, **1**, 429 (1970).
39. Barrall, E. M., *Thermochim. Acta*, **5**, 377 (1973).
40. Hall, P. G., *Trans. Faraday Soc.*, **67**, 556 (1971).
41. Rogers, R. N., and L. C. Smith, *Anal. Chem.*, **39**, 1024 (1967).
42. Rogers, R. N., and R. H. Dinegar, *Thermochim. Acta*, **3**, 367 (1972).
43. Murrill, E., and L. Breed, *Thermochim. Acta*, **1**, 239 (1970).
44. Murrill, E., and L. Breed, *Thermochim. Acta*, **1**, 409 (1970).
45. Murrill, E., M. E. Whitehead, and L. Breed, *Thermochim. Acta*, **3**, 111 (1972).
46. Ripmeester, J. A., and B. A. Dunell, *Canadian J. Chem.*, **49**, 2906 (1971).
47. Fauth, M. I., *Anal. Chem.*, **32**, 655 (1960).
48. Chiu, J., *Anal. Chem.*, **34**, 1841 (1962).
49. Dorko, E. A., R. S. Hughes, and C. R. Downs, *Anal. Chem.*, **42**, 253 (1970).
50. Krawetz, A. A., and T. Tovrog, *I & EC Prod. Res. Dev.*, **5**, 191 (1966).
51. Cross, C. K., *Am. Oil Chem. Soc.*, **47**, 229 (1970).
52. Bsharah, L. *I & EC Prod. Res. Dev.*, **6**, 246 (1969).
53. Vamos, E., *Schmierstoffe Schmierungstech.*, **1966**, 84.
54. Trzebowski, N., *Freiberger Forschungsh.*, **A367**, 257 (1955).
55. Noel, F., *J. Inst. Petroleum*, **57**, 357 (1971).
56. Noel, F., and L. W. Corbett, *J. Inst. Petroleum*, **56**, 261 (1970).
57. Gordon, S., *J. Chem. Educ.*, **40**, A87 (1963).
58. Kracek, F. C., *J. Phys. Chem.*, **33**, 1281 (1929).
59. Kracek, F. C., *J. Phys. Chem.*, **34**, 225 (1930).
60. Wendlandt, W. W., and J. A. Hoiberg, *Anal. Chim. Acta*, **28**, 506 (1963).
61. Wendlandt, W. W., and J. A. Hoibert, *Anal. Chim. Acta*, **29**, 539 (1963).
62. R. C. Mackenzie, ed., *Differential Thermal Analysis*, Academic, New York, 1970, Chaps. 7–15.
63. Berg, L. G., Ref. 62, Chap. 11.
64. Freeman, E. S., and W. K. Rudloff, Ref. 62, Chap. 12.
65. Dollimore, D., Ref. 62, Chap. 13.
66. Webb, T. L., and J. E. Kruger, Ref. 62, Chap. 10.
67. Mackenzie, R. C., Ref. 62, Chap. 9.
68. Webb, T. L., Ref. 62, Chap. 8.
69. Bollin, E. M., Ref. 62, Chap. 7.
70. Dollimore, D., Ref. 62, Chap. 14.
71. Wendlandt, W. W., and J. P. Smith, *Thermal Properties of Transition Metal Ammine Complexes*, Elsevier, Amsterdam, 1967.
72. Wendlandt, W. W., in *Chelates in Analytical Chemistry*, H. A. Flaschka and A.J. Barnard, eds., Dekker, New York, 1967, 107–143.
73. Mesmer, R. E., and R. R. Irani, *J. Chem. Eng. Data*, **8**, 530 (1963).
74. Ozawa, T., M. Momota, and H. Isozaki, *Bull. Chem. Soc. Japan*, **40**, 1583 (1967).
75. Ozawa, T., H. Isozaki, and A. Negishi, *Thermochim. Acta*, **1**, 545 (1970).

76. Kelley, K. K., *Bull. U. S. Bur. Mines*, **584**, (1960).
77. Sokolov, V. A., and N. E. Schmidt, *Izr. Sekt. Fiz. Khim. Anal.*, *Inst. Obsh. Neorg. Khim.*, *Akad. Nauk. SSR*, **27**, 217 (1956); *Chem. Absts.*, **50**, 15200b (1956).
78. Barrall, E. M., and L. B. Rogers, *Anal. Chem.*, **36**, 1405 (1964).
79. Wendlandt, W. W., *Thermochim. Acta*, **1**, 419 (1970).
80. Wendlandt, W. W., G. D'Ascenzo, and R. H. Gore, *Thermochim. Acta*, **1**, 488 (1970).
81. Wendlandt, W. W., G. D'Ascenzo, and R. H. Gore, *J. Inorg. Nucl. Chem.*, **32**, 3404 (1970).
82. Chiu, J., *Anal. Chem.*, **35**, 933 (1963).
83. DuPont DTA Apparatus Bulletin, DuPont Co.
84. Macak, J., and J. Malecha, *Anal. Chem.*, **41**, 442 (1969).
85. DuPont Application Brief, No. **900B31**, July, 1970.
86. Garn, P. D., *Anal. Chem.*, **41**, 447 (1969).
87. Ramachandran, V. S., *J. Thermal Anal.*, **3**, 181 (1971).
88. Ramachandran, V. S., *Materiaux et Constructions*, **4**, 3 (1971).
89. Beech, G., C. T. Mortimer, and E. G. Tyler, *J. Chem. Soc.*, (A) **1967**, 925.
90. Beech, G., S. J. Ashcroft, and C. T. Mortimer, Ref. 89, 929.
91. Beech, G., C. T. Mortimer, and E. G. Tyler, Ref. 89, 1111.
92. Ashcroft, S. J., *J. Chem. Soc.*, (A) **1970**, 1020.
93. Ashcroft, S. J., and C. T. Mortimer, *Thermochemistry of Transition Metal Complexes*, Academic, London, 1970.
94. Adams, J. J., and J. E. House, *Trans. Ill. Acad. Sci.*, **63**, 83 (1970).
95. Simchen, A. E., *Israel J. Chem.*, **9**, 613 (1971).
96. Block, J., *Anal. Chem.*, **37**, 1414 (1965).
97. Stone, R. L., *Anal. Chem.*, **32**, 1582 (1960).
98. Petricciani, J. C., S. E. Wimberley, W. H. Bauer, and T. W. Clapper, *J. Phys. Chem.*, **64**, 1309 (1960).
99. Wendlandt, W. W., T. D. George, and G. R. Horton, *J. Inorg. Nucl. Chem.*, **17**, 273 (1961).
100. Erdey, L., and F. Paulik, *Acta Chim. Acad. Sci. Hung.*, **7**, 27 (1955).
101. Dawson, J. B., and F. W. Wilburn, Ref. 62, Chap. 17.
102. Mackenzie, R. C., Ref. 62, Chap. 18.
103. Sudo, T., and S. Shimoda, Ref. 62, Chap. 19.
104. Vivaldi, J. L. M., and P. F. Hach-Ali, Ref. 62, Chap. 20.
105. Glasser, F. P., Ref. 62, Chap. 21.
106. Fisher DTA Instrument Bulletin, Fisher Scientific Co. Pittsburg, Pa.
107. Keith, M. L., and O. F. Tuttle, *Am. J. Sci.*, *Bowen Vol.*, 203 (1952).
108. *DuPont Thermal Analysis Application Brief*, No. 15, January 26, 1968.
109. Davis, C. E., and D. A. Holbridge, *Clay Minerals*, **8**, 193 (1969).
110. Cohen Arazi, A., and T. G. Krenkel, *Am. Mineralogist* **55**, 1329 (1970).
111. Reddick, K. L., *Sun Oil Quart.*, No. 3, 31 (1969).
112. Warne, S. S., and R. C. Mackenzie, *J. Thermal Anal.*, **3**, 49 (1971).
113. Garn, P. D., and S. S. Flaschen, *Anal. Chem.*, **29**, 271 (1957).
114. Lodding, W., and L. Hammell, *Anal. Chem.*, **32**, 657 (1960).
115. Berg, L. G., and I. S. Rassonskaya, *Dokl. Akad. Nauk SSSR*, **73**, 113 (1950).
116. Bollin, E. M., J. A. Dunne, and P. F. Kerr, *Science*, **131**, 661 (1960).
117. Bollin, E. M., and P. F. Kerr, *Am. Mineralogist*, **46**, 823 (1961).
118. Mitchell, B. D., and A. C. Birnie, in *Differential Thermal Analysis*, R. C. Mackenzie, ed., Academic, London, 1970, Chap. 24.

119. Pfeil, R. W., in *Proceedings of the Third Toronto Symposium on Thermal Analysis*, H. G. McAdie, ed., Chemical Institute of Canada, Toronto, 1969, p. 187.
120. Application Brief No. 20, DuPont Co., 1968.
121. Haighton, A. J., and J. Hannewijk, *J. Am. Oil Chem. Soc.*, **35**, 457 (1958).
122. Currell, B. R., and B. Robinson, *Talanta*, **14**, 421 (1967).
123. Garn, P. D., *Thermoanalytical Methods of Investigation*, Academic, New York, 1965, p. 124.
124. Edmonds, M. D., M. T. Core, A. Bauley, and R. F. Schwenker, *Tobacco Sci.*, **9**, 48 (1965).
125. Hoyer, H. W., and E. J. Barrett, *Anal. Biochem.*, **17**, 344 (1966).
126. Steim, J. M., *Perkin-Elmer Instru. News*, **19**, No. 2 (1968).
127. Puett, B., *Biopolymers*, **5**, 327 (1967).
128. Hoyer, H. W., *J. Am. Chem. Soc.*, **90**, 2480 (1968).
129. Hoyer, H. W., *Nature*, **216**, 997 (1967).
130. Moore, R., in *Thermal Analysis*, R. F. Schwenker and P. D. Garn, eds., Academic, New York, 1969, Vol. 1, p. 615.
131. Greaves, R., and J. Davies, *Ann. New York Acad. Sci.*, **125**, 548 (1965).
132. Luebke, H. W., and B. G. Breidenbach, *J. A. O. C. S.*, **46**, 60 (1969).
133. Cross, C. K., *J. A. O. C. S.*, **47**, 229 (1970).
134. Arseneau, D. F., *Can. J. Chem.*, **39**, 1915 (1961).
135. Pakulak, J. M., and G. W. Leonard, *Anal. Chem.*, **31**, 1037 (1959).
136. Mitchell, B. D., and Birnie, A. C., *Analyst*, **91**, 783 (1966).
137. Berlin, E., P. Kliman, and M. J. Pallansch, *Thermochim. Acta*, **4**, 11 (1972).
138. Labowitz, L. C., *Thermochim. Acta*, **3**, 419 (1972).
139. Berlin, E., P. G. Kliman, B. A. Anderson, and M. J. Pallansch, *Thermochim. Acta*, **2**, 143 (1971).
140. Williams, H. W., and B. L. Chamberland, *Anal. Chem.*, **41**, 2084 (1969).
141. Olafsson, P. G., and A. M. Bryan, *Mikrochim. Acta*, **1970**, 871.
142. Morita, H., *J. Am. Chem. Soc.*, **78**, 1397 (1956).
143. Morita, H., *Anal. Chem.*, **28**, 64 (1956).
144. Morita, H., *Anal. Chem.*, **29**, 1095 (1957).
145. Schulken, R. M., R. E. Roy, and R. H. Cox, *J. Polymer Sci.*, Part C, **6**, 1725 (1964).
146. Ke, B., ed., *Thermal Analysis of High Polymers*, Wiley-Interscience, New York, 1964.
147. Slade, P. E., and L. T. Jenkins, eds., *Techniques and Methods of Polymer Evaluation*, Marcel-Dekker, New York, 1966.
148. Schwenker, R. F., ed., *Thermoanalysis of Fibers and Fiber-Forming Polymers*, Wiley-Interscience, New York, 1966.
149. Porter, R. S., and J. F. Johnson, eds., *Analytical Calorimetry*, Plenum, New York, 1968.
150. Porter, R. S., and J. F. Johnson, eds., *Analytical Calorimetry*, Plenum, New York, Vol. 2, 1970.
151. Reich, L., and S. S. Stivala, *Elements of Polymer Degradation*, McGraw-Hill, New York, 1971.
152. Schwenker, R. F., and P. D. Garn, eds., *Thermal Analysis*, Academic, New York, 1969, Vol. 1, Section 2.
153. Murphy, C. B., in *Differential Thermal Analysis*, R. C. Mackenzie, ed., Academic, London, 1970, Chap. 23.
154. Chiu, J., *DuPont Thermogram*, **2**, No. 3, 9 (1965).

155. Gray, A. P., *Perkin-Elmer Instrument News*, **20**, No. 2, 8 (1969).
156. Gray, A. P., *Thermochim. Acta*, **1**, 563 (1970).
157. Dole, M., *J. Polymer Sci., Part C.*, **18**, 57 (1967).
158. Dole, M., *Fortschr. Hochpolym. Forsch.*, **2**, 221 (1960).
159. Duswalt, A. A., *Hercules Chemist*, No. **57**, 5 (1968).
160. Foltz, C. R., and P. V. McKinney, *Anal. Chem.*, **41**, 687 (1969).
161. Hoyer, H. W., A. V. Santoro, and E. J. Barrett, *J. Polymer Sci., Part A-1*, **6**, 1033 (1968).
162. Anderson, H. C., *Anal. Chem.*, **32**, 1592 (1960).
163. Murphy, C. B., J. A. Palm, C. D. Doyle, and E. M. Curtiss, *J. Polymer Sci.*, **28**, 447 (1958).
164. Murphy, C. B., J. A. Palm, C. D. Doyle, and E. M. Curtiss, *J. Polymer Sci.*, **28**, 453 (1958).
165. Clampitt, B. H., *Anal. Chem.*, **35**, 577 (1963).
166. Schwenker, R. F., and L. R. Beck, *Textile Res. J.*, **30**, 624 (1960).
167. Ke, B., *J. Polymer Sci.*, **42**, 15 (1960).
168. Ke, B., *J. Polymer Sci.*, **50**, 79 (1961).
169. Wunderlich, B., and W. H. Kashdan, *J. Polymer Sci.*, **50**, 71 (1961).
170. Scott, N. D., *Polymer*, **1**, 114 (1960).
171. Rudin, A., H. P. Schreiber, and M. H. Waldman, *Ind. Eng. Chem.*, **53**, 137 (1961).
172. Ke, B., and A. W. Sisko, *J. Polymer Sci.*, **50**, 87 (1961).
173. Murphy, C. B., and J. A. Hill, *Nucleonics*, **18**, 78 (1960).
174. *DuPont Thermal Analysis Application Brief*, No. 11, Jan. 15, 1968.
175. *Fisher DTA Instrument Bulletin*, Fisher Scientific Co.
176. Visser, M. J., and W. H. Wallace, *DuPont Thermogram*, **3**, No. 2, 9 (1966).
177. Kemme, H. R., and S. I. Kreps, *Anal. Chem.*, **41**, 1869 (1969).
178. Farritor, R. E., and L. C. Tao, *Thermochim. Acta*, **1**, 297 (1970).
179. Mita, I., I. Imai, and H. Kambe, *Thermochim. Acta*, **2**, 337 (1971).
180. Beech, G., and R. M. Lintonbon, *Thermochim. Acta*, **2**, 86 (1971).
181. Hoyer, H. W., A. V. Santoro, and E. J. Barrett, *J. Phys. Chem.*, **72**, 4312 (1968).
182. Gray, A. P., *Perkin-Elmer Instrument News*, **16**, No. 2 (1970).
183. Breger, I. A., and W. L. Whitehead, *Fuel*, **30**, 247 (1951).
184. Gamel, C. M., and W. J. Smothers, *Anal. Chim. Acta*, **6**, 442 (1952).

CHAPTER VIII

EVOLVED GAS DETECTION AND EVOLVED GAS ANALYSIS

A. Introduction

The detection or analysis of the gases evolved during a chemical reaction, as a function of temperature, constitute the techniques of thermal analysis called evolved gas detection (EGD) and evolved gas analysis (EGA), respectively. According to the ICTA nomenclature recommendations given in Chap. XIII, the technique of evolved gas detection is defined as the term covering any technique of detecting whether or not a volatile product is formed during thermal analysis. The technique of evolved gas analysis, likewise, is a technique of determining the nature and amount of volatile product or products formed during thermal analysis.

The detection and analysis technique used in EGD and EGA varies from a simple gas thermal conductivity detector to the more sophisticated and complex high-resolution mass spectrometer. In the EGD mode, the relative amount of evolved gases is continuously recorded as a function of temperature. The EGA mode may be continuous or intermittent in nature; using gas chromatography, the analysis is generally intermittent or stepwise, while mass spectrometry is used in the continuous mode of analysis. Both types of evolved gas sampling and analysis are widely used.

The area of EGD and EGA has been reviewed in a book by Lodding (76) and a book chapter by Kenyon (77).

B. Evolved Gas Detection

The technique of evolved gas detection (EGD) was probably envisioned by Stone (1) as a result of his dynamic gas atmosphere, enclosed furnace type of DTA apparatus. Unfortunately, he did not monitor the evolved gases from the furnace chamber but only recorded the DTA curves. In 1960, a thermal conductivity detector was placed on the gas outlets of a DTA furnace assembly by Lodding and Hammell (2), but no evolved gas curves were published. They included absorption tubes in the gas outlet system so that the gases could be analyzed as well as detected. In this same year, Rogers et al. (3) described a simple pyrolysis apparatus which was to give birth to the technique of evolved gas detection. Because of the thoroughness of this

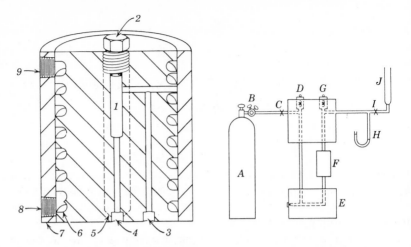

Figure VIII.1. Pyrolysis block and accessory apparatus as described by Rogers et al. (3). (*1*) Pyrolysis chamber; (*2*) nickel plug; (*3*) carrier gas inlet; (*4*) carrier gas outlet; (*5*) cartridge heater wells (*2*); (*6*) helical threads cut in inner body of block; (*7*) outer shell of block; (*8*) cooling jacket inlet; (*9*) cooling jacket outlet. (*A*) Carrier gas supply; (*B*) pressure regulator; (*C*) flow-control needle valve; (*D*) reference thermal conductivity; (*E*) pyrolysis chamber; (*F*) combustion tube; (*G*) active cell; (*H*) manometer; (*I*) pressure-control needle valve; (*J*) rotameter.

work, as far as the experimental parameters are concerned, Rogers et al. (3) are generally recognized as the founders of this technique.

The apparatus used by these investigators is illustrated in Figure VIII.1. It is based upon three primary sections: (*a*) an electrically heated pyrolysis chamber, (*b*) a combustion tube, and (*c*) a thermal conductivity cell. The pyrolysis chamber contained a rather long gas inlet tube which served as a preheater for the carrier gas. It was heated by two 240-watt cartridge heaters whose voltage input was controlled by a variable-voltage transformer. The combustion tube was a 5-in. segment of nickel tube which was heated by a resistance wire heater. It was filled with a mixture of fire brick and copper(II) oxide which was maintained at a temperature of 650 to 750°C. The purpose of this tube was to convert all volatile products into simple gases so that they would not condense in the tubes before reaching the detector. The thermal conductivity detector used model airplane glow plugs as detector elements; it was isolated from the pyrolysis chamber and combustion tube and maintained at room temperature. Voltage output from the detector was recorded on the *Y* axis of an *X–Y* recorder while the temperature, as detected by a Chromel-Alumel thermocouple, was recorded on the *X* axis.

The EGD curves obtained on this apparatus resemble the derivative of TG curves. Results are not normally comparable to those obtained with DTA where products are contained. The qualitative effects of the operating variables (flow rate, heating rate, sample mass, thermal conductivity cell sensitivity, pressure, and carrier gas) on the nature of the EGD curves were determined. These variables, with the exception of the variation in thermal conductivity cell sensitivity, affected the peak maximum temperatures and peak heights. In this case, only the peak heights were affected. The variation in EGD curves with sample mass is illustrated in Figure VIII.2. The peak maximum temperatures varied from 160°C for a 1.4-mg sample to 178°C for a 20.5-mg sample. The effect was even more pronounced with the EGD curve of TNT (trinitrotoulene). A 1.5-mg sample had a peak maximum temperature of 153°C, while a 20.9-mg sample shifted the peak maximum to 197°C. Thus, judicious control of the sample size appears to be necessary for reproducibility of peak temperatures.

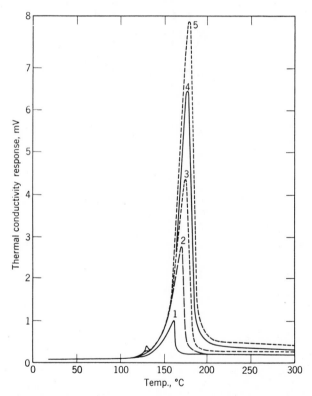

Figure VIII.2. Effect of sample size on EGA curve of PETN (3). (1) 1.4 mg PETN; maximum at 160°C; (2) 5.0 mg PETN; maximum at 167°C; (3) 9.7 mg PETN; maximum at 177°C; (4) 15.0 mg PETN; maximum at 178°C; (5) 20.5 mg PETN; maximum at 178°C.

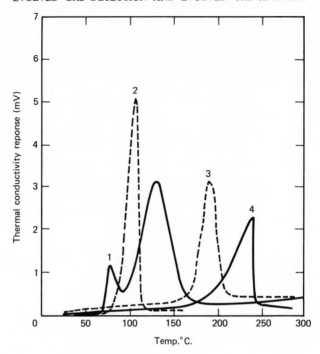

Figure VIII.3. EGD curves of various compounds (3); (*1*) 10 mg NaHCO$_3$ contaminated with Na$_2$CO$_3$·H$_2$O; (*2*) 10 mg CaSO$_4$·2H$_2$O; (*3*) 10 mg tetryl; (*4*) 10 mg hexanitroso-benzene.

Typical EGD curves of several representative compounds are shown in Figure VIII.3. In the case of a mixture of compounds with similar boiling points, resolution of the peaks was not affected. Melting points were, of course, not detected by this technique unless decomposition occurred at the same time. Quantitative as well as qualitative results can be obtained by this technique if, under identical conditions of pyrolysis, the areas under the EGD peaks are compared with standard substances.

A similar EGD apparatus was described by Vassallo (4) for the pyrolysis of various polymeric materials. The detection system consisted of a therm-istor thermal conductivity cell employing 100,000-Ω thermistors and operated at a bridge current of 10 mA. Located between the pyrolysis chamber and the detector cell was an 800°C CuO combustion chamber, similar to that previously described (3). A *X–Y* recorder was used to record the EGD curves instead of a strip-chart recorder. Contrary to previous comments, it was stated that EGD cannot replace TG; however, for comparing polymers of similar structure, EGD has more discrimination and is more rapid. A

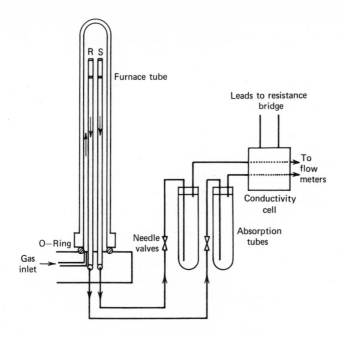

Figure VIII.4. Combined DTA-EGD apparatus of Lodding and Hamnell (2).

mass-loss of 2 mg can be detected at much faster heating rates than those used in TG.

The DTA–EGD apparatus used by Lodding and Hammell (2) is shown in Figure VIII.4   The furnace was constructed of recrystallized alumina and could be used at pressures up to 500 p.s.i.g. at a temperature of 1200°C. Gases introduced into the furnace chamber flowed through the sample and the reference material and exited the chamber separately.   It was stated that the gas detection and analysis system permitted determination of the origins of the DTA curve peaks.   No examples of these determinations were given, however.

The combination of EGD with other thermal techniques such as DTA and TG would be expected to yield especially significant results, as the advantages of each technique could be utilized.   Such has been the case with the EGD–DTA combination.   Measurements are made on a single sample at an identical furnace heating rate, furnace atmosphere, sample particle size, and other variable conditions.   A summary of EGD and DTA reactions for a specific chemical or physical change is given in Table VIII.1 (4).

One of the first systems for determining simultaneous EGD–DTA curves of a sample was that described by Ayres and Bens (5).   This apparatus, as

TABLE VIII.1
EGD and DTA Effects During a Chemical or Physical Change (5)

Chemical or physical change	DTA endotherm	EGD	
		Yes	No
Decomposition	x[a]	x	x
Fusion	x		x
Crystal transition	x[a]		x
Desorption	x	x	x
Vaporization			
Desolvation	x	x	
Ebullition	x	x[b]	
Sublimation	x	x[b]	

[a] May also be an exothermic reaction.
[b] Possible condensation before reaching gas detector.

shown in Figure VIII.5, permitted the continuous monitoring of the evolved gases and also selective sampling of any desired portion of the gases as well as providing for obtaining the DTA curve. The sample and reference holders were all constructed of Pyrex glass and were connected to a glass manifold by standard taper glass joints. Evolved gases were detected with a thermal conductivity cell. Glass beads (0.1 mm in diameter) were used as the sample diluent and inert reference material.

An interesting compound that was studied was nitroglycerin, the EGD–DTA curves of which appear in Figure VIII.6   The endothermic peaks in the 141–206°C temperature range are the result of vaporization of the nitroglycerin, boiling point 143°C, with some decomposition, as indicated by the EGD curve. A slight inflection corresponding to the maximum rate of vaporization (endothermic peak at 191°C) showed a change in rate of the decomposition gas reaching the detector, with a maximum rate of evolution being reached at 202°C. As the temperature was increased, more of the nitroglycerin previously distilled into the cool exit tube was decomposed, giving rise to further peaks in the EGD curve. The DTA curve showed no deviation in this region since all of the nitroglycerin had distilled away from the differential temperature probes.

Another application of this technique illustrating a phase transition and also a decomposition reaction is the thermal dissociation of ammonium perchlorate, as illustrated in Figure VIII.7. Ammonium perchlorate, a typical propellant oxidizer (5), had an endothermic phase transition at a peak maximum temperature of 204°C. This was due to the orthorhombic → cubic crystalline transition and gave little gas evolution as seen on the EGD curve. The subsequent exothermic peak and resultant EGD peak at 258°C

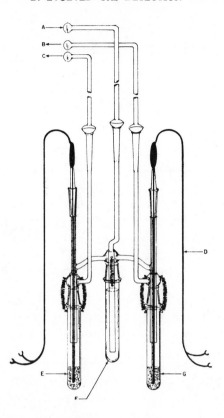

Figure VIII.5. Sample and reference cell used by Ayres and Bens (5). A, Sweeping gas inlet; B, gas train to detector, sample; C, gas train to detector, reference; D, thermocouple probe leads; E, reference cell with glass beads; F, sweeping gas preheater, with glass wool; G, sample cell, with sample and glass beads.

were attributed to the initial decomposition of NH_4ClO_4, with final decomposition beginning at 320°C. The results of this study were not in agreement with other DTA data (1, 6), due probably to differences in heating rate and diluents employed.

Ayres and Bens (5) pointed out that difficulties were observed in their system because of increased vaporization of the sample due to the gas flow and the occasional condensation of this vapor in the cool part of the exit tube. The vaporization of the sample below the boiling point made boiling-point determinations difficult, while the condensation of sample before passage through the gas detector presented difficulties, as later the gas stream may become hot enough to decompose the condensate.

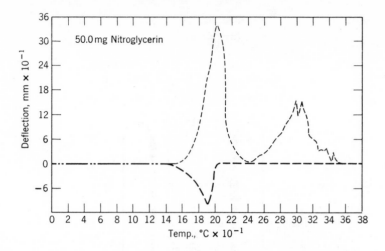

Figure VIII.6. Simultaneous EGD-DTA curves of nitroglycerin (5).

Wendlandt (7) has described a similar DTA–EGD apparatus, which was of a more robust construction that that previously described (5), and did not necessitate the use of a glass-bead diluent. The apparatus is illustrated schematically in Figure VIII.8.

The sample and reference materials were placed in small Inconel cups of 0.27-ml capacity which seated directly onto the differential thermocouple junctions. Helium was used as the carrier gas; the evolved decomposition

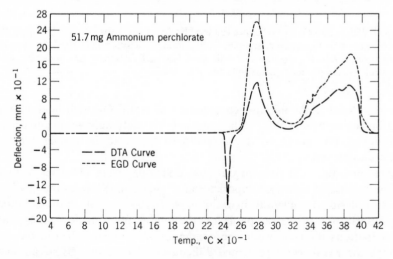

Figure VIII.7. Simultaneous EGD–DTA curves of ammonium perchlorate (5).

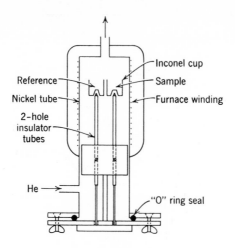

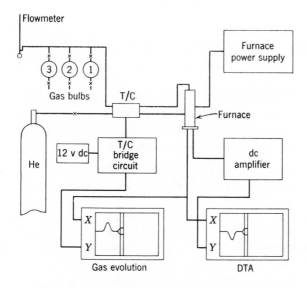

Figure VIII.8. Combined DTA–EGD apparatus (7).

gases were detected with a thermistor thermal conductivity cell of conventional design.   The thermistor bridge output and the ΔT amplified voltages were recorded on the Y axes of two different X–Y recorders; furnace temperature was recorded simultaneously on the two X axes.

The DTA–EGD curves for the dehydration of $CuSO_4 \cdot 5H_2O$ are illustrated in Figure VIII.9.   The origin of the various peaks was due to the following

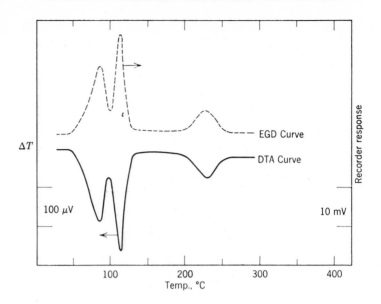

Figure VIII.9. Simultaneous DTA–EGD curves of $CuSO_4·5H_2O$ (7).

reactions:

$$85°C \text{ peak}: CuSO_4·5H_2O \text{ (s)} \rightarrow CuSO_4·3H_2O \text{ (s)} + 2H_2O \text{ (l)}$$
$$2H_2O \text{ (l)} \rightarrow 2H_2O \text{ (g) (from saturated solution)}$$
$$115°C \text{ peak}: CuSO_4·3H_2O \text{ (s)} \rightarrow CuSO_4·H_2O \text{ (s)} + 2H_2O \text{ (g)}$$
$$230°C \text{ peak}: CuSO_4·H_2O \text{ (s)} \rightarrow CuSO_4 \text{ (s)} + H_2O \text{ (g)}$$

From the EGD curve it can be seen that each of the DTA peaks was matched by a corresponding EGD peak showing that a volatile decomposition product, water vapor, was evolved.

Similarly, for the thermal dissociation of potassium ethylsulfate, as shown in Figure VIII.10, a crystalline phase transition indicated by the DTA curve did not give a peak in the EGD curve (8). The transition $\alpha \rightarrow \beta$ $KC_2H_5SO_4$ resulted in a DTA peak at 93°C. Further heating of the compound gave the fusion peaks, at 207°C, which again did not result in an EGD curve peak. The fusion peak was followed by the decomposition endothermic peak, which did give a corresponding EGD curve. The decomposition reaction was

$$\beta\text{-}KC_2H_5SO_4 \text{ (s)} \rightarrow K_2S_2O_7 \text{ (s)} + C_2H_4 \text{ (g)} + H_2O \text{ (g)}$$

A similar DTA–EGD set of curves was obtained for potassium methylsulfate.

A combined DTA–gas-sampling technique has been described by Murphy

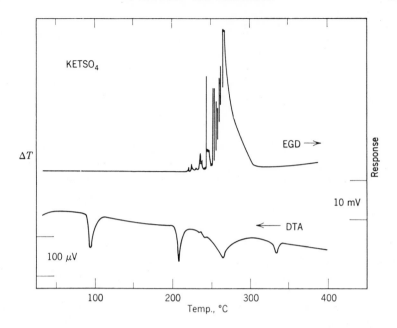

Figure VIII.10. Simultaneous DTA–EGD curves of potassium ethylsulfate (8).

et al. (9). The evolved gases were collected in a glass sampling bulb, from a system initially under a vacuum. Composition of the evolved gases was then determined by use of a mass spectrometer. This type of system would be useful only if one peak were involved in the DTA curve.

The potentialities of the combined DTA–EGD apparatus are readily apparent. The presence of a crystalline or other phase transition is easily detected, thus aiding in the interpretation of the DTA curve. By analysis of the evolved gases by some physical means, the stoichiometry of the reaction responsible for the peaks can be determined. It is also possible to study the thermal behavior of the sample under isothermal conditions as well as under various gaseous atmospheres and pressures.

As previously stated (5), the small amount of modification necessary to add an EGD apparatus to a regular DTA apparatus and the inexpensive equipment needed are far outweighed by the amount of additional information obtained from one sample. Several commercial firms, Columbia Scientific Industries, Technical Equipment Corporation (Deltatherm), the Perkin-Elmer Corporation, and others, now include this feature on certain types of their instruments.

An apparatus that permitted the study of dissociation reactions in a dynamic gas system was described by Thomas et al. (10). A platinum-rhodium sample tube was used to contain the sample during the dissociation and sintering reactions at elevated temperatures up to 1000°C. A thermal conductivity detector was used to monitor the evolved gases, the curve peak areas of which were integrated by a mechanical integrator attached to the recorder. The main purpose of the apparatus was to investigate sintering properties of various minerals and also to determine their surface areas while in the same apparatus.

Ware (11) developed a material having a thermal transition which could be used as a temperature calibration point during the gas evolution process. It was found that potassium sulfate containing a small amount of dissolved carbon dioxide released the latter substance in fusion so that the evolved gas peak coincided with the melting process and was independent of the heating rate and oxygen content of the atmosphere. The sample holder shown in Figure VIII.11 was used to contain the sample during the calibration run. It consisted of a platinum crucible placed in a alumina holder which was enclosed with a platinum metal cover. Carrier gas flows between the platinum cover and the alumina sample holder, up to the top of the sample holder, and impinges on the top of the sample. The gas then passes around the crucible and through a small hole in the center of the holder to a gas density detector. It was estimated that the time lag between the sample and detector for a gas flow rate of 30 ml/min was less than 6 sec. The preliminary results obtained using this system indicated that the gas evolution was not influenced by various furnace heating rates from 3 to 10°C/min, nor was it appreciably affected by oxygen present in the gas atmosphere. It should be considered as a possible system for EGD temperature calibration; however, a wider range of conditions are required before final acceptance is obtained.

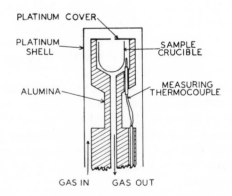

Figure VIII.11. EGD temperature calibration sample holder (11).

A simultaneous EGD–DRS sample holder was described by Wendlandt and co-workers (12, 13). This apparatus, described in Chap. IX, has been applied to the thermal dissociation of metal salt hydrates and coordination compounds.

Evolved gas detection techniques have been used to obtain kinetic and thermodynamic data by various investigators (3, 14, 15, 18). The methods developed are as useful as TG and DTA methods but have the advantages of less complicated instrumentation.

Critical evaluations of the EGD technique and related gas analysis problems have been made by Garn (16) and Findeis et al. (17).

C. Evolved Gas Analysis

The ability to determine the composition of the evolved gases greatly increases the usefulness of the evolved gas technique. The gases may be analyzed in an intermittant or continuous manner by the various techniques illustrated schematically in Figure VIII.12. Analytical techniques usually employed are mass spectrometry, gas or thin-layer chromatography, chemical

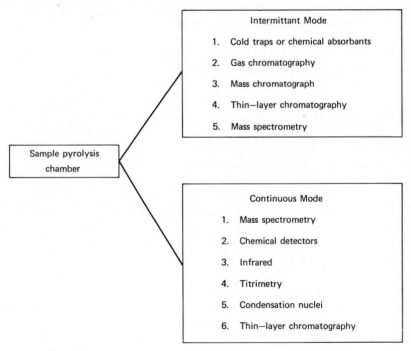

Fig. VIII.12. Analytical methods for the analysis of evolved gases.

detectors, and infrared spectroscopy. In most of the investigations, the continuous mode of analysis is preferred due to the decreased time and labor involved. However, these techniques are usually more expensive, especially when computer control is employed. The sample pyrolysis chamber may be a simple furnace system discussed in Section B of this chapter or it may be a DTA apparatus or thermobalance. Analysis of the radioactivity of a gas released during a chemical reaction will not be discussed here, as it is described in Chap. XI under emanation thermal analysis (ETA) (Section D).

1. GAS CHROMATOGRAPHY

The powerful analytical technique of gas chromatography has been used to determine the composition of the gases evolved during a chemical reaction. It is a fairly simple matter to attach a gas chromatograph to the pyrolysis chamber by means of a gas sampling valve. This valve can be operated manually (19) or automatically (20, 21). In some cases, the effluent gases from the gas chromatograph are introduced into a mass spectrometer (22) for positive identification of the gaseous products.

The gas train and valve arrangement for repetitive dual-sampling gas chromatography, as used by Garn (16), is shown in Figure VIII.13. The gaseous products from a pyrolysis chamber are passed through a pair of sampling valves. These valves operate at programmed intervals to withdraw samples of the gas and pass them through the columns and detectors. An adjustable cam timer can be connected to the valves so that the valves will open for 5 sec and then close. The oven temperature, flow rates, and sampling timer were chosen for each type of experiment. Three different columns were employed: fluoropak, silica gel, and molecular sieve.

A gas chromatograph was coupled to a thermobalance by several investigators (20–22). With this combination, the progress of the reaction can be monitored by TG, and at selected temperature intervals the evolved gases can be introduced into the gas chromatograph for analysis. The apparatus used by Wiedemann (21) is shown in Figure VIII.14. In order to keep the time delay in gas transfer to a minimum, the furnace volume was reduced to about 35 cm^3. Gases were introduced into the gas chromatograph by means of a gas sampling valve which could be opened at specific intervals. The results of this technique are illustrated in the TG and evolved gas curves shown in Figure VIII.15. During the decomposition of $CaC_2O_4 \cdot H_2O$, water, carbon monoxide, and carbon dioxide are evolved, but since the curves were obtained in an air atmosphere, the carbon monoxide was oxidized immediately to carbon dioxide. Wiedemann commented that compared to a mass spectrometer the gas chromatograph is slower, but as far as the quality of results is concerned the methods are about equal.

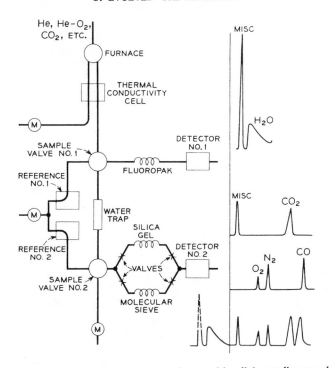

Fig. VIII. 13. Gas train and valve arrangement for repetitive dial-sampling gas chromatography (16).

Other gas-chromatography–evolved-gas systems have been described by Wist (23), Dorsey (24), Franc and Pour (25, 27), and Sarner et al. (26). The term "reaction thermal analysis" was used (25, 27) to describe the technique in which a sample is reacted with various reagents and the cleavage products are identified by gas chromatography.   Vapor phase pyrolysis, the controlled thermal fragmentation reactions of a sample (26), is used to provide unequivocal determination of the structure of GC effluents as well as for other basic studies.

An EGA-gas chromatograph was designed by Bollin (28) to be used to analyze the atmosphere and surface of the planet Mars.   The purpose of the chromatograph was to resolve the EGA peaks to determine if organic material has the ordered molecular-weight distribution characteristic of biologic origin, or if the distribution is random and therefore indicative of abiotic origin.   Of special interest were the specific gas detectors in which the desired sample gas reacted exothermically with the detector material.   The water detector employed calcium or lithium hydride, while the carbon dioxide detector may use sodium or lithium hydroxide.   The complete

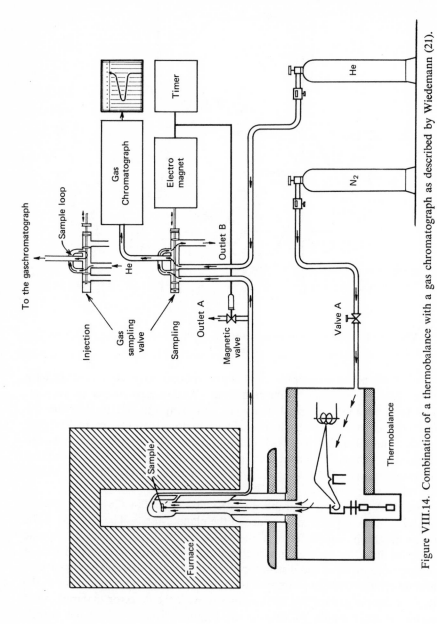

Figure VIII.14. Combination of a thermobalance with a gas chromatograph as described by Wiedemann (21).

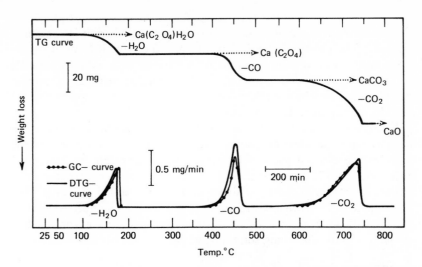

Figure VIII.15. TG and EGA curves for the decomposition of $CaC_2O_4 \cdot H_2O$ (21).

apparatus is illustrated in Figure VIII.16.   It should be noted that both the DTA–EGA instrument and the chromatograph are only one inch in diameter!

2. TITRIMETRIC METHODS

The amount of evolved gas can be determined by a continuous titration method.   A carrier gas removes the evolved gas from the furnace chamber

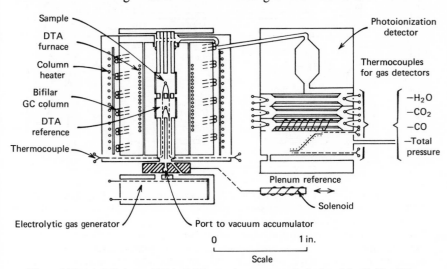

Figure VIII.16. Miniature DTA–EGA-gas chromatograph designed by Bollin (28).

and transports it to an aqueous absorbing solution where it is continuously titrated. The titrant used will depend upon the type of evolved gas to be determined. For example, ammonia is titrated with dilute hydrochloric acid, while water is determined by the Karl Fischer method. Compounds that can be determined include water, hydrogen chloride, ammonia, sulfur dioxide, carbon dioxide, and chlorine (29).

Paulik and co-workers (29–35) described the combination of titration analysis with the Derivatograph. By means of a continuous titrator, a curve of the evolved gas can be made, which they call the "thermo-gas-titrimetric" (TGT) curve, or the derivative of it, the "derivative thermo-gas-titrimetric" (DTGT) curve. This apparatus is illustrated in Figure VIII.17. The sample in the furnace is surrounded by a silica chamber which can be flushed with an inert carrier gas. Evolved gases are transported by the carrier gas to the absorbing solutions, where they are titrated with a suitable

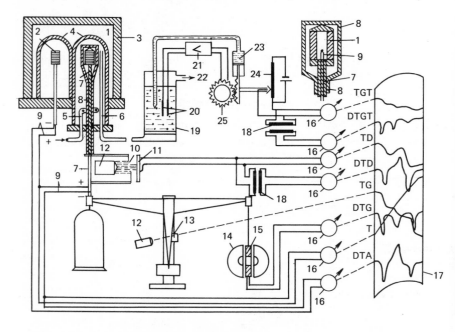

Figure VIII.17. Apparatus for the parallel recording of DTA, T, TG, DTG, TGT, and DTGT curves (35). 1, Compressed test piece; 2, compressed reference substance; 3, furnace; 4, silica bell; 5, inlet tube for carrier gas; 6, tube for gas extraction; 7, silica tube; 8, silica tube with stirrup-shaped end; 9, thermoelement; 10, diaphragms; 11, light cell; 12, lamps; 13, optical slit; 14, magnet; 15, coil; 16, galvanometer; 17, photographic paper; 18, deriving transformer; 19, absorber; 20, electrodes; 21, amplifier; 22, vacuum pump; 23, automatic burette; 24, potentiometer; 25, Servo-motor.

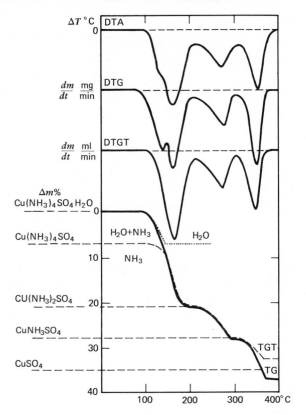

Figure VIII.18. Combined derivatographic and thermo-gas-titrimetric curves of [Cu(NH₃)₄]SO₄·H₂O (31).

reagent. The amount of titrant used is recorded as a function of time by means of a retransmitting potentiometer connected to the automatic buret.

The application of this technique to the analysis of the gaseous products formed during the thermal decomposition of $[Cu(NH_3)_4]SO_4 \cdot H_2O$ is shown in Figure VIII.18. Between 90 and 200°C, one mole of water and two moles of ammonia are evolved. The curve clearly shows that ammonia is evolved simultaneously with the water evolution, a process which would not be detected by TG and DTG alone. Ammonia is then released stepwise, one mole at a time, but during the release of the fourth mole, sulfur trioxide is also evolved due to the decomposition of copper(II) sulfate. Other applications of this technique include the thermal decomposition of $MnNH_4PO_4 \cdot H_2O$ (32), $[Ni(NH_3)_6]Cl_2$ (33), $ZnNH_4PO_4$ (33), $CuSO_4 \cdot 5H_2O$ (34), basic iron (III) acetate (34), glucose (34), fructose (34), lactose (34), and $MnCO_3 \cdot nH_2O$ (35).

3. INFRARED SPECTROSCOPY

Infrared spectroscopic techniques have long been used to analyze gas streams in industrial chemical processes.   Recently, with the advent of fast-scan infrared spectrometers, they have been used as gas chromatograph detectors.   One requirement of their use, needless to say, is that the compound must possess one or more infrared absorption band.   By means of a carrier gas, the evolved gas sample from a pyrolysis chamber can be readily passed through an infrared cell for analysis.   Infrared systems that can be employed include (*a*) nondispersive analyzers, (*b*) dispersion spectrometer, (*c*) band-pass filter-type instruments, and (*d*) interference spectrometers; all of these techniques have been adequately reviewed by Low (36).

Kiss (37, 38) coupled a Chevenard-type thermobalance with a type UR 10 spectrophotometer; the evolved gases from the thermobalance were passed through a 10-cm-long infrared cell.   A method was developed in which either the ammonia or the water (via C_2H_2 generation) content of the evolved gases could be determined.   Water was not measured directly because the presence of hydrogen bonding in the molecule greatly diminishes the sensitivity of the absorption measurement.   However, by passing the evolved

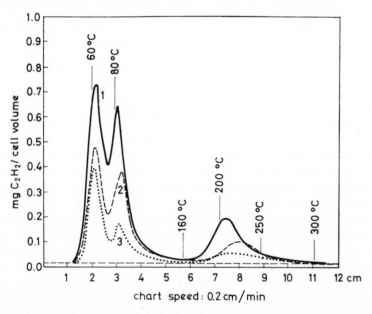

Figure VIII.19. Evolved water (C_2H_2) curves for the dehydration of $CuSO_4 \cdot 5H_2O$ (37); the amount of C_2H_2 is proportional to the water concentration. Sample sizes are: curve 1, 60 mg; curve 2, 40 mg; curve 3, 20 mg.

water over calcium carbide, acetylene was generated which could be detected by means of an absorption peak at 728.0 cm^{-1}. The water evolution curve from $CuSO_4 \cdot 5H_2O$ is shown in Figure VIII.19. Other compounds investigated include ammonium paramolybdate and ammonium paratungstate. In a later investigation, binary mixtures of ammonia and water (C_2H_2) were determined by this technique (38).

4. THERMOPARTICULATE ANALYSIS (TPA)

The technique of thermoparticulate analysis (TPA) consists of the detection of evolved particulate material in the evolved gases as a function of temperature. In the presence of supersaturated water vapor, these particles provide condensation sites for water, and hence can be detected by light-scattering techniques. Water droplets grow very rapidly on the particulate matter (condensation nuclei) until they are of a sufficient size to scatter light. The scattered light, as detected by a phototube in a dark-field optical system, is proportional to the number of condensation nuclei initially present. It is an extremely sensitive measurement, with the capability of detecting one part of material in 10^{15} parts of air. The technique was first employed by Doyle (39) and has been reviewed by Murphy (40, 41).

A schematic diagram of the apparatus for TPA measurements is shown in Figure VIII.20. The polymer samples, $1 \times 0.125 \times 0.014$ in., were heated in a copper tube furnace which was programmed for a temperature rise of 50°C per hr. Hydrogen was passed over the sample at a 3-ml-per-sec flow rate, but because the condensation nuclei counter required a gas flow of 100 ml per sec, additional amounts of nitrogen were added beyond the furnace and heat exchangers. Some materials studied by this technique are shown in Table VIII.2.

Murphy (40) has also described a converter which will generate condensation nuclei from a reactant which would normally not be detected by TPA. For example, ammonia can be detected by passing it through a flask containing a small amount of hydrochloric acid. The gaseous HCl above the solution reacts with the ammonia to form condensation nuclei of ammonium chloride. Other gases and the conversion processes employed are shown in Table VIII.3. The TPA method is continuous, capable of gas analysis through conversion techniques, and able to detect condensation nuclei. It is the only thermal technique with the last-named capability.

5. THIN LAYER CHROMATOGRAPHY (TLC)

Although not a very widely used technique in this application, thin layer chromatography (TLC) has been used to analyze evolved gaseous products

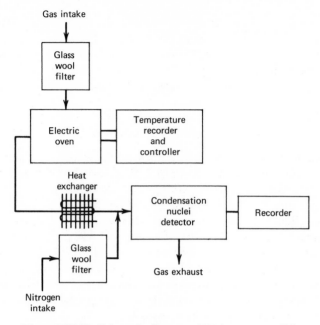

Figure VIII.20. Schematic diagram of TPA apparatus (41).

TABLE VIII.2

Decomposition Temperatures of Polymeric Materials by Thermoparticulate Analysis
(40)

Material	Atm.	Heating rate, °C/hr	Decomposition temp., °C
Poly(methyl methacrylate)	H_2	50	310–320
	Air	50	242
	Air	180	237
Polystyrene	H_2	50	337
Polystyrene filled with SiO_2	H_2	50	337
Phenol-formaldehyde resin	H_2	50	432
Polychlorotrifluoroethylene	N_2	180	340
Copper phthalocyanine	N_2	180	230
Zinc 4,4′-bis′thiopicolin- amidodibenzophenone	N_2	180	255
	Air	180	350

TABLE VIII.3
Gases Detected by Condensation Nuclei Techniques (40).

Compound	Conversion process	Easily detected concn., ppm
Ammonia	Acid–base	0.005
Benzene	Photochemical	2
Carbon dioxide	Electrochemical	5
Carbon monoxide	Chemical	1
Chlorine	Chemical	1
Ethyl alcohol	Reverse photochemical	5
Freon 12-21	Pyrolysis	2
Hydrogen chloride	Acid–base	0.5
Hydrocarbon	Photochemical	0.1
Methyl mercaptan	Oxidation-photochemical	0.01
Ethylamine	Acid–base	0.5
Mercury	Photochemical	0.001
Nitrogen dioxide	Hydrolysis	0.5

and also for kinetics studies. Permanent-type gases, of course, cannot be handled by this technique, but high-molecular-weight compounds, which may be difficult to identify by other methods, can be separated and characterized. Also, the equipment required for TLC is much less expensive than that required for any of the other methods.

The technique used by Rogers (42, 43) involved the heating of the sample under a flowing carrier-gas stream at a programmed rate (or isothermally) while impinging the carrier-gas stream containing the decomposition products onto the surface of an activated TLC plate. The plate is transported across the orifice of the pyrolysis stream as a function of sample temperature. Therefore, any position along the zone of application of the plate corresponds to a specific sample temperature. The plate can then be developed by the usual chromatographic techniques to separate the individual products of the reaction. The final plate yields two types of data: R_F data in the direction of development, and temperature data along the zone of application.

A developed TLC plate for TNT (42) with a DTA curve superimposed on it is shown in Figure VIII.21. The TNT and volatile impurities begin to vaporize and appear on the TLC plate between 125 and 135°C, corresponding to the first appearance of gas in the pyrolysis curve. Most of the TNT vaporizes and is collected undecomposed as it is a relatively stable compound thermally. Within the temperature range where TNT dissociates exothermally, as indicated by the DTA curve, the following products appear: 2,4,6-trinitrobenzyl alcohol (TNB-OH); 4,6-dinitroanthranil (DNA); 1,3,5-trinitrobenzene (TNB); 2,4,6-trinitrobenzoic acid (TNB-a); and a

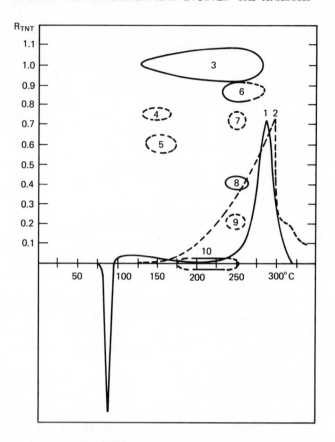

Figure VIII.21. Graphical compilation of thermal data for TNT (42). 1, DTA curve, 2, pyrolysis curve, 3, TNT zone, 4, 2,6- and 3,5-DNT zone, 5, 2,4-DNT zone, 6, TNB zone, 7, DNA zone, 8, TNB-OH zone, 9, unidentified zone, 10, TNB-a zone Sample 0.284 mg, heating rate 11°C/min, air carrier gas.

trace of an unidentified compound.   The combination of the precision of the TLC method with the characteristic colors of the spray reagent make it relatively simple to identify all of the major components found.

The TLC technique has also been applied to the study of reaction kinetics by Rogers and Smith (43).

Stahl (44) described a pyrolysis procedure which also used TLC to identify the decomposition products.   The sample was introduced into a glass cartridge with a conical tip, and heated rapidly for a short period of time. The emerging gases were deposited as a spot on a TLC plate and developed with suitable reagents.   He described numerous variations of the standard procedure and demonstrated its application to the analysis of drugs, food additives, and residues, and to phytochemistry. Rogers (42) concluded his investigation of this technique by stating that it will probably be most useful for the study of polymeric materials. It should be possible to study large repeating units rather than just the simplest ultimate pyrolysis products. Because no thermal conductivity, flame ionization, or other detector is used, it is not necessary to use specific carrier gases for best results. Mechanisms and interactions can be studied in any gas or vapor system, for example, oxygen, air, nitrogen, helium, hydrogen, carbon dioxide, or mixtures containing water vapor, acids, ammonia, and so on.

6. FLAME IONIZATION DETECTION

A flame ionization detector has been used by Eggertsen and co-workers (45–49) to detect carbon-containing compounds in gaseous pyrolysis products. The apparatus consists of a small sample furnace coupled to a high-temperature flame ionization detector. A dynamic carrier gas transports the evolved gases from the pyrolysis furnance to the detector. The essential components of the apparatus are shown in Figure VIII.22. The sample furnace and detector jet are constructed of Vycor tubing as a single unit. Due to the unitized construction and the high temperature of the flame ionization detector (FID), normally 500°C, complete recovery of the volatile products is possible. The furnace is heated with a resistance wire heater element which is controlled by a precision temperature programmer. The

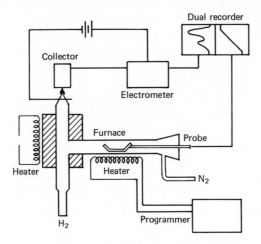

Figure VIII.22. Schematic diagram of flame ionization detector system (49).

TABLE VIII.4
Comparison of FID Sensitivity with TG

	TG	FID
Sample size	5–100 mg	0.1–5 mg
Sensitivity	10–40 mg	0.01 μg/min
Polyethylene		
T_{50}	456°C	458°C
T_{10}	430°C	424°C
T_5	418°C	419°
%/min		
0.1	—	370°
0.02	—	330°
0.01	—	300°

flame jet is heated by a stainless-steel block. Provision is made for injecting standard gas samples from a syringe, via a septum inlet, for convenient calibration of the detector response. The sample boat is made of aluminum, gold, or platinum and is held in a stainless-steel wire frame attached to a sheathed thermocouple.

The FID method possesses high sensitivity, as seen by its comparison with TG in Table VIII.4. The FID method uses smaller samples than TG, can routinely detect decomposition rates of 0.01%/min, and can detect rates down to 0.001%/min if desired. The lowered limit of detection is of the order of 1×10^{-4} μg of carbon per min. However, as found in previous work on the thermal stability of polymers (46), it is hardly feasible to utilize this high sensitivity because of the heating blank obtained even with a well-cleaned system (burnout with air at 550°C). The high sensitivity of the detector makes it possible to obtain meaningful thermal data curves with as little as a few micrograms of material. Vapor pressure measurements can be made below the level of 1 mTorr (45).

A typical curve obtained from the FID system is shown in Figure VIII.23. The polyvinyl chloride polymer formulation studied yielded four components, as seen from the curve peaks. Compounds identified included a pesticide (79°), a plasticizer (179°), a stabilizer (275°), and the polyvinyl chloride (438°).

Trace amounts of organic carbon compounds in water can also be determined by this technique (48). The sample is heated in a nitrogen carrier gas in two stages to determine volatile (<150°C) and nonvolatile (150–550°C) organic carbon. The water evaporated in the first stage changes the detector response to some extent, but this can be taken care of by proper calibration technique. The lower limit of detection is about 0.2 ppm. Results of some

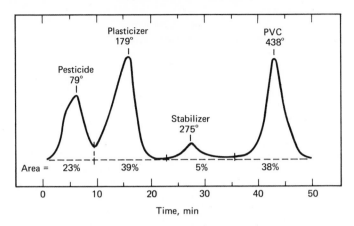

Figure VIII.23. FID curve for a polyvinyl chloride formulation; 9.4-mg sample heated at 10°C/min (49).

of the analyses for organic carbon content of various substances are shown in Table VIII.5. The method is especially effective for determining trace organic material that is volatile or steam-distillable under the conditions of the analysis. Results for the nonvolatile portions are sometimes low, particularly for natural organic material such as carbohydrates and proteins. In some cases, however, this can be compensated for by proper calibration techniques.

7. MASS SPECTROMETRY

Perhaps the most widely used of all of the analytical techniques employed in the analysis of evolved gases is the mass spectrometer. This technique can give positive identification of the gases evolved during the thermal reaction and in many cases can give the amount of each component present. Coupled with other thermal techniques such as TG and DTA, it provides perhaps the ultimate in technique for determining the nature of the reacting system under investigation.

The use of mass spectrometry in thermal analysis systems is shown schematically in Figure VIII.24. Evolved gases from a pyrolysis chamber, DTA apparatus, or thermobalance can be introduced directly into the mass spectrometer. If the evolved gas consists of a complex mixture of compounds, a gas chromatograph may be employed for preliminary separation and then introduction into the mass spectrometer. Because of the volume and complexity of data obtained from the mass spectrometer, data reduction by a large digital computer may be necessary, or a small digital computer may be employed to control the data output and presentation from the instrument.

TABLE VIII.5
Analysis of Dilute Solutions for Carbon Content (48).

Sample, ppm	Dilution factor[a]	Carbon, ppm	
		Found	Calcd.[b]
1. Acetic acid, 428	0	90	—
	10	10.3	9.0
	50	2.1	1.8
2. Heptanoic acid, 91	0	53	—
	5	10.0	10.6
3. NEODOL 25-12, 213	0	72	—
	10	5.9	7.2
	50	1.4	1.4
4. Cat-cracked gas oil	0	24	—
	10	2.7	2.4
	50	0.8, 0.6	0.5
5. Starch, 158	0	14	—
	5	1.7	2.8
6. Gelatin, 90	0	14	—
	5	3.4	2.8

[a] Dilutions were carried out in the sample boat with distilled water.
[b] Values were calculated from amount found undiluted and the dilution factor.

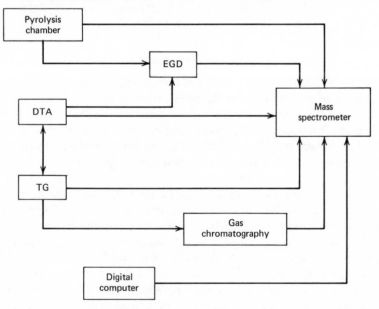

Figure VIII.24. Thermal analysis systems employing mass spectrometry.

Only recently has the latter mode been employed in thermal analysis. The entire subject of mass spectrometry in thermal analysis has been reviewed by Friedman (50) and others (51).

a. EGA–MS

Perhaps the simplest of the EGA–MS techniques is the linear programmed heating of the sample in the vacuum chamber of the mass spectrometer. The evolved gases are analyzed continuously as the mass spectrometer scans the mass range of interest. A plot of ion current versus time or temperature for each ionic mass observed may be made, or a single ionic mass may be monitored as a function of time or temperature of the pyrolysis. Both methods are commonly employed (50).

The first modern use of the EGA–MS technique was by Langer and Gohlke (52) in 1963. They heated a sample in the vacuum chamber of a time-of-flight mass spectrometer by means of a small furnace and recorded the mass spectra of the decomposition products at selected intervals. Compounds studied included $BeSO_4 \cdot 4H_2O$, germanium(IV) EDTA 2-hydrate, $CaSO_4 \cdot 2H_2O$, and $CuSO_4 \cdot 5H_2O$.

Friedman and co-workers (50) originally planned to use EGA–MS to aid in the kinetic analysis of TG data, but found the former to be of such good quality as to render the results of the latter useless for complex gas evolution. They used remote pyrolysis, about a meter away from the electron beam of the mass spectrometer, by tube furnace, and eventually introduced the capability to pyrolyze the sample after the furnace tube had been baked out at very high temperatures. Most of the compounds investigated were polymers such as polybenzimidazoles, polyesters, polyamides, and others.

Shulman and co-workers (53–55) used a system similar to that of Friedman in which a time-of-flight mass spectrometer was used to analyze the gases evolved from a Knudsen cell. Other coupled EGA–MS systems include those by Austin et al. (56, 57), Gray et al. (58), and others (59).

b. Simultaneous EGD–MS

Wendlandt and Southern (60) described an apparatus which was capable of recording simultaneously the EGD and MS curves of a sample. This apparatus, as shown in Figure VIII.25, permitted heating of the sample at atmospheric pressure in an inert or other atmosphere instead of the *in vacuo* conditions of the usual mass spectrometric technique (52). The pyrolysis chamber consisted of a tube furnace which was heated by a temperature programmer. Evolved gases were swept from the pyrolysis chamber by a carrier gas, usually helium, and detected in the flowing gas stream by a

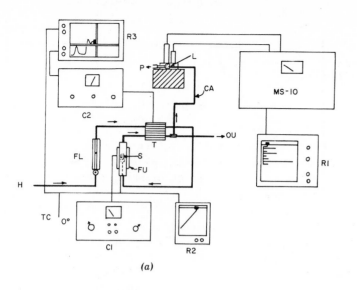

(a)

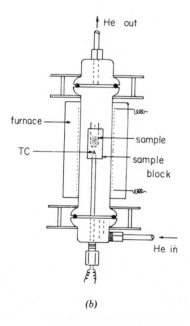

(b)

Figure VIII.25. Simultaneous EGA–MS apparatus of Wendlandt and Southern (60). (a) Complete apparatus; (b) pyrolysis chamber.

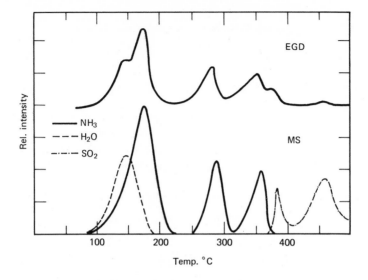

Figure VIII.26. EGD and MS curves of [Cu(NH$_3$)$_4$]SO$_4$·H$_2$O (60).

thermal conductivity cell.   A short length of stainless-steel capillary tubing transferred part of the evolved gases to the mass spectrometer for analysis. The mass spectrometer could be used to monitor one m/e value as a function of temperature, or a complete mass spectrum could be recorded at various time intervals.

Use of the apparatus is illustrated by the thermal dissociation of

$$[Cu(NH_3)_4]SO_4·H_2O,$$

as shown in Figure VIII.26.   An early question concerned whether the water or ammonia was evolved first in the initial thermal dissociation step.   From the slight difference in molecular masses involved, 18 versus 17, TG could not be expected to settle the question.   With the use of mass spectrometry, the question was easily answered.   As seen from the EGD curve, six maxima are present, at peak temperatures of 145, 175, 285, 380, and 460°C, respectively.   The MS curve enables an interpretation to be made of the EGD peaks, as shown in the following equations:

$$[Cu(NH_3)_4]SO_4·H_2O \text{ (s)} \rightarrow [Cu(NH_3)_4]SO_4 \text{ (s)} + H_2O \text{ (g)}$$
$$[Cu(NH_3)_4]SO_4 \text{ (s)} \rightarrow Cu(NH_3)_2SO_4 \text{ (s)} + 2NH_3 \text{ (g)}$$
$$Cu(NH_3)_2SO_4 \text{ (s)} \rightarrow Cu(NH_3)SO_4 \text{ (s)} + NH_3 \text{ (g)}$$
$$Cu(NH_3)SO_4 \text{ (s)} \rightarrow CuSO_4 \text{ (s)} + NH_3 \text{ (g)}$$
$$CuSO_4 \text{ (s)} \rightarrow CuO·CuSO_4 \text{ (s)} + SO_2 \text{ (g)} + \tfrac{1}{2}O_2 \text{ (g)}$$

c. Simultaneous DTA–EGD–MS

The addition of another thermal parameter, DTA, increases the utility of the mass spectrometric technique. Langer and Gohlke (61, 62) first used a modified DTA cell from a commercial instrument to provide the DTA data. At various time intervals the evolved gases were introduced into the mass spectrometer for analysis. This system was later modified by the addition of an internal DTA cell (63) for high-vacuum operation.

Wendlandt et al. (64) modified the pyrolysis chamber of their EGD–MS system so that DTA measurements could also be made simultaneously with the other two measurements.

Other DTA–EGD–MS systems have been described by Redfern et al. (65) and Gaulin et al. (59).

d. Simultaneous TG–EGD–MS or TG–DTA–EGD–MS

The direct combination of a thermobalance and a mass spectrometer was first reported by Zitomer (66), who coupled a DuPont thermobalance with a time-of-flight mass spectrometer. He applied this technique to the study of

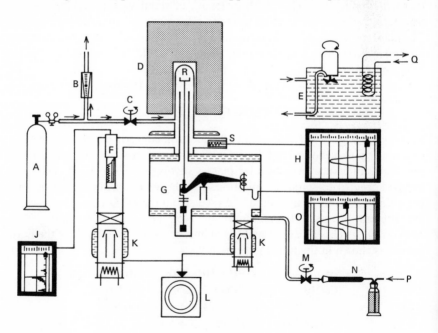

Figure VIII.27. Mettler thermobalance-quadrapole mass spectrometer system (68, 69).

the thermal decomposition of polymethylene sulfide and maleic hydrazide-methyl vinyl ether copolymer. Wilson and Hamaker (67) reported the coupling of an ultrahigh-vacuum thermobalance to a quadrupole mass spectrometer. For a heat source, they used quartz lamp heaters programmed at a heating rate of 10°C/min to a maximum temperature of 435°C. Studies were made of the thermal decomposition of polymethyl-methacrylate, poly-oxymethylene, polystyrene, and other polymers.

Perhaps the most elaborate type of TG–DTA–EGD–MS system is that of the Mettler thermobalance–quadrapole mass spectrometer system described by Wiedemann (68) and others (69). This apparatus is illustrated in Figure VIII.27. The sample may be studied under vacuum ($\sim 10^{-6}$Torr) or under higher pressures to one atmosphere. The reaction chamber, R, is surrounded by the furnace and is separated from the balance by a diffusion baffle. The evolved gases pass directly to the mass analyzer, F, which is connected to a recorder, J, through the mass spectrometer control panel. Total pressure is determined by an ionization gage, S, which also permits the

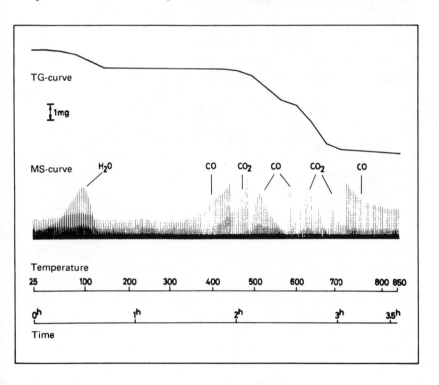

Figure VIII.28. Thermal dissociation of $CaC_2O_4 \cdot H_2O$ as described by Wiedemann (68).

recording of the EGD curve (in this case, due to the pressure change in the system). The relation between measured total pressure and the ion current of the calibration gas permits calibration of the mass spectrometer in absolute partial pressure units or amp/Torr.

Application of the apparatus is illustrated by the TG and MS curves of $CaC_2O_4 \cdot H_2O$, as illustrated in Figure VIII.28. The sample was heated under high vacuum, during which the partial pressure was continuously monitored with the mass spectrometer. Mass values of 18 (H_2O^+), 28 (CO^+ and N_2^+), and 44 (CO_2^+) were cyclically registered on the recorder. A sensitivity change was necessary during the run because of the sharp increase of CO and CO_2 evolution.

Gibson and co-workers (69–72) used the above apparatus to study gas evolution from geochemical and lunar soil samples. An example of these investigations is the TG–MS curves of an Apollo 15 rille soil, as shown in Figure VIII.29. All calculations of mass spectral data were made by a dedicated mini-computer system.

Smith and Johnson (73, 74) described a multipurpose apparatus which permitted the determination of the TG, DTA, EGD, and MS curves of a sample. It was applied to oil-shale research problems. Chang and Mead (75) described a TG–gas chromatograph–high-resolution mass spectrometer system and its application to the degradation of polymers. Application of the (75) apparatus to polystyrene foam is shown by the three curves in Figure VIII.30. In (a), a TG curve is presented for a 3.0-mg sample heated at 15°C/ min in a helium carrier-gas atmosphere. The gas chromatogram of the evolved gases is shown in curve (b), while the low-resolution mass spectrum of peak no. 15 is shown in curve (c). The evolved gases consisted of a mixture containing about 20 components, of which 16 were readily identified by the high-resolution mass spectrometer.

8. THERMOBAROGRAVIMETRY (TBG)

The technique of thermobarogravimetric analysis was introduced in 1965 by Bancroft and Gesser (78), who developed an apparatus for the continuous measurement of mass, pressure, and temperature. This apparatus, which was most useful at low initial pressures and at temperatures up to 800°C, was applied to the study of the thermal dissociation of metal bromates.

A more elaborate apparatus for thermobarogravimetry was described by Maycock and Verneker (79), who measured the pressure changes in a Mettler thermoanalyzer system using a thermocouple gauge or a M. K. S. Baratron pressure gauge. This apparatus is illustrated schematically in Figure VIII.31. The sensitivity of the Baratron gauge is 1×10^{-3} Torr and can be used to a pressure of 760 Torr.

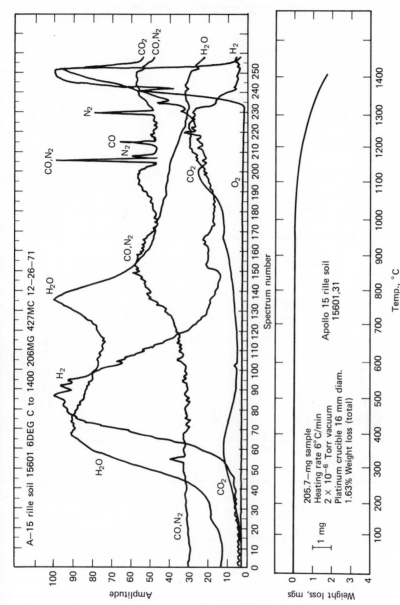

Figure VIII.29. TG–MS data for an Apollo 15 rille soil (72).

353

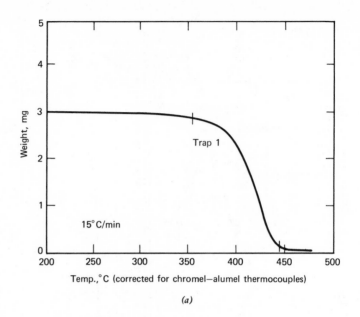

(a)

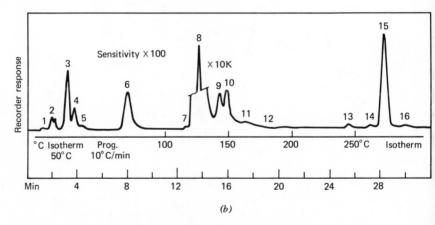

(b)

Figure VIII.30. TG–GC–MS curves of a polystyrene foam sample (75); (*a*) TG curve; (*b*) GC curve; and (*c*) mass spectrum of peak no. 15 in GC curve.

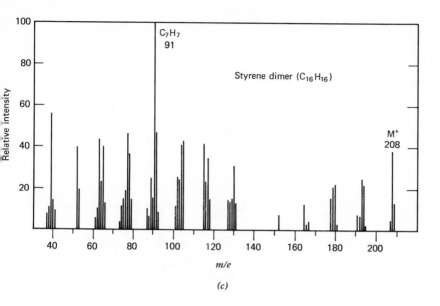

Figure VIII.30.

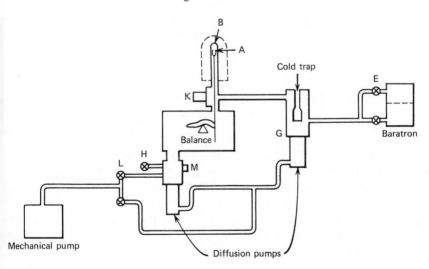

Figure VIII.31. Apparatus for thermobarogravimetry as described by Maycock and Pai Verneker (79).

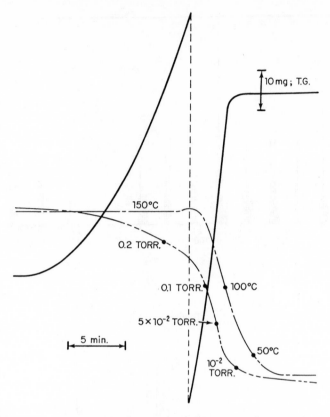

Figure VIII.32. Simultaneous mass-loss, pressure increase, and temperature profile for nitronium perchlorate (79).

An example of the use of this technique is shown in Figure VIII.32. Simultaneous measurements of sample mass-loss and accumulatory dissociation gas pressure were made at 10° intervals in the temperature range from 80 to 150°C. From these data, information concerning the sublimation and decomposition behavior of nitronium perchlorate could be obtained. Other applications of this technique by Maycock and co-workers have been to explosives (80) and ammonium perchlorate (81).

References

1. Stone, R. L., *Anal. Chem.*, **32,** 1582 (1960).
2. Lodding, W., and L. Hammell, *Anal. Chem.*, **32,** 657 (1960).

3. Rogers, R. N., S. K. Yasuda, and J. Zinn, *Anal. Chem.*, **32**, 672 (1960).
4. Vassallo, O. A., *Anal. Chem.*, **33**, 1823 (1961).
5. Ayres, W. M., and E. M. Bens, *Anal. Chem.*, **33**, 568 (1961).
6. Gordon, S., and C. Campbell, *Anal. Chem.*, **27**, 309 (1955).
7. Wendlandt, W. W., *Anal. Chim. Acta*, **27**, 309 (1962).
8. Wendlandt, W. W., and E. Sturm, *J. Inorg. Nucl. Chem.*, **25**, 535 (1963).
9. Murphy, C. B., J. A. Hill, and G. P. Schacher, *Anal. Chem.*, **32**, 1374 (1960).
10. Thomas, J., G. M. Hieftje, and D. E. Orlopp, *Anal. Chem.*, **37**, 762 (1965).
11. Ware, R. K., *Thermochim. Acta*, **3**, 49 (1972).
12. Wendlandt, W. W., and E. L. Dosch, *Thermochim Acta*, **1**, 103 (1970).
13. Wendlandt, W. W., and W. S. Bradley, *Thermochim. Acta*, **1**, 143 (1970).
14. Ingraham, T. R., in *Proceedings of the Second Toronto Symposium on Thermal Analysis*, H. G. McAdie, ed., Chemical Institute of Canada, Toronto, 1967, p. 21.
15. Wist, A. O., in *Thermal Analysis*, R. F. Schwenker and P. D. Garn, eds., Academic, New York, 1969, Vol. 2, p. 1095.
16. Garn, P. D., *Talanta*, **11**, 1417 (1964).
17 Findeis, A F., K. D. W. Rosinski, P. P. Petro, and R. E. W. Earp, *Thermochim. Acta*, **1**, 383 (1970).
18. Ingraham, T. R., and D. Fraser, in *Proceedings of the Third Toronto Symposium on Thermal Analysis*, A. G. McAdie, ed., Chemical Institute of Canada, Toronto, 1969, p. 101.
19. Chiu, J., *Anal. Chem.*, **40**, 1516 (1968).
20. Chiu, J., *Thermochem, Acta*, **1**, 231 (1970).
21. Wiedemann, H. G., in *Thermal Analysis*, R. F. Schwenker and P. D. Garn, eds., Academic, New York, 1959, Vol. 1, p. 229.
22. Chang, T. L., and T. E. Mead, *Anal. Chem.*, **43**, 534 (1971).
23. Wist, A. O., *J. Gas Chromatog.*, March, 157 (1967).
24. Dorsey, G. A., *Anal. Chem.*, **41**, 350 (1969).
25. Franc, J., and J. Pour, *Anal. Chim. Acta*, **48**, 129 (1969).
26. Sarner, S. F., G. D. Pruder, and E. J. Levy, *Am. Lab.*, Oct., 57 (1971).
27. Franc, J., and J. Pour, *J. Chromatog.*, **32**, 2 (1968).
28. Bollin, E. M., in *Thermal Analysis*, R. F. Schwenker and P. D. Garn, eds., Academic, New York, 1969, p. 255.
29. Paulik, J., and F. Paulik, *Thermochim. Acta*, **4**, 189 (1972).
30. Paulik, J., F. Paulik, and L. Erdey, *Mikrochim. Acta*, **1966**, 886.
31. Paulik, J., and F. Paulik, *Talanta*, **17**, 1224 (1970).
32. Paulik, J., and F. Paulik, *Proceedings of the Third Analytical Chemistry Conference*, Budapest, 1970, p. 225.
33. Ref. 32, p. 231.
34. Gal, S., J. Simon, and L. Erdey, Ref. 32, p. 243.
35. Paulik, F., and J. Paulik, *Thermochim. Acta*, **3**, 13, 17 (1971).
36. Low, M. J. D., in *Gas Effluent Analysis*, W. Lodding, ed., Marcel-Dekker, New York, 1967, Chap. 6.
37. Kiss, A. B., *Acta Chim. Acad. Sci. Hung.*, **61**, 207 (1969).
38. Kiss, A. B., *Acta Chim. Acad. Sci. Hung.*, **63**, 243 (1970).
39. Doyle, C. D., WADD Tech. Rept. **60-283**, U.S. Air Force, Wright-Patterson Air Force Base, Ohio, May, 1960.
40. Murphy, C. C., in *Gas Effluent Analysis*, W. Lodding, ed., Marcel-Dekker, New York, 1967, Chap. 7.

41. Murphy, C. B., F. W. Van Luik, and A. C. Pitsas, *Plastics Design and Processing*, July, 16 (1964).
42. Rogers, R. N., *Anal. Chem.*, **39**, 730 (1967).
43. Rogers, R. N., and L. C. Smith, *J. Chromatog.*, **48**, 268 (1970).
44. Stahl, E., *Analyst*, **1969**, 723.
45. Eggertsen, F. T., and F. H. Stross, *J. Appl. Polymer Sci.*, **10**, 1171 (1966).
46. Eggertsen, F. T., H. M. Joki, and F. H. Stross, in *Thermal Analysis*, R. F. Schwenker and P. D. Garn, eds., Academic, New York, 1969, p. 341.
47. Eggertsen, F. T., E. E. Seibert, and F. H. Stross, *Anal. Chem.*, **41**, 1175 (1969).
48. Eggertsen, F. T., and F. H. Stross, *Anal. Chem.*, **44**, 709 (1972).
49. Eggertsen, F. T., and F. H. Stross, *Thermochim. Acta*, **1**, 451 (1970).
50. Friedman, H. L., *Thermochim. Acta*, **1**, 199 (1970).
51. Langer, H. G., and R. S. Gohlke, in *Gas Effluent Analysis*, W. Lodding, ed., Marcel-Dekker, New York, 1967, Chap. 3.
52. Langer, H. G., and R. S. Gohlke, *Anal. Chem.*, **35**, 1301 (1963).
53. Shulman, G. P., *Polymer Letters*, **3**, 911 (1965).
54. Schulman, G. P., and H. W. Lochte, *J. Appl. Polymer Sci.*, **10**, 619 (1966).
55. Schulman, G. P., and H. W. Lochte, *J. Macromol. Sci.* (*Chem.*), **A1**, 413 (1967).
56. Austin, F., J. Dollimore, and B. Harrison, in *Thermal Analysis*, R. F. Schwenker and P. D. Garn, eds., Academic, New York, Vol. 1, p. 311.
57. Brown, J. G., J. Dollimore, C. M. Freedman, and B. H. Harrison, *Thermochim. Acta*, **1**, 499 (1970).
58. Gray, D. N., G. P. Shulman, and R. T. Conley, *J. Macromol. Sci.* (*Chem.*), **A1**, 395 (1967).
59. Gaulin, C. A., F. Wachi, and T. H. Johnston, in *Thermal Analysis*, R. F. Schwenker and P. D. Garn, eds., Academic, New York, 1969, Vol. 2, p. 1453.
60. Wendlandt, W. W., and T. M. Southern, *Anal. Chim. Acta*, **32**, 405 (1965).
61. Langer, H. G., R. S. Gohlke, and D. H. Smith, *Anal. Chem.*, **37**, 433 (1965).
62. Gohlke, R. S., and H. G. Langer, *Anal. Chem.*, **38**, 530 (1966).
63. Langer, H. G., and T. P. Brady, in *Thermal Analysis*, R. F. Schwenker and P. D. Garn, eds., Academic, New York, 1969, Vol. 1, p. 295.
64. Wendlandt, W. W., T. M. Southern, and J. R. Williams, *Anal. Chim. Acta*, **35**, 254 (1966).
65. Redfern, J. P., B. L. Treherne, M. L. Aspimal, and W. A. Wolstenholme, *17th Conference on Mass Spectrometry and Allied Topics*, Dallas, Texas, May, 1969.
66. Zitomer, F., *Anal. Chem.*, **40**, 1091 (1968).
67. Wilson, D. E., and F. M. Hamaker, in *Thermal Analysis*, R. F. Schwenker and P. D. Garn, eds., Academic, New York, 1969, Vol. 1, p. 517.
68. Wiedemann, H. G., Ref. 67, p. 229.
69. Gibson, E. K., and S. M. Johnson, *Thermochim. Acta*, **4**, 49 (1972).
70. Gibson, E. K., *Thermochim. Acta*, **5**, 243 (1973).
71. Gibson, E. K., and S. M. Johnson, *Proceedings of the Second Lunar Science Conference*, M.I.T. Press, Cambridge, Mass., 1971, Vol. 2, p. 1351.
72. Gibson, E. K., and G. W. Moore, *Thermochim. Acta*, in press.
73. Smith, J. W., and D. R. Johnson, in *Thermal Analysis*, R. F. Schwenker and P. D. Garn, eds., Academic, New York, 1969, Vol. 2, p. 1251.
74. Smith, J. W., and D. R. Johnson, *Am. Lab.*, Jan., 8 (1971).
75. Chang, T. L., and T. E. Mead, *Anal. Chem.*, **43**, 534 (1971).
76. Lodding, W., *Gas Effluent Analysis*, Marcel-Dekker, New York, 1967.

77. Kenyon, A. S., in *Techniques and Methods of Polymer Evaluation*, Marcel-Dekker, New York, 1966, Chap. 5.
78. Bancroft, G. M., and H. D. Gesser, *J. Inorg. Nucl. Chem.*, **27**, 1537 (1965).
79. Maycock, J. N., and V. R. Pai Verneker, *Anal. Chem.*, **40**, 1935 (1968).
80. Maycock, J. N., and V. R. Pai Vernecker, *Thermochim. Acta*, **1**, 191 (1970).
81. Pai Vernecker, V. R., M. McCarty, and J. N. Maycock, *Thermochim. Acta*, **3**, 37 (1971).

CHAPTER IX

SPECTROSCOPIC, PHOTOMETRIC, AND OPTICAL THERMAL TECHNIQUES

A. High-Temperature Reflectance Spectroscopy and Dynamic Reflectance Spectroscopy

1. INTRODUCTION

The measurement of the radiation reflected from a mat surface constitutes the area of spectroscopy known as diffuse reflectance spectroscopy. The difference between this technique and transmittance spectroscopy is shown in Figure IX.1. The reflected radiation is expressed as the ratio of I_i/I_r, where I_i is the incident and I_r the reflected radiation. The reflected radiation may be in the ultraviolet, the visible, or the infrared region of the electromagnetic spectrum. From a mat surface, the total reflected radiation, R_T, consists in general of two components: a regular reflectance component R (sometimes referred to as surface or mirror reflection), and a diffuse reflection component R_∞ (1, 2). The former component is due to the reflection at the surface of single crystallites, while the latter arises from the radiation penetrating into the interior of the solid and reemerging to the surface after being scattered numerous times.

According to the Kubelka and Munk theory (3), the diffuse reflection component for 1–3 mm-thick layers of a powdered sample (an increase in thickness beyond this point has no effect on the reflectance) at a given wavelength is equal to

$$R_\infty = \frac{I}{I_0} = \frac{1 - [k/(k + 2s)]^{1/2}}{1 + [k/(k + 2s)]^{1/2}} \tag{IX.1}$$

where I is the reflected radiation, I_0 is the incident radiation, k is the absorption coefficient, and s the scattering coefficient. The absorption coefficient is the same as that given by the familiar Beer-Lambert Law, $T = e^{-kd}$. The regular reflection component is governed by Fresnel's equation

$$R = \frac{I}{I_0} = \frac{(n - 1)^2 + n^2K^2}{(n + 1)^2 + n^2K^2} \tag{IX.2}$$

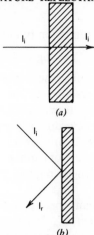

(a)

(b)

Figure IX.1. Reflectance and transmittance spectroscopy. (a) Transmittance mode; (b) reflectance mode.

where n is the refractive index and K is the absorption index, defined through Lambert's law,

$$I = I_0 \exp\left(\frac{-4\pi K d}{\lambda_0}\right) \tag{IX.3}$$

The λ_0 denotes the wavelength of the radiation in vacuum, and d is the layer thickness.

With some algebraic manipulation, equation (IX.1) can be rewritten into the more familiar form

$$\frac{(1 - R_\infty)^2}{2R_\infty} = \frac{k}{s} \tag{IX.4}$$

The left-hand side of the equation is commonly called the *remission function* or the *Kubelka-Munk function* and is frequently denoted by $f(R_\infty)$. Experimentally, one seldom measures the absolute diffuse reflecting power of a sample, but rather the relative reflecting power of the sample compared to a suitable white standard. In that case, $k = 0$ in the spectral region of interest, $R_{\infty \text{ std}} = 1$ [from equation (IX.4)], and one determines the ratio

$$\frac{R_{\infty \text{ sample}}}{R_{\infty \text{ std}}} = r_\infty \tag{IX.5}$$

from which the ratio k/s can be obtained using the remission function

$$f(r_\infty) = \frac{(1 - r_\infty)^2}{2r_\infty} = \frac{k}{s} \tag{IX.6}$$

Taking the logarithm of the remission function gives

$$\log f(r_\infty) = \log k - \log s \qquad \text{(IX.7)}$$

Thus, if $\log f(r_\infty)$ is plotted against the wavelength or wave number for a sample, the curve should correspond to the absorption spectrum of the compound (as determined by transmission measurements) except for the displacement by $-\log s$ in the ordinate direction. The curves obtained by such reflectance measurements are generally called *characteristic color curves* or *typical color curves*. Sometimes there is a small systematic deviation in the shorter-wavelength regions due to the slight increase in the scattering coefficient.

By use of modern double-beam spectrophotometers equipped with some type of a reflectance attachment, r_∞ is automatically plotted against the wavelength. Many investigators replot the data as *percent reflectance* ($\%R$), or plot by use of a remission-function table (4) $f(r_\infty)$ or k/s as a function of wavelength or wavenumber. The most common method is probably the former above.

The above brief introduction to reflectance spectroscopy outlines the most elementary principles of the technique. As would be exected, the technique is widely used for the study of solid or powdered solid samples, although it can be used for liquids or pastelike materials as well. The technique is a rapid one for the determination of the "color" of a sample, and is generally convenient to use since commercial instrumentation is readily available. Since only the surface of the sample is responsible for the reflection and absorption of the incident radiation, it is widely used in the study of the chemistry and physics of surfaces (5).

2. HIGH-TEMPERATURE REFLECTANCE SPECTROSCOPY

Practically all of the studies in reflectance spectroscopy (it should be noted that the term reflectance spectroscopy used here will denote diffuse reflectance spectroscopy only) have been carried out at ambient temperatures, or in some cases at subambient temperatures. The latter would most probably be used in single-crystal studies for the elucidation of "hot bands," that is, transitions which originate from vibrationally excited ground states. However, in many cases, a great amount of additional information on a chemical system can be obtained if the reflectance spectrum of a compound is obtained at *elevated* temperatures. Normally, temperatures in the range from 100 to 300°C have been used, although there is no reason why higher temperatures could not be employed.

Two modes of investigation are used for high-temperature reflectance studies. The first is the measurement of the sample spectra at various fixed

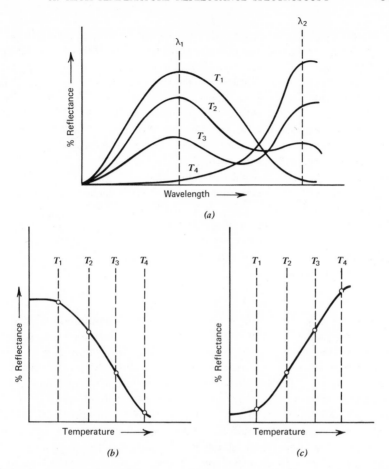

Figure IX.2. (*a*) High-temperature reflectance spectra curves; (*b*) and (*c*) dynamic reflectance spectra curves, at λ_1 and λ_2, respectively (1).

or isothermal temperatures; the second is the measurement of the change in reflectance of the sample as a function of the increasing temperature. The first procedure will be called the static method or *high-temperature reflectance spectroscopy* (HTRS) (6); the second is a dynamic method and has been termed *dynamic reflectance spectroscopy* (DRS) (7, 8). The two methods are illustrated in Figure IX.2. In (*a*), the HTRS curves, the spectra of the sample is recorded at increasing fixed temperatures, T_1 to T_4. As can be seen, the curve maximum at wavelength λ_1 decreases with increasing temperature while a new curve maximum is formed at λ_2. By measuring the spectra at small temperature increments, the minimum temperature at which the sample begins to undergo a thermal transition can be determined. By use of the dynamic technique, these transition temperatures can be determined in a more precise manner, as shown in (b) and (c). Plotting reflectance of the sample versus temperature (b) as the sample temperature is increased at a slow fixed rate, at fixed wavelength λ_1, the reflectance is seen to decrease with an increase in temperature. Using fixed wavelength λ_2, the DRS curve in (c) is obtained which shows the increase in reflectance of the sample with increasing temperature. These *isolambdic* curves reveal the temperatures at which sample thermal transitions begin and end, and also permit the investigation of only a single thermal transition; mass-loss and enthalpic effects do not interfere with the measurements. The DRS technique is useful for determining the thermal stability of a substance, and also sample structural changes which are a function of temperature. Indeed, the technique shows great promise as a complementary method to other thermal techniques such as thermogravimetry, differential thermal analysis, high-temperature X-ray diffraction, and others (45).

3. INSTRUMENTATION

The use of a heated sample holder to contain the compound under investigation has been described by several investigators. Asmussen and Anderson (9) studied the reflectance spectra of several $M_2[HgI_4]$ complexes at various elevated temperatures in order to investigate their thermochromic $\beta \to \alpha$ form transitions. The heated sample container consisted of a nickel-plated brass block, 60 mm in diameter by 85 mm in height, the lower end of which contained a chamber in which a small light bulb was mounted. Regulation of the current through the bulb filament permitted temperature regulation of the block. The upper end of the block contained the sample chamber, which was 35 mm in diameter by 0.5 mm deep. A copper-Constantan thermocouple, embedded in the powdered sample, was used to detect the sample's temperature.

Kortum (2) measured the reflectance spectrum of mercury(II) iodide at 140° but did not describe the heated sample block or other experimental details. Another heated block assembly was described by Hatfield et al. (10). It consisted of a metal block in which a heating element was embedded. No other details are available, such as temperature detection or sample thickness.

In 1963, Wendlandt et al. (11) described the first of their heated sample holders for high-temperature reflectance spectroscopy. The main body of the sample container was 60 mm in diameter by 11 mm thick and was machined from aluminum. The sample itself was contained in a circular indentation, 25 mm in diameter by 1 mm deep, machined on the external face of the cell. Two circular ridges were cut at regular intervals on the indentation to increase the surface area of the holder and to prevent the compacted-powdered sample from falling out of the holder when it was in a vertical position. The sample holder was heated by coils of Nichrome wire wound spirally on an asbestos board and then covered with a thin layer of asbestos paper. Enough wire to provide about 15 Ω of resistance was used. The temperature of the sample was detected by a Chromel-Alumel thermocouple contained in a two-holed ceramic insulator tube. The thermocouple junction made contact with the aluminum block directly behind the sample indentation. To prevent heat transfer from the sample holder to the integrated sphere, a thermal spacer was constructed from a loop of 0.25-in. aluminum tubing and wet shredded asbestos. After drying, the thermal spacer was cemented to the sphere and the sample holder attached to it by a spring-loaded metal clip.

A modification of the above sample holder was described by Wendlandt and George (12) and by Wendlandt (13). The circular aluminum disk of the holder was heated by means of a cartridge heater element inserted directly behind the sample well. Two Chromel-Alumel thermocouples were placed in the block, one adjacent to the heater and the other in the bottom of the sample well so as to be in intimate contact with the compacted sample. The block thermocouple was used to control the temperature programmer, while the sample thermocouple was used to detect the sample temperature.

Still another heated sample holder was described by Wendlandt and Hecht (1). It consisted of a block of aluminum, 50 mm in diameter by 25 mm thick, into which was machined a 25-mm by 1-mm deep sample well A 35-watt stainless-steel sheathed heater cartridge embedded in the main block of the holder was used as the heater. The same two-thermocouple systems, one for the temperature programmer and the other for sample temperature, was employed. For samples which evolved gaseous products, a Pyrex or quartz cover glass was used to prevent contamination of the integrating sphere.

A heated sample holder, based upon the design of Frei and Frodyma (14), which could be used for studying small samples, was recently described by Wendlandt (15). The sample is placed as a thin layer on glass fiber cloth which is secured to the heated aluminum metal block by a metal clamp and a cover glass. Dimensions of the aluminum block are 4.0 by 5.0 cm. The block is heated by a circular heater element contained within the holder. Electrical connections to the heater and to the thermocouple are made by means of the terminal strip mounted at the top of the assembly. Both the aluminum block and the terminal strip are mounted on a 5.0- by 5.0-cm transite block.

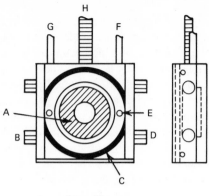

Front side

Figure IX.3. Schematic illustration of the heated sample holder (16). A, Silver sample block and heater; B, glass or quartz cover plate; C, "0" ring; D, thumb screw (one of four); E, gas inlet to sample chamber; F, gas outlet tube; G, gas inlet tube; and H, connecting cable.

Generally, in the previously described heated sample holders, few attempts were made to control the atmosphere surrounding the sample as it was heated. A cover plate of Pyrex glass or quartz was employed but its main purpose was to prevent the sample from accidentally falling into the integrating sphere of the spectroreflectometer. In order to control the sample atmosphere, the sample holder shown in Figure IX.3 was constructed by Wendlandt and Dosch (16).

The sample is contained in a 1- by 10-mm diameter indentation machined in the surface of a silver heater block. The circular block is 25 mm in diameter and it is heated by two 2.6-Ω Nichrome wire heaters. It is contained in an enclosure, 55 mm square and 13 mm thick. The heater is thermally insulated from the main body of the sample holder by a thin layer of ceramic fiber insulation. The sample side of the holder is enclosed by a quartz plate, 50 mm on an edge by 2 mm thick, which is held firmly in place by two metal strips. Each metal strip is fastened to the holder by two thumb screws; they (and hence the cover plate) can easily be removed and to facilitate sample loading and removal. A gas-tight seal between the cover plate and the sample holder is provided by a 44-mm-ID "O" ring. Two 0.125-in.-diameter aluminum tubes, located at the top of the holder, are used to control the gas inlet and outlet to the sample chamber.

With the controlled-atmosphere heated sample holder, it was a simple matter to connect a thermistor-type thermal conductivity cell to the system and, by means of an external multichannel recorder, record the DRS and the

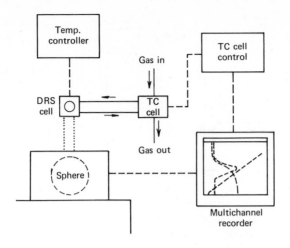

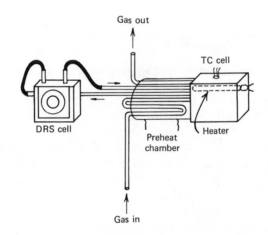

Figure IX.4. Schematic diagrams of DRS–EGD system (17).

evolved gas detection (EGD) curves simultaneously (17).   This modification
of the apparatus is shown in Figure IX.4.   The cell was connected to a Carle
Model 1000 Micro-Detector system by means of metal and rubber tubing.
The thermal conductivity cell was enclosed by an aluminum block which was
heated to 100°C by means of a cartridge heater.   The block was connected
to a preheat chamber, also operated at 100°C, which was used to preheat
the helium gas stream before it entered the detector.   The output from the
detector bridge was led into one channel of a four-channel 0–5 mV Leeds and

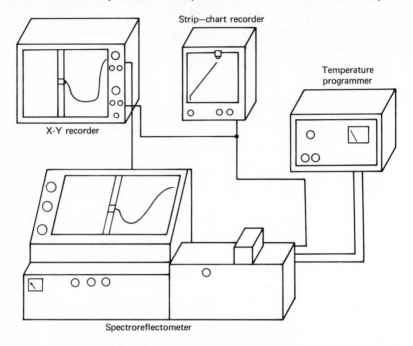

Figure IX.5. Schematic diagram of the HTRS–DRS system (1).

Northrup multipoint strip-chart potentiometric recorder. The temperature programmer from a Deltatherm III DTA instrument was used to control the temperature rise of the DRS cell. Output from the Beckman Model DK-2A spectroreflectometer was also led into the multichannel recorder, as was the output from a thermocouple located in the DRS heater block.

The complete HTRS-DRS system is illustrated in Figure IX.5. The high-temperature sample holder is used in conjunction with a temperature programmer and two recorders. One recorder recorded the sample temperature versus time; the other, an $X-Y$ recorder, was used to record reflectance versus temperature, as required for the DRS studies. A Beckman Model DK-2A or a Bausch & Lomb Spectronic 505 spectroreflectometer was employed for the spectral measurements.

4. APPLICATIONS OF HTRS AND DRS TO INORGANIC COMPOUNDS

a. The Octahedral → Tetrahedral Transition in $Co(py)_2Cl_2$

The *octahedral* → *tetrahedral* structure transition of bis(pyridine) cobalt(II) chloride, $Co(py)_2Cl_2$, has been the subject of a number of investigations (12, 22). Wendlandt and George (22) studied the transition using the techniques of high-temperature reflectance spectroscopy (HTRS) and dynamic

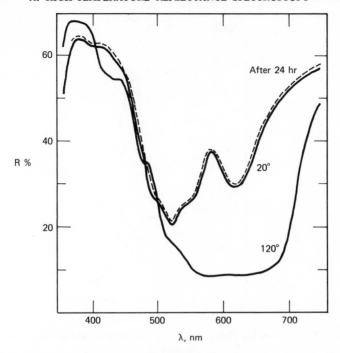

Figure IX.6.   HTRS curves of Co(py)$_2$Cl$_2$ at various temperatures (18).

reflectance spectroscopy (DRS).   The thermal transition was found to begin at about 100° and was completed at about 135°.   The color change reported for the transition was from violet (*octahedral*) to a dark blue (*tetrahedral*) color; it was stated that the change was nonreversible on cooling to room temperature.

Recently, Wendlandt (18) reported that the structural change was actually reversible.   The blue form, after standing at room temperature for 24 hours, reverted to the original violet compound.   The HTRS curves of Co(py)$_2$Cl$_2$ at 20 and 120°C are given in Figure IX.6.   At 20°C, reflectance minima were observed in the curve at 520 and 620 nm, with shoulders at 500 and 550 nm, respectively.   The compound reflected rather strongly in the 350–450-nm region and at 580 nm.   The blue *tetrahedral* form, at 120°, absorbed very strongly in the 500–700-nm region, with shoulders at 425, 480, and 510 nm, respectively.   After standing 24 hours at room temperature, the reflectance curve of the blue form was again recorded at 20°.   As can be seen, the curve obtained was almost identical with that of the original compound having the *octahedral* structure.   Thus it is seen that the *tetrahedral* → *octahedral* transition takes place rather slowly on standing; it does not revert back to the *octahedral* form immediately upon cooling to room temperature.

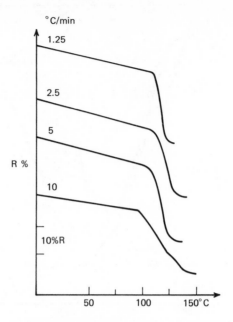

Figure IX.7. DRS curves of $Co(py)_2Cl_2$ at various heating rates; curves recorded at 675 nm (18).

The effect of heating rate on the *octahedral* → *tetrahedral* transition is shown by the DRS curves in Figure IX.7. The heating rate varied from 1.25°/min to 10°/min; the latter value is considered to be rather high for DRS studies. Surprisingly, the procedural transition temperature was greatest (107°) for the 1.25°/min rate and lowest (95°) for the highest rate studied. However, on increasing the heating rate, the reaction temperature interval increased, from 95–145° for the 10°/min rate to 107–123° for the slowest heating rate. This is just the opposite to that observed in dissociation reactions involving volatile products.

The spectra of $Co(py)_2Cl_2$ in the visible and near-infrared regions (15) are shown in Figure IX.8. At room temperature, reflectance minima were found at 1140, 1670, 2150, and 2440 nm, respectively, for the α form. On heating to 125°, all of these minima disappear except for a small minimum at 2440 nm. The β-form curve is practically identical to the curve obtained for *tetrahedral* $Co(py)_2Br_2$.

b. $[Cu(en)(H_2O)_2]SO_4$

The deaquation of $[Cu(en)(H_2O)_2]SO_4$ was studied using HTRS and DRS by Wendlandt (15). This reaction, which takes place between 75 and 150°C,

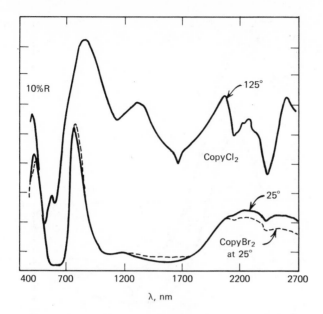

Figure IX.8. HTRS curves of α-Co(py)$_2$Cl$_2$ in the visible and near-infrared wavelength region (15).

follows the equation

$$[Cu(en)(H_2O)_2]SO_4 \text{ (s)} \rightarrow Cu(en)SO_4 \text{ (s)} + 2H_2O \text{ (g)}$$

The HTRS curves, from 25 to 180°C, are shown in Figure IX.9. Two sets of curves are shown, one set at 25° and 75°, and the other at 150° and 180°. The first set has a peak minimum at about 625 nm (corresponding to maximum absorption) and corresponds to the curves for the initial compound, while the second set has a peak minimum at 575 nm and corresponds to the curves for the deaquated compound, Cu(en)SO$_4$. Thus, the deaquation reaction must have occurred between 75 and 150°.

    To obtain the transition temperature for the deaquation reaction, the DRS technique was employed, as shown in Figure IX.10. The transition temperature dependence upon the sample heating rate is readily seen; it varied from 115° at 6.7°C/min to 165° at 45.8°C/min. This behavior is not unexpected, because it occurs with practically all of the other thermal techniques where some physical property of the sample is measured as a dynamic function of temperature (13). In all DRS studies, the sample heating rate obviously must be specified.

    Since the Beckman DK-2A spectroreflectometer is capable of recording the sample spectra in the near-infrared wavelength region, the HTRS curves

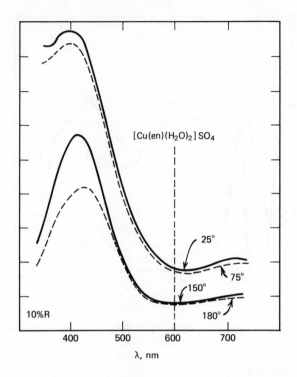

Figure IX.9. HTRS curves of $[Cu(en)(H_2O)_2]SO_4$ (15).

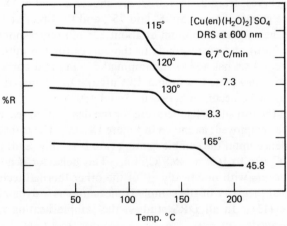

Figure IX.10. DRS curves of $[Cu(en)(H_2O)_2]SO_4$ at 600 nm (15).

372

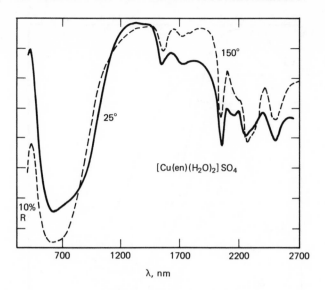

Figure IX.11. Visible and near-infrared reflectance spectra of [Cu(en)(H₂O)₂]SO₄ (15).

were recorded to 2700 nm, as shown in Figure IX.11. At room temperature, the reflectance curve contained minima (absorption bands) at 1560,
1725, 2050, 2150, 2260, and 2500 nm, respectively. On heating the sample
to 150°C, the bands at 1560, 1725, and 2050 nm remained unchanged, while
the 2150 nm disappeared. The 2260-nm band shifted to 2280 nm, and the
2510-nm band shifted to 2525 nm. There were rather pronounced changes
in intensity for all of the bands discussed, which may be due to the sample-
particle size changes.

c. $CuSO_4 \cdot 5H_2O$

The HTRS curves in the visible and near-infrared regions are given in
Figures IX.12 and IX.13, while the DRS curve, at 625 nm, is given in Figure
IX.14.

As in the case of [Cu(en)(H₂O)₂]SO₄, two sets of curves are shown for
$CuSO_4 \cdot 5H_2O$ in Figure IX.12. At room temperature the reflectance minimum occurs at about 680 nm; on heating to 135° the minimum shifts to
715 nm. In this temperature range, the compositional change of the compound is that due to deaquation from $CuSO_4 \cdot 5H_2O$ to $CuSO_4 \cdot H_2O$. At still
higher temperatures, such as 250°C, the last mole of water per mole of copper
sulfate is evolved to give the anhydrous salt. In the near-infrared region at
room temperature, reflectance minima were observed at 1510, 1675, and
2000 nm, respectively. At 150°, the first two bands had disappeared while the
2000-nm band has shifted to 2060 nm. A new band, at 2400 nm, was

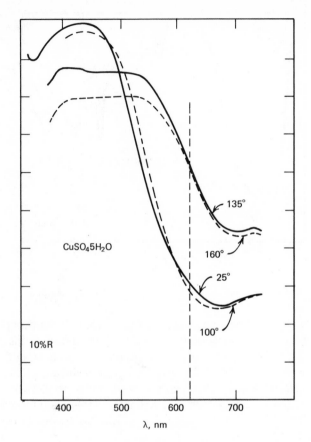

Figure IX.12. HTRS curves of $CuSO_4 \cdot 5H_2O$ in the visible wavelength region (15).

observed at the higher temperature. At still higher temperatures, 200°, all of the bands in this region were absent.

The DRS curve in Figure IX.14, showed that a major increase in sample reflectance began at about 105°, although the reflectance was gradually increased from room temperature up to 100° (about 0.1 a % R unit). At about 125°, the reflectance of the compound again decreased gradually until a maximum temperature of about 200° was attained.

d. $CoCl_2 \cdot 6H_2O$

In an earlier investigation, Wendlandt and Cathers (26) studied the HTRS and DRS of the reaction between $CoCl_2 \cdot 6H_2O$ and KCl. The DRS curve

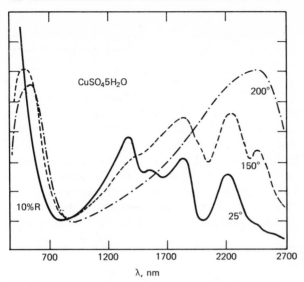

Figure IX.13. HTRS curves of $CuSO_4 \cdot 5H_2O$ in the visible and near-infrared wavelength region (15).

revealed that the following structural and compositional changes occurred: *octahedral*–$CoCl_2 \cdot 6H_2O$ → *tetrahedral*–$CoCl_4^{2-}$ → *octahedral*–$CoCl_2 \cdot 2H_2O$ → *tetrahedral*–$CoCl_4^{2-}$. The above reactions took place, of course, in the presence of an excess of chloride ion; hence, the final product was K_2CoCl_4 rather than anhydrous cobalt(II) chloride.

More recently, the deaquation of $CoCl_2 \cdot 6H_2O$ was investigated in the

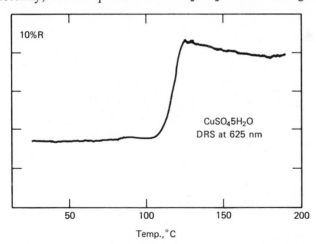

Figure IX.14. DRS curve of $CuSO_4 \cdot 5H_2O$ at 625 nm and a heating rate of 6.7°C/min (15).

absence of potassium chloride.   This compound is a rather difficult one to study because it fuses at about 50°, and since the heated sample holder is mounted in a vertical position on the spectroreflectometer, the liquid $CoCl_2 \cdot 6H_2O$ is impossible to retain on the sample holder.   This problem was solved, however, by placing a thin layer of the powdered sample on a 25-mm-diameter round cover glass which was then retained on the glass fiber cloth by the rectangular cover glass.   The viscous nature of the melt prevented the compound from leaving the sample area.

The HTRS and DRS curves of $CoCl_2 \cdot 6H_2O$ are shown in Figures IX.15 and IX.16, respectively.   The HTRS curves reveal a rather interesting series of structural changes, both in the liquid and solid states.   At 25°, solid $CoCl_2 \cdot 6H_2O$ has an *octahedral* structure with a reflectance minimum at 535 nm and shoulder minima at 460 and 500 nm, respectively.   On heating

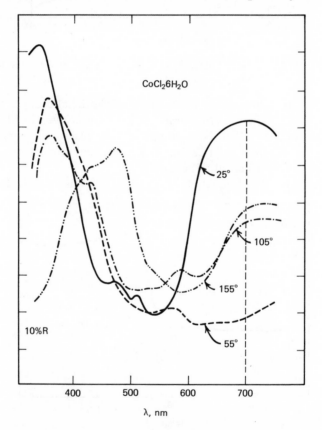

Figure IX.15.   HTRS curves of $CoCl_2 \cdot 6H_2O$ (15).

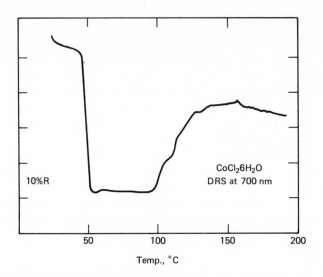

Figure IX.16. DRS curve of $CoCl_2 \cdot 6H_2O$ at 700 nm and a heating rate of 6.7°C/min (15).

the compound to 55°, it fused and gave a reflectance curve which had one minimum at 525 nm and a rather broad minimum between 600 and 700 nm. This latter curve is similar to the one previously observed for a mixture of *octahedral–* and *tetrahedral–*cobalt(II) complexes by Simmons and Wendlandt (27). Thus, a possible interpretation would be that the 55° curve is probably a mixture of *octahedral–*$CoCl_2 \cdot 6H_2O$ and *tetrahedral–*$Co[CoCl_4]$. On further heating, the mixture underwent further deaquation and gave, at 155°, anhydrous *octahedral–*$CoCl_2$. This latter curve contained a peak minimum at 590 nm, with a shoulder minimum at 535 nm.

The DRS curve, Figure X.16, showed a pronounced decrease in reflectance at 45° which was due to the formation of the *octahedral–tetrahedral* mixture. At 100°, the reflectance of the mixture began to increase, reaching a maximum value at about 150°, then decreasing slightly above this temperature. The curve reflects the various structural changes that have been discussed previously.

e. $Ni(py)_4Cl_2$

Yang (28) studied the HTRS of the deamination of $Ni(py)_4Cl_2$, the curves of which are illustrated in Figure IX.17. The spectrum at 25°C is that for the initial compound, $Ni(py)_4Cl_2$. From 125 to 175°C, two moles of pyridine per mole of complex are lost, so that the spectrum at 175°C is that for the complex $(Ni(py)_2Cl_2$. From 175 to 275°C, another pyridine is evolved, so that the 275°C spectrum is that for $Ni(py)Cl_2$. The loss of pyridine and

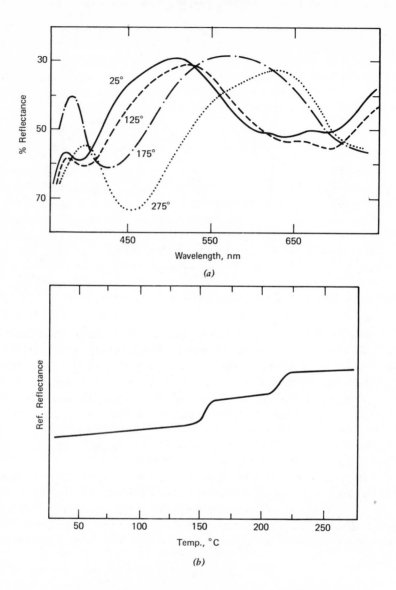

Figure IX.17. (a) HTRS of $Ni(py)_4Cl_2$ in an Al_2O_3 (50%) matrix. (b) DRS curve of $Ni(py)_4Cl_2$ in an Al_2O_3 (50%) matrix recorded at 450 nm (28).

the changes in the reflectance of the initial complex are shown in the 450-nm DRS curve in Figure IX.17. The transition, $Ni(py)_4Cl_2 \rightarrow Ni(py)_2Cl_2$, began at 145° and was completed at 160°C; the loss of an additional pyridine began at 210°C and was completed at 220°C. The increase in slope throughout the DRS curve was due to the increasing sample temperature.

f. Thermochromism of $Ag_2[HgI_4]$

The thermochromism of $Ag_2[HgI_4]$ has been of great interest since its first preparation by Caventou and Willm (29) in 1870. The transition was first investigated in a thorough manner by Ketelaar using specific heat, X-ray, and electrical conductivity techniques (30–33). Additional information concerning the color changes (34, 36, 39), dilatometry (35), crystal structure (37, 38), magnetic susceptibility (9), electrical conductivity (39, 40), and thermal stability (41) of the compound has been reported. The compound has been proposed as a temperature indicator (36, 42) and as a pigment for temperature-indicator paints (42–44).

The thermochromism of $Ag_2[HgI_4]$ is due to an order-disorder transition which involves no less than three phases. According to Ketalaar (33), both the yellow low-temperature β modification and the red high-temperature α form contain iodide ions which are cubic close-packed, while the silver and mercury ions occupy some of the tetrahedral holes. The β form has tetragonal symmetry, with the mercury ion situated at the corners of a cubic unit cell and the silver ions at the midpoints of the vertical faces. As the temperature is increased it becomes possible for the silver and mercury ions to occupy each other's lattice sites and also the two extra lattice sites (top and bottom face centers of the unit cube) which were unoccupied at lower temperatures. Above 52°C, the mercury and silver ions are completely disordered. The α modification has, therefore, averaged face-centered cubic symmetry. More recently, magnetic (39) and dielectric polarization (37, 39) measurements confirm the presence of a third phase, the β' modification. With an increase in temperature, the silver ions become disordered, occupying at random $\frac{2}{3}$ of the face-centered positions of the unit cube during the $\beta \rightarrow \beta'$ transition. During the $\beta' \rightarrow \alpha$ transition, the silver and mercury ions become further disordered at random $\frac{3}{4}$ of the corners plus face centers of the unit cell (37) corresponding to two cubic (but not isotropic) cells stacked one on top of the other. The Patterson function suggests that a portion of the silver atoms are disordered, having left sites surrounded tetrahedrally by iodide ions and appearing in interstitial (octahedrally occupied) sites. The interstitial silver ions would be expected to be rather labile, since the octahedral holes are large compared to those at the tetrahedral sites. This is also apparent from

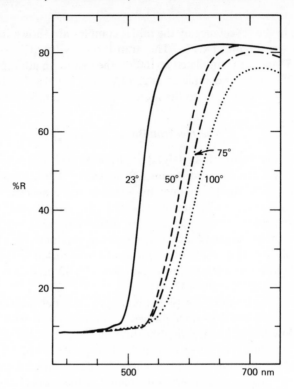

Figure IX.18. HTRS curves of $Ag_2[HgI_4]$ (45).

the low activation energy obtained (37) for the conduction process in β-β-$Ag_2[HgI_4]$, 12 kcal/mole below 20°C.

The reflectance curves for $Ag_2[HgI_4]$ from 23 to 100°C are shown in Figure IX.18. The yellow β form reflects rather strongly above 500 nm, with the maximum shifting to higher wavelengths during the transition to the red α form. The change in color is dependent upon the rate of heating. At 2.5°C/min the transition is completed at a somewhat lower temperature than at the 10°C/min heating rate. This heating rate is extremely rapid compared to the temperature rise of 5°C/day used by Neubert and Nichols (39) in their magnetic studies. The transition temperature found here was not very well defined in that the color change appeared to take place over the temperature range from 30 to about 60°C. Reported transition temperatures include 50.7 ± 0.2, 51.2, 51, 50.5, and 52°C.

The DRS curves of a number of $M_n[HgI_4]$ complexes (M = Pb, Cu, Hg, Ag, and Tl) are given in Figure IX.19. All the compounds exhibit rather sharp thermochromic transition, with the exception of $Tl_2[HgI_4]$. The latter

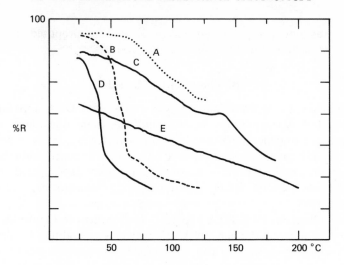

Figure IX.19. DRS curves of complexes (5°C/min). A, Pb[hgI$_4$], 585 nm; B, Cu$_2$[HgI$_4$], 650 nm; C, Hg[HgI$_4$], 600 nm; D, Ag$_2$[HgI$_4$], 575 nm; E, Tl$_2$[HgI$_4$], 550 nm (45).

compound is reported to have a transition at 116.5°C (9); however, it is not evident from the DRS curve. The change in reflectance of the compound appears to decrease linearly with temperature.

g. Thermal Matrix Reactions

The techniques of HTRS and DRS were used by Wendlandt and co-workers (46–49) in the investigation of reactions between chromium(III) and cobalt(III) ammine complexes and ammonium salts (thermal matrix reactions). Such a reaction is illustrated by (46)

$$[Cr(en)_3]Cl_3 + NH_4Cl(excess) \rightarrow cis\text{-}[Cr(en)_2Cl_2]Cl + \text{other products} \qquad (IX.9)$$

A 1:1 mass ratio of [Cr(en)$_3$]Cl$_3$ and NH$_4$X (X = fluoride, chloride, bromide, iodide, and thiocyanate) were heated up to 200°C in a high-temperature sample holder. The HTRS mode was used to identify the reaction products, while the DRS mode was used to determine the temperature range at which the reaction took place. As a result of these studies, new synthetic procedures were developed for the preparation of cis-[Cr(en)$_2$X$_2$]X, cis-[Cr(pn)$_2$X$_2$]X, and cis-[Cr(tm)$_2$X$_2$]X complexes (pn = 1, 2-propanediamine and tm = 1,3-propanediamine).

Chang and Wendlandt (47–49) investigated several series of compounds of the types (a) cis- and trans-[Co(en)$_2$(H$_2$O)$_2$](NO$_3$)$_2$ + NH$_4$X and (b) cis- and trans-[Co(NH$_3$)$_4$(H$_2$O)$_2$](NO$_3$)$_3$ + NH$_4$X. In all cases, a trans reaction product was obtained. Various mechanisms were proposed as well as synthetic procedures for the *trans* isomeric products.

B. Photometric Methods

A simultaneous photothermal analysis (PTA) and DTA apparatus has been described by David (50). It was found after examining a large number inorganic and organic compounds that the two techniques provided complimentary information but that the PTA curves contained features which were not present in the DTA curves. Compounds investigated included limestone, clay, $CuSO_4·5H_2O$, polystyrene, polyvinylchloride, and several other polymers.

The simultaneous PTA-DTA apparatus is illustrated in Figure IX.20. It consisted of a Stone-Premco DTA cell which was modified by drilling a 0.275-in. opening in the cell cap and furnace chamber to permit sample

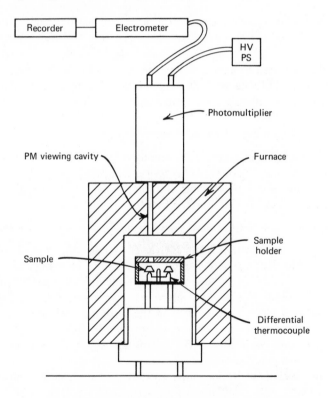

Figure IX.20. Simultaneous PTA–DTA apparatus (50).

viewing by the photomultiplier tube. With an "end-on" photomultiplier tube mounted on the top of the furnace, the sample in the ring thermocouple container could be viewed directly. Using a two-channel recorder, one channel recorded the output from the phototube and the other recorded the DTA curve.

In the case of several limestone samples that were examined, the DTA curves revealed small, broad exothermic peaks in the temperature range from 300 to 375°C, which coincided with the major glow peaks in the PTA curves. David concluded that DTA appeared to be as sensitive as or perhaps more sensitive than PTA for examining these types of materials and perhaps would be better suited to studies involving geological dating than thermoluminescent analysis. The lack of sensitivity in the PTA curves may be due to the slow heating rate that was employed (10°C/min). Typical thermoluminescent curve determinations are made at heating rates from 10°C/min to 16°C/sec. In the case of polymeric samples, David (50) found that melting and/or glass transitions were not detected by PTA but decomposition reactions were observed. Thus, the DTA-PTA technique permitted differentiation between these types of transitions. A PTA curve peak began at approximately the temperature where oxidation began, much like that observed in oxyluminescence. This behavior was observed even though the samples were heated in a flowing nitrogen atmosphere.

A somewhat more sophisticated apparatus was described by Rupert (51) for observing phase transitions of incandescent materials. In this apparatus, a photomultiplier tube was used to follow the temperature changes of the heated sample. The phototube responds to the luminosity of the sample, which is proportional to the temperature. It is not necessary to know the exact relationship between the output of the phototube and the temperature of the sample because temperature calibration is accomplished by using a calibrated optical pyrometer which receives part of the light from the sample. The sample was contained in a crucible located within the current concentrator. The latter receives power from an induction heater whose output is controlled by an induction heater controller. Light from the sample emerges through a 0.070-in.-diameter hole in the top of the crucible, and travels upward through a Pyrex or quartz window into the lower end of a beam splitter. Part of the light is reflected at approximately 90° to the axis of the beam splitter, by a partially aluminized mirror, to the optical pyrometer used to measure the sample temperature. Typical cooling curves studied by this technique include the freezing of a molybdenum carbide–carbon mixture at 2540°C, a cooling curve of the freezing of zirconium carbide–carbon eutectic mixture at 2855°C, and a solid-state transition of uranium dicarbide at 1800°C.

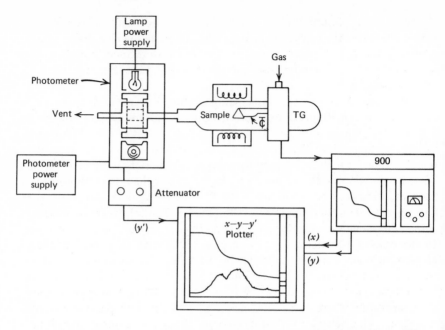

Figure IX.21. Simultaneous TG-smoke density apparatus of Loehr and Levy (52).

An apparatus to measure simultaneously smoke density and TG was de-scribed by Loehr and Levy (52).   The apparatus, as shown in Figure IX.21, consisted of a DuPont 950 thermogravimetric analyzer coupled to a DuPont 410 precision photometer.   The latter employed a split-beam (sample versus reference) system, with the light transmission being detected by a S-5 type photocell coupled to a tungsten-iodine light source.   These com-ponents were chosen to give the closest facsimile to human eye response. Output from the photocell was attenuated to read linearly in optical density by the photometer control console.   The smoke detection chamber con-sisted of a 1-ml stainless-steel enclosure with sapphire windows having a dead volume of 0.59 cubic inches.

Two different flame retardants in the same urethane foam were evaluated to determine the amount of smoke from each, as well as their flame retarding abilities.   Simultaneous TG–smoke-density curves for these two samples, in an air atmosphere, are shown in Figure IX.22.   Sample 1 decomposed leaving 10% residue at 450°C, with a maximum optical density at the same temperature.   Under the same conditions, sample 2 had a residue of 25% with an increase in optical density of ~20% over that of sample 1.   The latter sample is indicated to be a better flame retardant in that the amount

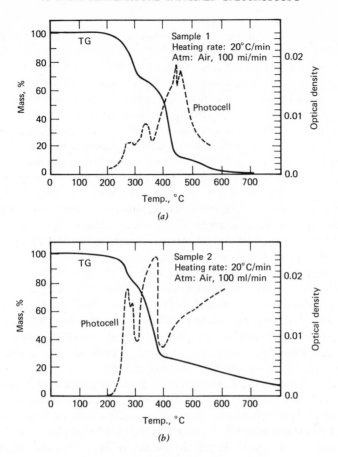

Figure IX.22. Simultaneous TG-smoke density curves of two urethane samples (52); (a) Sample 1 in air; (b) Sample 2 in air.

of residue at any given temperature is greater. However, it produces a greater amount of smoke at lower temperatures.

C. High-Temperature Infrared Spectroscopy

Although the KBr disk technique in infrared spectroscopy is well known, few quantitative kinetic studies of solid-state chemical reactions in this medium have been reported. Hisatsune and co-workers (53–58) found that many chemical reactions were initiated by heating the disks to elevated temperatures and that the kinetics of these reactions could be conveniently followed by this technique. The disks were placed in an oven heated at the desired temperature for a specific period of time, and the IR spectrum of the disk was recorded after it was quenched to room temperature. A typical disk weighing about 0.5 g cooled from about 600°C to room temperature in less than a minute. Disks prepared from potassium salts could be heated in air to about 600°C, but above this temperature appreciable sublimation of the matrix salt occurred. Initial heating usually produced the greatest change in the appearance of the disks. They turned opaque, expanded, and often showed blisters on the surface when gaseous products were formed by the decomposition of the solutes. In some cases, the transparency of the disk could be restored by breaking it into small pieces and repressing. For quantitative studies the rim of the expanded disk was sanded off until it fitted the die cavity. Studies reported included the trapping of the BO_2^- ion (54), the carbon dioxide anion (CO_2^-) free radical (55), the carbonate anion (CO_3^-) free radical (56), the formate ion from the acetate ion (57), and the decomposition of the perchlorate ion (58).

The thermal decomposition kinetics of silver carbonate using the disk technique was reported by Wydenen and Leban (59). Continuous, *in situ* quantitative analysis of infrared active reactants and products of the decomposition reactions was made possible by use of a heated cell. The cell was constructed of stainless steel and could be heated to 500°C with the KRS-5 cell windows maintained at room temperature by cooling water. A similar approach was used by Wendlandt (60).

A heated temperature-programmable IR cell was used by Tanaka et al. (61) to study the thermal decomposition of a number of cobalt(III) ammine complexes. The disk matrix material was either KCl or KBr, and it was reported that they frequently became opaque to infrared radiation at elevated temperatures. A heated IR cell for use up to 200°C was also described by LeRoux and Montano (62).

D. Thermal Optical Microscopy Techniques

1. FUSION MICROSCOPY

The term "fusion microscopy" includes the methods and procedures that involve the heating of a compound or a mixture of compounds on a microscope slide (64). It includes all observations made during the heating of the preparation (description, sublimation, decomposition, melting, etc.) on the melt itself (refractive index, boiling point, critical solution temperature, etc.), the solidification of the melt (crystal angles, birefringence, rate of growth, etc.), and the cooling (polymorphic transformations, orthoscopic and conoscopic observations, composition diagrams, etc.). The most important applications appear to be (*a*) characterization and identification of pure compounds, (*b*) determination of purity, (*c*) analysis of binary mixtures, (*d*) determination of composition diagrams for binary and ternary systems, (*e*) elucidation of phase diagrams, and studies of (*f*) polymorphism, (*g*) crystal growth kinetics, and (*h*) crystal-lattice strain (63, 64). Identification of a fusible compound by this technique is very rapid, consumes only small quantities of material, and requires relatively little specialized training or equipment. The purity of a fusible compound is, in many cases, very quickly determined by fusion methods since impurities usually are visible as high-melting material, as liquid eutectic, or as early-melting material. A mixed fusion gives a rapid and dependable means of determining whether two given samples are the same compound. Complete binary and ternary composition diagrams can be determined usually in a few hours' time. Recent reviews on this subject include those by Vaughan (65), Smith (66), Sommer (67, 68), and McCrone (63).

The various hot stages employed have been reviewed by McCrone (63, 64); the most successful appears to be that described by Kofler. It was first described in the 1930s and became commercially available in 1940. This instrument was successful mainly because it was reliable; temperatures indicated by the thermometer could be dependably associated with the temperatures of the preparation under investigation. A much more sophisticated instrument, the Mettler FP-2 hot stage, which was introduced in 1968, is schematically shown in Figure IX.23. In the control system, the temperature is detected by a platinum resistance thermometer placed in very close proximity to the sample. Power to the heating elements is proportionally controlled by the difference between the resistance of the resistance thermometer and the resistance of a motor-driven potentiometer. As long as the actual temperature and the required program temperature are in agreement with each other, the rotation of the program signal-generator motor will be a linear function of the temperature. Digital temperature readout is presented by connecting the program signal motor to a series of counter wheels. A choice of three heating rates may be preselected: 0.2, 2.0, and 10.0°C/min; the temperature range is from $-20°$ to 300°C. Temperature measurement and control accuracy is $\pm 0.1°C$ below 100°C and $\pm 0.1\%$ above this temperature. The hot stage, which can be fitted to any standard microscope, consists of an adjustable sample carrier, two metal heating plates in which heating wires and the platinum thermometer are embedded, and a compact blower for circulating air within the chamber. The design of the hot stage is such that the sample is heated simultaneously from above and below and is thus maintained in a uniform-temperature field.

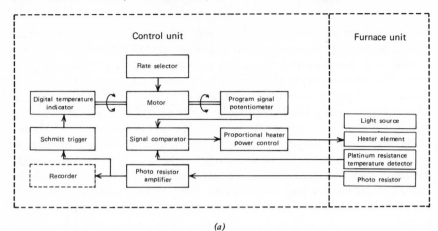

(a)

Figure IX.23. The Mettler FP-2 hot stage. (a) Schematic diagram of the control system; and (b) schematic diagram of hot stage.

For detailed application of fusion microscopy to chemical and physical problems, the work of McCrone (63, 64) should be consulted.

2. DEPOLARIZED LIGHT INTENSITY AND PHOTOMETRIC THERMAL MICROSCOPY

In the usual procedures used in fusion microscopy, thermal changes in the sample are visually observed; photomicroscopic techniques may be employed to obtain a permanent record of the thermal events. Because the eye is not particularly sensitive to subtle changes taking place over a period of several minutes, automatic devices have been used to detect these changes. One method is to use some type of photocell to measure changes in the sample light transmission or intensity of birefringence (45, 65, 69–78). Another method is to use an infrared scanning camera such as described by Hyzer (79). The former is primarily used for the accurate determination of melting points, polymorphic transitions, and crystallization rates, while the latter more sophisticated method can reveal the minute differences in temperature in the area under investigation.

The apparatus used for recording the changes in light transmission of a sample usually consists of the following components (65, 69, 73): (a) photo-detector such as a photocell, photomultiplier tube or photo-resistor; (b) hot stage; (c) temperature programmer; (d) microscope; (e) some type of sample holder; and (f) recorder (usually an X-Y type). The output of the photodetector is recorded as a function of sample temperature as the sample is being heated at rates from 0.2 to 30°C/min. The sample preparation (69) for light-transmission methods consisted of placing it either melted between cover slips or mounted in a silicone oil and then covering with a cover slip. The oil cuts down light refraction and permits improved birefringence intensity measurements. For crystalline or partially crystalline polymers, an embedding technique may be used. The polymers are embedded in a thin layer (0.5 mm) of high-viscosity polydimethylsiloxane polymer spread over the microscope slide. Still another technique (65) is to use glass capillary tubes to contain the sample. Powdered samples are normally packed to a height of 3–4 mass in the bottom of the tube, while fats, oils, and waxes can be easily loaded into the tubes in their melted state by means of a long-needled syringe.

The photometric heating and cooling curves of ammonium nitrate (73) are illustrated in Figure IX.24. The four polymorphic transitions between 25 and 200°C are clearly indicated. The *orthorhombic → monoclinic* transition at 42° is observed as an increase in light intensity, while the *monoclinic → tetragonal* transition at 78°C is accompanied by a decrease in intensity. A sharp decrease in intensity indicates the *tetragonal → cubic* transition and then the fusion of the cubic form at 169°C. On cooling, the melt crystallizes to the cubic form at 175°C. Several large changes of light intensity are shown as the ammonium nitrate undergoes the various phase transitions. All of these measurements were made under polarized light.

The technique of depolarized light intensity (DLI) microscopy was introduced by Magill (80) in 1960. Basic elements of the apparatus were a light source, polarizers, a sample holder, an analyzer, and a suitable recording system. Barrall and Johnson (74) and Miller (75, 76) have described applications of this technique to polymeric samples. Miller (75) prefers to call this technique *thermal polarization analysis* (TPA).

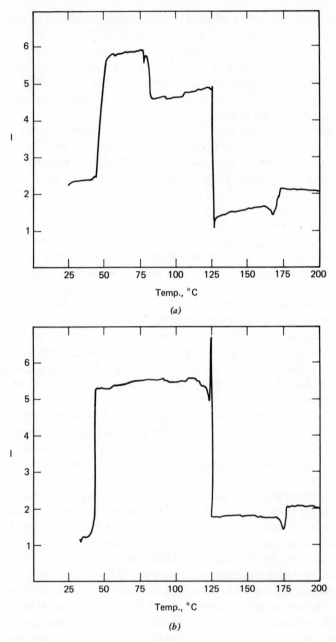

Figure IX.24. Photometric curves of ammonium nitrate (73); (a) Heating curve; (b) cooling curve.

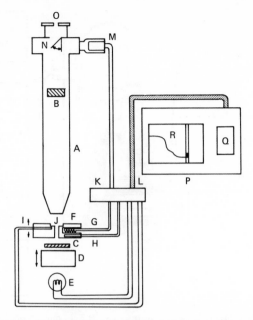

Figure IX.25. Schematic diagram of depolarized light intensity apparatus (74). A, Unitron MPS Microscope; B, Analyzer Sheet; C, Polarizer Sheet; D, Movable Abbe Condenser; E, Tungsten Lamp; F, Movable Hot Stage; G, Heater; H, Program Thermocouple; I, Sample Temperature Thermocouple; J, Sample Well; K, Pin Plug Terminal Attached to Microscope Stand; L, Shielded Cable to Du Pont Module; M, Photocell; N, Movable Mirror; O, Ocular; P, Du Pont 900 Differential Thermal Analyzer Module; Q, Programmer; R, X–Y Recorder.

The DLI apparatus used by Barrall and Johnson (74) is illustrated in Figure IX.25. They found it convenient to use the recording system from a DuPont 900 DTA module. The intensity of the depolarized light was measured with a photoconductive cell used in conjunction with a Wheatstone bridge circuit. A hot stage, constructed from a copper block, could be used in the temperature range from −40 to 600°C at a heating rate of about 5°C/min or in an isothermal operational mode. For faster heating rates, a low-thermal-mass furnace was used which contained a platinum heater and was usable up to 800°C.

Wendlandt (45) used a microscopic method for the determination of the reflectance of the sample. The apparatus, as shown in Figure IX.26, consisted of a low-power (100×, generally) reflection-type microscope, A, which is illuminated by means of a monochromator, B. The reflected radiation is detected by a photomultiplier tube, C, and amplifier, D, and recorded either on an $X–Y$ recorder, E, or a strip-chart recorder, F. In order to heat the

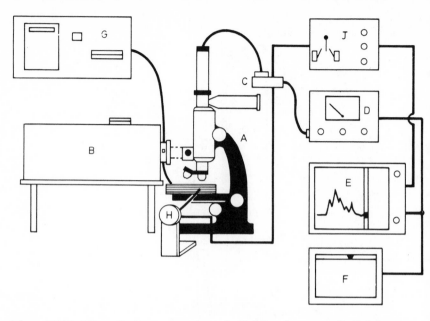

Figure IX.26. Microreflectance apparatus. A, B & L microscope; B, B & L mono-chromator and lamp; C, photomultiplier tube; D, amplifier and power supply; E, X-Y recorder; F, strip-chart recorder; G, Mettler hot stage; H, reversible motor; J, relay and timer (45).

sample to 250°C, a Mettler Model FP-2 hot stage, G, is employed. Either isothermal ($\pm 1°$C) or dynamic sample temperatures may be attained by this device. The sample is moved through the illuminated optical field by means of the reversible motor, H. The motor is reversed at preset intervals by a relay circuit and timer, J. Thus, it is possible to scan the reflectance from the sample, which may consist of a single crystal or a powdered mixture. Powdered samples may be placed directly on the heated microscope slide or else placed in 0.0–1.1-mm-ID glass capillary tubes. In the latter case, it is possible to obtain the reflectance curve of a sample contained in a sealed tube.

Two modes of operation of the apparatus are possible: (*a*) The scanning mode in which the sample surface reflectance can be recorded as a function of scanning distance at ambient room temperature or at elevated temperatures. (*b*) The change in reflectance of the sample as a function of temperature can be recorded. The former mode is called *high-temperature scanning micro-reflectance* (SMR), while the latter is called *dynamic microreflectance* (DMR). The use of these two modes is illustrated by the deaquation of $CuSO_4\cdot5H_2O$.

The scanning microreflectance curves of a single crystal of $CuSO_4\cdot5H_2O$ (78) at room temperature are shown in Figure IX.27. The curves represent

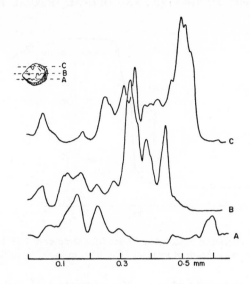

Figure IX.27. Scanning microreflectance curves of a single crystal of $CuSO_4 \cdot 5H_2O$ at a wavelength of 450 nm and at a temperature of 30°C (100× magnification) (78).

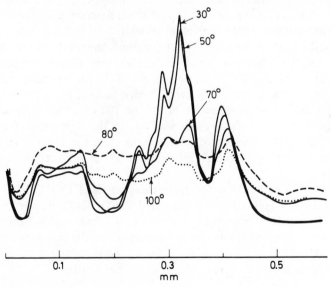

Figure IX.28. Scanning microreflectance curves of a single crystal of $CuSO_4 \cdot 5H_2O$ at various temperatures; wavelength of incident light 450 nm, 100× magnification (78).

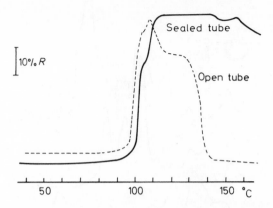

Figure IX.29. Microreflectance of sealed and open tube samples of $CuSO_4 \cdot 5H_2O$; heating rate of 10°C/min; wavelength of incident light, 450 nm; 80× magnification (78).

the reflectance of the crystal surface at scans at points A, B, and C. Since the reflectance geometry of 90°/90° was used, the curve maxima represent maximum specular reflectance from the crystal surface. Thus, surfaces perpendicular to the incident beam reflect the strongest, giving the curve peak maxima. The curves are not very reproducible from crystal to crystal due to the different surfaces of the individual crystals.

The SMR curves of the same crystal at various temperatures are illustrated in Figure IX.28. The curves changed little on increasing the temperature of the crystal from 30–50°C. However, at 70°C, the specular reflectance maxima all showed a general decrease, which became more pronounced as the temperature was increased from 80° to 100°C. The decrease in the specular reflectance of the crystals was due to the formation of a surface layer of $CuSO_4 \cdot 3H_2O$ which is more opaque than the original compound. Thus, the formation of the former can easily be followed by the SMR technique.

The evolution of liquid water in the deaquation of $CuSO_4 \cdot 5H_2O$ is illustrated by the microreflectance of powdered $CuSO_4 \cdot 5H_2O$ in sealed and open glass capillary tubes, as shown in Figure IX.29. In an open capillary tube, the sample reflectance of the samples decreased as the temperature was increased. However, since the liquid water was confined to the capillary tube, the reflectance did not increase again on further heating.

References

1. Wendlandt, W. W., and H. G. Hecht, *Reflectance Spectroscopy*, Interscience, New York, 1966, Chaps. 3 and 4.
2. Kortum, G., *Trans. Faraday Soc.*, **58**, 1624 (1962).

3. Kubelka, P., and F. Munk, *Z. Techn. Physik*, **12,** 593 (1931).
4. Ref. 1, pp. 275–279.
5. Ref. 1, Chap. 8.
6. Wendlandt, W. W., P. H. Franke, and J. P. Smith, *Anal. Chem.*, **35,** 105 (1963).
7. Wendlandt, W. W., *Science*, **140,** 1085 (1963).
8. Anon., *Chem. Eng. News*, April 15, 1963, p. 62.
9. Asmussen, R. W., and P. Anderson, *Acta Chem. Scand.*, **12,** 939 (1958).
10. Hatfield, W. E., T. S. Piper, and U. Klabunde, *Inorg. Chem.*, **2,** 629 (1963).
11. Wendlandt, W. W., P. H. Franke, and J. P. Smith, *Anal. Chem.*, **35,** 105 (1963).
12. Wendlandt, W. W., and T. D. George, *Chemist-Analyst*, **53,** 100 (1964).
13. Wendlandt, W. W., *Thermal Methods of Analysis*, Interscience, New York, 1964, Chap. 10.
14. Frei, R. W. and M. M. Frodyma, *Anal. Chim. Acta*, **32,** 501 (1965).
15. Wendlandt, W. W., in *Modern Aspects of Reflectance Spectroscopy*, W. W. Wendlandt, ed., Plenum, New York, 1968.
16. Wendlandt, W. W., and E. L. Dosch, *Thermochim. Acta*, **1,** 103 (1970).
17. Wendlandt, W. W., and W. S. Bradley, *Thermochim. Acta*, **1,** 143 (1970).
18. Wendlandt, W. W., *J. Thermal Anal.*, **1,** 469 (1970).
19. Cox, E. G. A. J. Shorter, W. Wardlaw, and W. J. R. Way, *J. Chem. Soc.*, **1937,** 1556.
20. Dunitz, J. D., *Acta Cryst.*, **10,** 307 (1957).
21. Mellor, D. P. and C. D. Coryell, *J. Am. Chem. Soc.*, **60,** 1786 (1938).
22. Wendlandt, W. W., and T. D. George, *Chemist-Anal.*, **53,** 71 (1964).
23. Ocone, L. R., J. R. Soulen, and B. P. Block, *J. Inorg. Nucl. Chem.*, **15,** 76 (1960).
24. Beech, G., C. T. Mortimer, and E. G. Tyler, *J. Chem. Soc.*, **1967,** 925.
25. Murgulescu, I. G., E. Segal, and D. Fatu, *J. Inorg. Nucl. Chem.*, **27,** 2677 (1965).
26. Wendlandt, W. W., and R. E. Cathers, *Chemist-Analyst*, **53,** 110 (1964).
27. Simmons, E. L., and W. W. Wendlandt, *J. Inorg. Nucl. Chem.*, **28,** 2187 (1966).
28. Yang, W. Y., unpublished results.
29. Caventou E., and E. Willm, *Bull. Soc. Chim. Fr.*, **13,** 194 (1870).
30. Ketelaar, J. A. A., *Z. Krist.*, **87,** 436 (1934).
31. Ketelaar, J. A. A., *Z. Physik. Chem.*, **B26,** 327 (1935).
32. Ketelaar, J. A. A., *Z. Physik. Chem.*, **B30,** 35 (1938).
33. Ketelaar, J. A. A., *Trans. Faraday Soc.*, **34,** 874 (1938).
34. Suchow, L. and P. H. Keck, *J. Am. Chem. Soc.*, **75,** 518 (1953).
35. Thomas, D. G., L. A. K. Staveley, and A. F. Cullis, *J. Chem. Soc.*, **1952,** 1727.
36. Bachman, C. H., and J. B. Maginnis, *Am. J. Phys.*, **19,** 424 (1951).
37. Olsen, C. E., and P. M. Harris, *Phys. Rev.*, **86,** 651 (1952); U.S. Department of Commerce, Office, Tech. Ser., PB Dept. 156, 106, 61 pages (1959).
38. Hahn, H., G. Frank, and W. Klinger, *Z. Anorg. Allgem. Chem.*, **279,** 271 (1955).
39. Neubert, T. J., and G. M. Nichols, *J. Am. Chem. Soc.*, **80,** 2619 (1958).
40. Rothstein, J., *Phys. Rev.*, **98,** 271 (1955).
41. Heintz, E. A., *J. Inorg. Nucl. Chem.*, **21,** 64 (1961).
42. Andrews, W. S., *Gen. Elec. Rev.*, **29,** 521 (1926).
43. Perez, H. G., *Quim. Ind. Sao Paulo*, **4,** 137 (1936).
44. Horiguchi, Y. T. Funayama, and T. Nakanishi, *Sci. Papers Inst. Phys. Chem. Res. Tokyo*, **53,** 274 (1959).
45. Wendlandt, W. W., *Pure and Appl. Chem.*, **25,** 826 (1971).
46. Wendlandt, W. W., and C. H. Stembridge, *J. Inorg. Nucl. Chem.*, **27,** 575 (1965).
47. Chang, F. C., and W. W. Wendlandt, *J. Inorg. Nucl. Chem.*, **32,** 3535 (1970).

48. Chang, F. C., and W. W. Wendlandt, *Thermochim. Acta*, **2**, 293 (1970).
49. Chang, F. C., and W. W. Wendlandt, *Thermochim. Acta*, **3**, 69 (1971).
50. David, D. J., *Thermochim. Acta*, **3**, 277 (1972).
51. Rupert, G. N., *Rev. Sci. Instr.*, **34**, 1183 (1963).
52. Loehr, A. A., and P. F. Levy, *Am. Lab.*, Jan., 11 (1972).
53. Hisatsune, I. C., *Perkin-Elmer Instrument News*, **16**, No. 2, p. 2 (1965).
54. Hisatsune, I. C., and N. Haddock Suarez, *Inorg. Chem.*, **3**, 168 (1964).
55. Hartman, K. O., and I. C. Hisatsune, *J. Chem. Phys.*, **44**, 1913 (1966).
56. Hisatsune, I. C., T. Adl, E. C. Beahm, and R. J. Kempf, *J. Phys. Chem.*, **74**, 3225 (1970).
57. Hisatsune, I. C., E. C. Beahm, and R. J. Kempf, *J. Phys. Chem.*, **74**, 3444 (1970).
58. Hisatsune, I. C., and D. G. Linnehan, *J. Phys. Chem.*, **74**, 4091 (1970).
59. Wydeven, T., and M. Leban, *Anal. Chem.*, **40**, 363 (1968).
60. Wendlandt, W. W., and J. P. Smith, *Thermal Properties of Transition Metal Ammine Complexes*, Elsevier, Amsterdam, 1967, p. 35.
61. Tanaka, N., M. Sato, and M. Nanjo, *Sci. Reports Tohoku Univ.*, **48**, 1 (1964).
62. LeRoux, J. H., and J. J. Montano, *Anal. Chem.*, **38**, 1808 (1966).
63. McCrone, W. C., *Mettler Technical Information Bulletin*, No. 3003, 1968.
64. McCrone, W. C., *Fusion Methods in Chemical Microscopy*, Interscience, New York, 1957.
65. Vaughan, H. P., *Thermochim. Acta*, **1**, 111 (1970).
66. Smith, R. V., *Am. Lab.*, Sept., 85 (1969).
67. Sommer, G., *Instr. Tech. Southern Africa*, **2**, 10 (1965).
68. Sommer, G., and P. F. Jochens, *Minerals Sci. Eng.*, **3**, 3 (1971).
69. Kolb, A. K., C. L. Lee, and R. M. Trail, *Anal. Chem.*, **39**, 1206 (1967).
70. Vaughan, H. P., *Microscope*, **17**, 71 (1969).
71. Reese, D. R., P. N. Nordberg, S. P. Ericksen, and J. V. Swintosky, *J. Pharm. Sci.*, **50**, 177 (1961).
72. Hock, C. W., and J. F. Arbogast, *Anal. Chem.*, **33**, 462 (1961).
73. Faubion, B. D., *Anal. Chem.*, **43**, 241 (1971).
74. Barrall, E. M., and J. F. Johnson, *Thermochim. Acta*, **5**, 41 (1972).
75. Miller, G. W., *Thermochim. Acta*, **3**, 467 (1972).
76. Miller, G. W., in *Analytical Calorimetry*, R. S. Porter and J. F. Johnson, eds., Vol. 2, Plenum, New York, 1970, p. 397.
77. Bruckner, H. P., and K. Heide, *Z. für Chemie*, **10**, 125 (1970).
78. Wendlandt, W. W., *Thermochim. Acta*, **1**, 419 (1970).
79. Hyzer, W. G., *Research/Development*, Feb., 61 (1972).
80. Magill, J. H., *Nature*, **187**, 770 (1960).

CHAPTER X

CRYOSCOPIC PURITY DETERMINATION

A. Introduction

The purity of organic and inorganic compounds can be determined by a number of techniques ranging from simple physical methods of boiling- and melting-point determinations to more sophisticated instrumental methods such as absorption or emission spectroscopy. An attempt is made to summarize the principal instrumental techniques for purity (or impurity) determinations in Table X.1. The figures cited are not very accurate and may vary widely, depending on the main component as well as the impurities present. The first six methods may each show a number of contaminants in one single experiment and permit the determination of each of them. Electrical conductance permits the estimation of ions in aqueous or non-aqueous solutions as well as the ionic components in semiconductors. Although the latter is a rather limited technique, it does approach the optimum purity-control method for group contaminants.

The method offering the widest potential for the determination of the purity of a substance is *thermal analysis* (1). It is applicable to all substances which are sufficiently stable at their melting points and permits the determination of the total quantity of impurity not soluble in the solid phase. Thermal analysis may be defined as a method for the determination of the amount of contaminant(s) in a substance from an analysis of thd temperature-time or temperature-heat content curves at its melting point. Glasgow and Ross (38) prefer to use the broad term of "cryoscopy," which they define as the science of the determination of temperatures, from solid–liquid equilibria, of the freezing points of liquids and of the melting points of solids, and the uses of such measurements for analytical purposes. The terms "freezing point" and "melting point" are commonly accepted (38) as referring to the temperature where an infinitesimal amount of solid is in equilibrium with liquid when the measurements are performed in equilibrium with air at one atmosphere.

Various methods have been used to determine the temperature–time or temperature–heat content curves of a substance. They include the following:

(a) *Thermometric methods* (not to be confused with thermometric titrations) in which temperature–time curves are obtained at various intervals. Heat evolution or absorption occurs continuously and preferably at a constant rate. The amount of heat supplied per unit time is not measured

TABLE X.1
Methods of Impurity Determination of Chemical Compounds (1)

Method	Sensitivity[a]	Substance	
		Organic	Inorganic
Emission and X-ray spectroscopy	10^{-4}–10^{-5}		x
Activation analysis	10^{-1}–10^{-8}	x	x
Polarography	10^{-6}	x	x
Mass spectroscopy	10^{-3}–10^{-5}	x	x
Chromatography	10^{-5}	x	x
Absorption spectroscopy	10^{-3}	x	
Electrical conductance	10^{-8}		x
Thermal analysis	10^{-5}	x	x

[a] Smallest fraction of impurity still detectable.

directly but may be calculated as a fraction of the total heat of melting of the substance.

(b) *Calorimetric methods* in which temperature–heat-content curves are obtained. An adiabatic calorimeter or a differential scanning calorimeter may be employed. The latter instrument is much more convenient to use and is capable of almost the same accuracy and precision as the former technique.

(c) *Dilatometry* (volume-temperature curves) and *dielectric constant*. The latter method appears to be as accurate as the other techniques, and, by virtue of use of an extensive property, it is not influenced by the amount of material used. These methods will not be discussed here.

Various reviews on the subject of thermal analysis as a means of purity determination of organic compounds have been published by Sturtevant (2), Cines (3), Mathieu (4), and Smit (1, 5), Glasgow and Ross (38), Skau and Arthur (39), and others (40–42).

B. Theory

The treatment of temperature–heat-content curves from a theoretical viewpoint has been carried out by a number of investigators, starting with White (6) in 1920. Other early papers on the subject are by Andrews et al. (7), Skau (8), Mair et al. (9), Glasgow et al. (10), Malotaux and Straub (11), and Thomas and Parks (12). More recent treatments have been given by Rossini (13), Mastrangelo and Dornte (14), Badley (15), and Smit (1). The reader is referred to the above for a more comprehensive presentation than that given here.

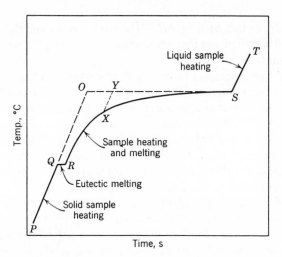

Figure X.1. Melting-temperature curve for two component system; ———, actual equilibrium curve; – – – –, idealized process, heating without melting (specific-heat component) (16).

To analyze the temperature versus time curve, the curves in Figure X.1 will be discussed. The basic analysis of this curve has been described by White (6) and modified by Carleton (16). The resulting temperature versus time curve is based on the linear relationship of heat input to time and to the equation

$$N_2 = \frac{\Delta H}{RT^2}(T_0 - T) = A(T_0 - T) \qquad (X.1)$$

where N_2 is the mole fraction of solute, ΔH is the heat of fusion of the solvent, T_0 is the freezing point of pure solvent, and T is the equilibrium temperature. This equation is restricted to those examples which are nearly pure and in which, on freezing, the pure major component solidifies, leaving the impurity in solution.

The linear relationship of heat input to time follows from the maintenance of a constant temperature interval between the sample and the bath. During melting, this heat input has two components: a specific-heat component which raises the temperature of the sample and the thermometer bulb, and a melting component. The separation of these two components on a time scale is illustrated in Figure X.1. The initial straight-line portion, PQ, represents the heating of the solid sample and thermometer bulb. In the case of a two-component system, there is a flattening of the curve at the eutectic temperature, QR. Above this eutectic, melting and heating occur simultaneously, as indicated by RS. The curve arches and flattens out until all of the solid is melted and then the slope changes abruptly, where heating of the liquid sample begins, at ST.

The dashed line represents an idealized process in which all solid is heated to the freezing point and then all the sample is melted isothermally. The two lines, QO and OS, are thus separate specific-heat and melting components for the actual process. The flat line, OS, represents the melting of the two substances, solvent and solute, with different heats of fusion. However, for most substances studied, the amount of solute is always small so that on the central part of the curve, which is used for analysis, the fraction of material melted is proportional to the distance along OS.

Analysis of a temperature versus time curve depends on the construction of several projections, such as XY, from the actual curve to the ideal flat line, or, the sample slope as that of PQ could be used with negligible error. However, a run usually begins at a temperature at which some melting is already under way, so that the slope of the separate specific-heat component is not known. When the properties of the sample are known, a slope for XY may be approximated from the dimensions of the apparatus, the specific heat, the heat of fusion of the sample, and a rough estimate of the impurity, assuming ideal solution behavior. Usually, however, these properties will not be known, but either the properties or the slope may be estimated with sufficient accuracy to be useful.

If x represents the mole fraction of impurity in the original sample and T_1 the freezing point of the impure sample (temperature at point S), then equation (X.1) gives

$$N_2 = A(T_0 - T) \tag{X.2}$$

$$x = A(T_0 - T_1) \tag{X.3}$$

whence

$$N_2 = x + A(T_1 - T) \tag{X.4}$$

If f is the mole fraction of the solvent frozen at temperature T,

$$f = \frac{N_2 - x}{N_2} = \frac{A(T_1 - T)}{x + A(T_1 - T)} = \frac{A\,\Delta T}{x + A\,\Delta T} \tag{X.5}$$

If Δt represents the time difference from the point of complete melting, YS, and Δt_1 the total time represented by the ideal-melting flat line, OS, then

$$f = \frac{\Delta t}{\Delta t_1} \tag{X.6}$$

Equations (X.5) and (X.6) are rearranged to

$$\Delta t = \Delta t_1 - \frac{x}{A}\left(\frac{\Delta t}{\Delta T}\right) \tag{X.7}$$

Thus, the theory predicts that a plot of Δt against $\Delta t/\Delta T$ will give a straight line whose slope is $-x/A$. The term A is equal to

$$A = \frac{\Delta H}{RT^2} \tag{X.8}$$

and is a characteristic property of the major component in the sample. When ΔH is not known, A may be determined from an additional run on the sample containing a known mole fraction of added solute. Alternatively, ΔH may be estimated by comparing the curve for the sample with a curve for a reference substance of known heat of fusion, obtained in the same apparatus. Then x is the product of A and x/A determined from the slope of the curve.

For the temperature–heat-content curves, a similar expression can be derived. Rossini (13) has shown that the thermodynamic relation for equilibrium between a liquid phase, of the major and minor components, and a crystalline phase of the major component alone is given by

$$\ln N_1 = \ln (1 - N_2) = -A(T_0 - T)[1 + B(T_0 - T) + \cdots] \tag{X.9}$$

where N_1 and N_2 are the mole fractions of the major and minor components, respectively, in the liquid phase. The temperature T_0 is the freezing point of the pure major component ($N_1 = 1$), and T is the equilibrium temperature for the mixture. The quantity A is given by equation (X.8), while B is the other cryoscopic constant

$$B = \left(\frac{1}{T_0}\right) - \left(\frac{\Delta C_p}{2\,\Delta H}\right) \tag{X.10}$$

where C_p is the molar heat capacity of the liquid less that of the solid. For highly purified samples, T approaches T_0 and N_2 approaches zero, so that equation (X.9) can be written as

$$N_2 = A(T_0 - T) \tag{X.11}$$

If N_2 is the mole fraction of impurity in the liquid phase for a fraction F of the sample liquid, then

$$N_2 = N_2^* \frac{1}{F} \tag{X.12}$$

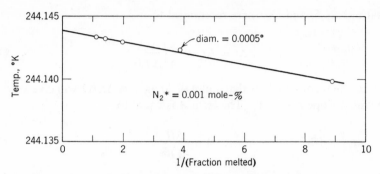

Figure X.2. The melting curve of benzotrifluoride (17).

where $N_2{}^*$ is the mole fraction of impurity in the sample. Combining equations (X.11) and (X.12) gives

$$T = T_0 - \left(\frac{N_2{}^*}{A}\right)\left(\frac{1}{F}\right) \qquad (X.13)$$

A plot of T versus $(1/F)$ will give a straight line of slope $-(N_2{}^*/A)$ and the intercept at $(1/F) = 0$ will be T_0. Thus, from the slope of the line, the purity of the sample can be determined. Such a curve of $(1/F)$ versus temperature for benzotrifluoride is given in Figure X.2.

The above procedure is based upon the assumptions that (17) (a) the values of T are thermodynamic equilibrium temperatures, (b) an ideal solution is formed in the liquid phase, (c) the impurity is insoluble in the solid phase, and (d) $N_2{}^*$ is very much less than one. Departure from linearity in a plot of T versus $(1/F)$ may be taken as an indication that one or more of these basic assumptions is not met fully.

Gunn (37) has proposed for quantitative purity determination another method, which can also be used for the estimation of heat capacities and heats of fusion of the sample.

Assuming Newton's law, the heat transfer to the sample is

$$\frac{dH}{dt} = k(T_b - T_s) \qquad (X.14)$$

where T_b is the temperature of the block, T_s is the temperature of the sample, and k is the heat-transfer coefficient. The rate of temperature change of the sample is

$$\frac{dT_s}{dt} = \frac{k(T_b - T_s)}{nC_s + C_g} \qquad (X.15)$$

where n is the number of moles of sample, C_s is its molar heat capacity, and C_g is the heat capacity of the glass bulb, sample well packing, and part of the thermocouple. If T_b is increased at a constant rate, r, T_s will approach asymptotically and follow a parallel time-versus-temperature line such that

$$\frac{dT_s}{dt} = \frac{dT_b}{dt} = r \qquad (X.16)$$

displaced in temperature at a given time by a thermal heat, h:

$$h = T_b - T_s = r(nC_s + C_g)/k \qquad (X.17)$$

and displaced in time at a given temperature by a lag, l:

$$l = \frac{T_b - T_s}{r} = \frac{nC_s - C_g}{k} \qquad (X.18)$$

Thus, l is a function only of the heat capacity and k, but h is also a function of heating rate, r:

$$h = rl \qquad (X.19)$$

The treatment for purity determinations used by Gunn (37) assumes that k is a constant, but that its value need not be known; likewise, the values of n, C_s, and C_g need not be known.

In Figure X.3 is illustrated an idealized melting curve to be described by this treatment. Curve AB represents the block temperature, increasing at an approximately constant rate, r. Curve CGF represents the sample temperature, t, being selected before this temperature departs from a line parallel to AB. From equation (X.14), it follows that the heat transferred to the sample in warming it from T_i to T_f is

$$H_f - H_i = k\int_{t_i}^{t_f} (T_b - T_s)\, dt = k(ABFGC) \qquad (X.20)$$

If no latent heat were associated with the fusion, the sample would warm along path $CDEF$, where T_0 is the melting point and the curves CD and EF are separated from AB by the different lags, l_c and l_1, which reflect the different heat capacities of the solid and liquid, that of the solid generally being lower. The absorption of heat would be $k(ABFEDC)$; hence, the molar heat of fusion, ΔH, of the sample is

$$n\Delta H = k(ABFGC) - k(ABFEDC) = k(CDEFG) \qquad (X.21)$$

The area, $CDEFG$, will be called Z; in practice, it is evaluated by graphical integration, that is, by dividing the area into several easily measured triangles which cover an area judged visually to be equal.

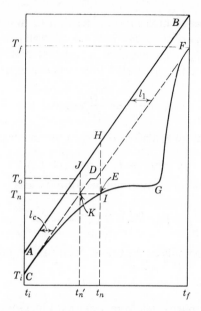

Figure X.3. Idealized melting curve (37).

The heat transferred to the sample to warm it from T_i to T_n is $k(AHIC)$, and is denoted as W.   The heat required to warm the solid from T_i to T_n in the absence of melting would be $k(AJKC)$, and is denoted as X.   Instead of integrating X graphically, it is noted that

$$X = rl_c(t_n' - t_i) = h_c(t_n' - t_i) \qquad (X.22)$$

The amount of heat which has been used to melt part of the sample at time t_n is $k(W - X)$; the quantity $W - X$ is denoted as Y.   The reciprocal of the fraction of the sample melted, F^{-1}, is

$$F^{-1} = \frac{Z}{Y} \qquad (X.23)$$

where F^{-1} may be calculated for as many points as desired on the melting curve.

For the ideal or sufficiently dilute solutions, the van't Hoff law of freezing-point lowering has the form

$$T = T_0 - \frac{N_2 F^{-1} R T_0^2}{\Delta H} \qquad (X.24)$$

where T_0 is the melting point of the pure material and N_2 is the mole fraction of impurity.   Hence, the values of T_n plotted against F^{-1} should lie on a straight line whose slope multiplied by the cryoscopic constant, $RT^2/\Delta H$, is equal to N_2.

C. Experimental Techniques

For thermal analyses by the static method, a precise adiabatic calorimeter is required. Although many adiabatic calorimeters have been described in the literature, Glasgow et al. (18) have described a calorimeter which was used to determine the purity of benzene and other substances in the temperature range from 10 to 300°K. The calorimeter is illustrated schematically in Figure X.4.

The sample container, suspended in the calorimeter by a small tube, was constructed of copper and had a capacity of about 106 ml. Tinned copper vanes were arranged radially from the central reentrant well, containing a heater and a platinum resistance thermometer, to the outer wall of the container. The vanes were held in place by means of a thin coating of tin. A thin copper thermal shell was attached to the upper periphery of the container to obtain a nearly isothermal surface. The outer surface of the container, the inner and outer surfaces of the shell, and the inner surface of the adiabatic shield were gold plated and polished to minimize heat transfer by radiation. A high vacuum, 10^{-6} Torr, was maintained in the space surrounding the sample container and the adiabatic shield.

The resistance of the platinum thermometer was measured by means of a Mueller bridge. The electrical input energy was determined from the measurements of the current and potential across a 100 Ω Constantan wire heater and the time interval of heating. The heater current and potential were measured by means of a Wenner potentiometer in conjunction with a resistor and a volt box. The time interval of heating was measured by means of a precise interval timer.

The calculations involved in the determination of the specific heat of a sample have been described by Stull (19). During a heat input, an electric current of I amperes flowed through the sample heater because of a voltage e impressed on the heater terminals for t seconds. The heat in calories, H, is then

$$H = \frac{Iet}{4.1840} \tag{X.25}$$

This heat input caused the temperature of the sample to go from its initial state, T_i, before the heat was applied, to T_f, the final temperature of the sample after the sample had reached a constant temperature. Thus, $T_f - T_i = \Delta T$, the rise in temperature due to H, and $\frac{1}{2}(T_f - T_i) = T_a$, the average temperature of the space heat input.

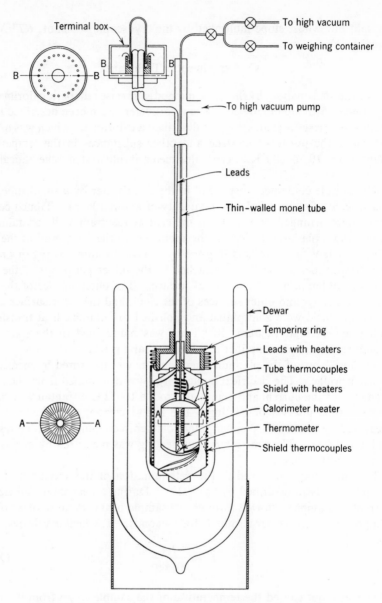

Figure X.4. Adiabatic calorimeter for volatile compounds (18).

Now heat was absorbed by the sample container of weight w grams and specific heat C_{pc} at T_a, as well as by the sample of W grams and specific heat C_{ps} at T_a. Expressed mathematically,

$$H = [wC_{pc} + WC_{ps}] \Delta T \qquad (X.26)$$

and combining equations (X.14) and (X.15),

$$\frac{Iet}{4.1840WT} - \frac{W}{W} C_{pc} = C_{ps} \qquad (X.27)$$

Equation (X.27) is the basic equation used to calculate the heat capacity of the sample. By slight modification of the equation, the calculations can be made by an electronic digital computer.

Other calorimeters that have been used for melting determinations have been described by Clarke et al. (20), Aston and Fink (21), Pilcher (22), Mazee (23), and Ruehrwein and Huffman (24).

In the case of the dynamic method, the constant heat supply to the sample is obtained by maintaining a constant thermal head between the sample and its surroundings. This may be done by two different methods: (*a*) by a constant-wall apparatus, and (*b*) by an adapted-wall apparatus (1). A constant-wall apparatus maintains a constant temperature between the wall of the sample container and the sample. In an adapted-wall apparatus, a constant temperature between the wall of the sample container and the sample. In an adapted-wall apparatus, a constant heat supply to the sample is also maintained when the sample is surrounded by a mantle and its temperature is continuously adapted to the temperature of the sample in such a way that the difference between both temperatures remains constant.

The various constant- and adapted-wall apparatuses have been summarized by Smit (1). The former type have been built by White (6) and by Rossini and co-workers (9, 10). Instruments of the latter type have been described by Thomas and Parks (12), Malotaux and Straub (11), Carleton (16), Smit and Kateman (25), Smith (26, 27), Glasgow and Tenenbaum (28), Glasgow et al. (18), Handley (29), and Barnard-Smith and White (30).

The applications of the constant-wall instruments are mainly for the determination of cooling or freezing curves, and not for heating or melting curves. This is probably because when heat must be transported to the sample, the outer wall of the apparatus, and thus the isolating mantle, must be at a temperature much higher than when heat must be transported from the sample (1). Since due to radiation the isolating power of a vacuum jacket decreases rapidly at increasing temperatures, it is clear that the thermal head for heating a sample at a permissable rate will be lower than the opposite thermal head for cooling the sample at the same rate.

Depending upon the temperature range to be covered, the wall of the adapted-wall instruments consists of a glass bulb immersed in a liquid bath or a thick cylindrical mantle made of metal. The temperature of the bath or of the metal mantle is adapted to the temperature of the sample so that the difference of the two temperatures remains constant. Between the wall and the measuring vessel containing the sample, there is an air space which provides the necessary insulation. The thermal gradient or difference usually amounts to about 2°C, and the rate of heating of the sample is quite low, about 0.1–0.3 °C/min. This type of apparatus is not stirred.

A simple apparatus of the adapted-wall type has been described by Carleton (16) and is a modification of the apparatus described by Smit (26). The apparatus is schematically illustrated in Figure X.5.

The enclosure of the sample is in the form of a thin, uniform film surrounding the bulb of a 0.1°C graduated mercury thermometer. The thermometer was positioned by means of a bored cork in a glass sample tube drawn to the proper dimensions in the portion surrounding the thermometer bulb. To reduce the effects of temperature fluctuations, the sample tube was jacketed with a slightly larger tube retained by a plastic ring. The entire sample assembly was placed in a 300-ml round-bottomed flask, in such a position that the thermometer bulb was at the approximate center of the flask. The flask was immersed to the neck in a suitable heating bath which was provided with a stirrer, thermostat, and thermometer. The volume of

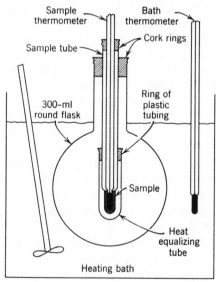

Figure X.5. Apparatus for determination of melting temperatures (16).

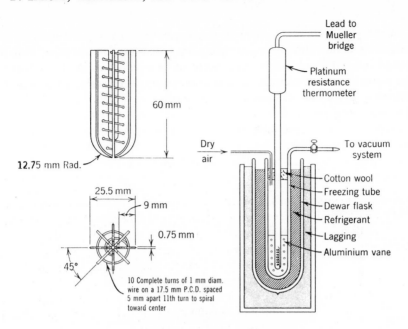

Figure X.6. Freezing-point apparatus (30).

sample required for a determination was about 0.3 ml. The outside bath was heated at a rate of 0.3°C per 100 s or per minute. A plot of sample temperature versus time was started at 15 to 20°C below the melting point of the substance.

Another apparatus which was similar to that described previously by Glasgow et al. (10), and modified by Barnard-Smith and White (30), is schematically illustrated in Figure X.6. The sample, usually about 25 ml, was frozen and melted in a double-walled tube, the rate of heat transfer from the refrigerant or heating bath to the sample being controlled by the vacuum between the walls of the tube. A rotating stirrer was used and, for smaller samples, an aluminum tube was inserted to reduce the volume of the sample chamber. The temperature of the sample was measured by a platinum resistance thermometer and a Mueller bridge. The instrument could also be used for heat-of-fusion measurements by insertion of a series of aluminum vanes. These vanes assisted in the even distribution of heat throughout the sample.

D. Errors, Limitations, and Other Factors Affecting Results

The errors in the determination of temperature versus heat content or time curves have been discussed in detail by Smit (31) and McCullough and

Waddington (17). The former discussed the qualitative consideration concerning the rates of phase transitions, the rates of diffusion, and the temperature differences occurring with the "thin-film" method. The latter were concerned with the limitations of the calorimetric method based on the results of more than 125 melting-point studies.

1. LIMITATIONS OF THE DYNAMIC METHOD

a. Solid–Liquid Transitions

When heat is supplied to a system, its temperature will increase until the net rate of melting equals the rate of heat supply. When heat is withdrawn from the system, its temperature will decrease until the rate of heat production by crystallization equals the rate of cooling. This is only possible below a temperature T_m, the temperature at which thermodynamic equilibrium exists. However, in this region, the rate of heat production may be low and may increase only slightly at decreasing temperatures. Consequently, the temperature obtained may differ appreciably from the equilibrium temperature. The temperature finally obtained remains constant so long as the rate of heat production of the system equals the rate of cooling. This is illustrated by the heating and cooling curves of antipyrine containing 9.1-mole-% acetanilide, as shown in Figure X.7. The heating curve of pure antipyrine showed a range of constant temperature at 110.45°C, which appeared to be independent of the rate of heating. The cooling curve determined at a

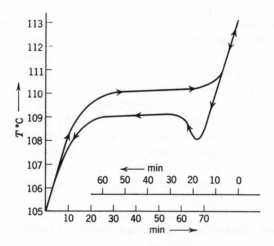

Figure 7. Heating and cooling curves of a sample of antipyrine containing 0.1 mole-% acetanilide. Upper curve is the heating curve (31).

comparable rate of cooling showed a small undercooling peak and then the temperature raised to a maximum at about 109°C. The height of the maximum was dependent on the rate of cooling. The curves obtained by heating and cooling the antipyrine and 0.1-mole-% acetanilide mixture were comparable to those obtained with the pure antipyrine.

Similar analogous behavior has been observed with azobenzene, benzyl benzoate, and p-xylene, and slight differences with naphthalene.

b. Solid-Phase Transitions

A heating curve usually shows the existence of *solid → solid* or enantiotropic transitions. When melting occurs before the *solid → solid* transition is completed, the melting curve will obviously be unreliable. Smit (31) recommended that the sample be stored for a period of time at a temperature above the transition temperature before determination of the melting curve.

c. Rates of Diffusion

When the solid and liquid of a multicomponent system are in thermodynamic equilibrium, the composition of the solid will usually differ from that of the liquid. When the system is submitted to further melting or crystallization, the composition of at least one of the phases will change in the vicinity of the contact surface. Diffusion tends to equalize the concentration differences occurring both in the solid and in the liquid phases and should, therefore, be promoted.

d. Effect of Stirring

Stirring promotes the homogeneity of the liquid phase only and does not affect the inhomogeneities occurring in the solid phase. Thus, even when stirring is applied, thick layers may be disadvantageous. Stirring is an advantage at times in that it may cause disintegration of solid particles which may promote the bulk rate of crystallization. The advantage is rather dubious, according to Smit (31), because stirring can only be applied over a limited range of solid–liquid ratios.

e. Rate of Heat Transport

When heat is supplied or withdrawn from a calorimeter, temperature differences will occur throughout all parts of the calorimetric system, including the wall, the sample, and even the thermometer. These differences constitute a source of errors, the magnitude of which depend on the rate of heating, the sizes of the system components, and the heat conductivities of the construction materials.   The magnitude of these errors has been calculated by Smit (31).

f. Temperature Differences During Melting

Melting, of course, starts at the inner wall of the sample container and subsequently proceeds to the thermometer bulb.   As soon as melting starts, the flow of heat to the thermometer decreases appreciably.   It is not reduced to zero, however, because the temperature of the thermometer is below the temperature of the melting zone.   The difference between the temperature indicated by the thermometer and the temperature of the melting zone constitutes an error which gradually decreases with time.   The deviations will be large at the start of the melting process and gradually approach zero as the curve is continued.   It is important to know within what time this error has decreased to a value not exceeding the limit of accuracy of the determination.   An attempt has been made by Smit (31) to calculate this exact time.

g. Influence of Contact Between Layers

Contact is never perfect between the glass wall, the sample, and the thermometer bulb.   This imperfect contact can give rise to extra temperature differences.   Heat can flow from the environment along the stem to the bulb of the thermometer and subsequently to the sample.   The temperature of the thermometer will be high when imperfect contact exists between the sample and the bulb.

2. LIMITATIONS OF THE STATIC METHOD

The limitations of the static method undoubtedly apply to a greater or lesser extent to any melting-point method.   These limitations are as follows (17).

a. Uncertainty of Impurity Values

Inhomogeneous distribution of impurity in the liquid phase may result in low values of N_2^* because the slope of the melting curve is usually decreased by this effect.   A more important source of error, formation of solid solutions has long been recognized as a possible limitation of all melting-point purity methods, but it has not been realized that the phenomenon is so common.

b. Evidence of Solid-Solution Formation

It is not unreasonable to expect that solid solutions may be formed in highly purified samples, for the impurities may often be isomeric with the main component.   About half of the melting curves observed by McCullough and Waddington (17) showed moderate to pronounced deviation from linearity of the T versus $1/F$ plots, indicative of solid-solution formation. In fact, linear melting curves over the entire range of fractions melted are rare.   Both the formation of solid solutions and inhomogeneous distribution melting-point studies to be too low.   The calculations of impurity values from the slope of the melting curve at high fractions melted will minimize errors in most cases.

c. Application of Solid-Solution Theory

If a melting curve shows evidence of appreciable solid-solution formation, it may require application of a solid-solution treatment (14, 15) to give an accurate impurity value, although Smit (1) has criticized one of the treatments (14).   Unfortunately, the method often has failed to give an adequate representation of observed melting curves.   In some instances, the solid-solution treatment has given an excellent representation of experimental data, but the high sensitivity of the method to small thermometric errors makes the calculated impurity values unreliable.   For example, the difference in temperatures observed with 70% and 90% of a sample melted may easily be in error by $\pm0.0005°C$.   For the solid-solution treatment, such an error would correspond to an uncertainty of 500% in the impurity value for very pure compounds with normal cryoscopic constants, whereas the same $0.0005°C$ error corresponds to 150% uncertainty if solid insolubility is assumed.

3. Comparison of Results Obtained by the Static and Dynamic Methods

It is rather interesting to note that the impurity values determined by static methods are systematically lower than those determined on the same sample by dynamic methods (3, 4).   However, an extremely careful study by Glasgow et al. (18) on a sample of very pure benzene contaminated by known amounts of n-heptane showed that the divergence between the two methods of determination was not as large as was formerly obtained.   The results of this study are given in Table X.2.   It is suggested that the difference in values may be due to chemisorbed water as a source of contamination.

4. Recommendations

The following recommendations have been suggested by Smit (31) for thermal analysis:

(1) When a static method is used, each period of heat supply to the

TABLE X.2

Comparison of the Results from Dynamic and Static Methods (18)

Sample	Computed from contamination	Dynamic	Static
		Purity, mole-%	
A	100[a]	99.994 ± 0.002	99.9937 ± 0.0010
B	99.9964	99.970 ± 0.004	99.958 ± 0.005
C	99.9610	99.940 ± 0.002	99.947 ± 0.005

[a] The "pure" sample was assumed to be pure beyond the sensitivity of the methods of analysis employed.

substance should be followed by a period of "adiabatic conditions" of sufficient length so as to approach equilibrium to a desired extent.

(2) For the dynamic method, heating curves are preferred to cooling curves.

(3) Before starting a measurement of a heating curve, the sample should be kept at a temperature slightly below the initial melting point for at least one hour.

(4) The stirring method for determining heating curves is not recommended.

(5) The rate of heating of samples with small heats of fusion should be decreased as far as practical.

(6) Subject each curve to an internal check and also select a reliable part of the curve for purity determination. Besides experimental checks on the technique, the curve should be checked to see if it obeys the equation (31)

$$T_y = T_a - \frac{C_f C_m}{C_f + Y(C_m - C_f)}\, p \qquad (X.28)$$

where T_y is the temperature at which a fraction Y of the sample has melted, T_a is the melting point of the absolutely pure substance, C_f and C_m are constants, and p is the mole-% of contamination present in the sample.

Equations (X.17) can be rearranged to give

$$T_y = C_1 - \frac{C_2}{C_3 + Y} \qquad (X.29)$$

where C_1, C_2, and C_3 are constants which can be resolved algebraically by selecting three pairs of corresponding values of T_y and Y. If the T_y values of the melting curve are plotted as a function of $1/(C_3 + Y)$, the plot should be a straight line with slope C_2.

With all of the above distressing sources of error and limitations, thermal analysis has several incomparable advantages (1).   Being a physical method, it may be applied without any knowledge concerning the chemical properties of the main component or the contaminants of the sample.   It is sensitive, although not equally sensitive, to all types of contaminants.   When the sample may be considered as a binary system, it certainly permits quantitative determination of its content of contaminants.

E. Applications to Impurity Determinations and Other Problems

The impurities in synthetic mixtures of naphthalene with anthrazene or diphenyl were determined by the melting-curve method of Carleton (16). Melting curves for pure naphthalene alone and for naphthalene containing 1.45-mole-% diphenyl are given in Figure X.8.   The ideal-melting flat lines extend across from T_0, and diagonal lines representing heating without melting (specific-heat effect) are drawn in at selected values of T.   Slopes of these lines are obtained by resolving the slope of the equilibrium curve at 70°C into separate specific-heat and melting components, calculated from the dimensions of the apparatus and the properties of naphthalene.

The calculation was as follows for pure naphthalene: The quantities T_0 and x were estimated as 79.7°C and 0.003, respectively.   In the 10°C temperature interval from 65 to 75°C, the change in fraction melted, $(1 - f)$, was calculated from

$$(1 - f) = \frac{x}{N_2} = \frac{x}{x + A(T_0 - T)} \qquad (X.30)$$

Because A is equal to 0.0184 for naphthalene, $(1 - f)$ is equal to 0.0226. The melting component for the 65 to 75°C temperature range is the product of 0.0226 and 35.6 cal/g (ΔH of naphthalene), or 0.81 cal/g.   From the dimensions of the apparatus employed, the total value of the specific-heat component was 11.1 cal/g.

A triangle was constructed with the slope of the equilibrium curve at 70°C forming the long side as shown in Figure X.8a.   The other two sides represent the melting and specific-heat components, whose ratio was 0.81 to 11.1.   The slope of the specific-heat component was used in the analysis. Lines having the slope of the specific-heat component were then drawn for selected values of ΔT, and the intersections with the ideal-melting flat line gave the corresponding values of ΔT, as shown in Figure X.8b.

Figure X.9 shows a plot of Δt against $\Delta t/\Delta T$ for the naphthalene–biphenyl mixture.   The best straight line drawn through these points had a slope of -0.95; hence $x = 0.0175 = 1.75$ mole-% contaminant which includes the added biphenyl and the original impurity.   From a similar analysis of the

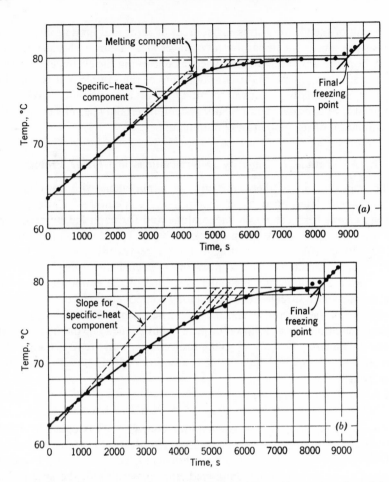

Figure X.8. (a) Melting curve of pure naphthalene; (b) melting curve of naphthalene containing 1.45 mole-% diphenyl (16).

naphthalene by itself, $x = 0.34$ mole-% impurities. Thus, the mole-% biphenyl found experimentally was 1.41 compared to the 1.45 actually added.

In the method employed by Schwab and Wichers (32), the amount of contaminant originally present in a sample can be obtained by determining the freezing curve of the original sample and also the curve of the original sample plus a known amount of contaminant. The above comparative method is said to be applicable even if the fraction frozen does not vary linearly with time. Herington (33) has also described the use of this method using a similar experimental apparatus as previously described above.

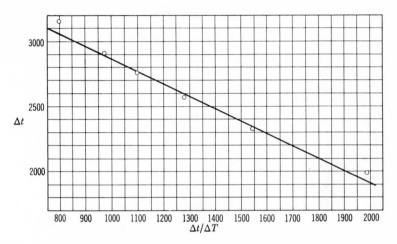

Figure X.9. Derived line of Δt versus $\Delta t/\Delta T$ for naphthalene $+$ 1.45 mole-% diphenyl (16).

A set of freezing curves used in the comparative method is given in Figure X.10. The difference, ΔT, between the initial freezing temperature and the temperature at a time equal to half that required for complete freezing is found, as shown in curve 1. A known amount of impurity, x_1 mole-%, is then added and another freezing curve is obtained using the same rate of cooling as previously employed. A value, ΔT_1, for the difference between the new initial freezing temperature and the temperature at a time equal to half that required for complete freezing is thus obtained. The same procedure is carried out after the addition of a second amount of impurity, x_2, and another ΔT_2 value is obtained.

From the relationship

$$\frac{\Delta T_1 - \Delta T}{x_1} = \frac{\Delta T_2 - \Delta T}{x_2} \qquad (X.31)$$

the amount of contaminant originally present, x, can be given by

$$x = \frac{x_1 \Delta T}{\Delta T_1 - \Delta T} \qquad (X.32)$$

The experiments are carried out in duplicate or triplicate and the standard error computed. The standard error was found to vary from sample to sample, but the mean of several results indicated a value of approximately $\pm x/8$ for this quantity.

If this technique is used, it is important to choose a suitable substance to add to the system. In general, the melting point of this material should not be

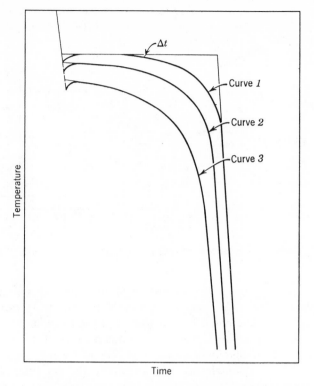

Figure X.10. Temperature–time curves obtained in freezing experiments (33).

higher than the melting point of the main material and should have a lower melting point. The chosen material should not form solid solutions, nor should it form a compound with the main component.

When a freezing curve is obtained, the values may vary at times, due to the nonlinearity of the temperature in the entire system, defects in the temperature detection, and so on. The determination of an actual curve to fit the experimental data presents a difficult problem. Various techniques, such as the use of a flexible spline, have been employed to draw this curve. An optical method, using a lantern projector, has been employed successfully by Saylor (34). Another method that has been suggested is that given by Kienitz (35). A hyperbola is constructed from certain values of time and temperature which best represent the measured curve. In this way, the freezing point of a sample is better obtained than with the analytical or geometrical methods of evaluation via three points on the equilibrium curve.

The purity of a *n*-pentane sample was determined by a calorimetric method by Clarke et al. (20). The results obtained for pure *n*-pentane and for a

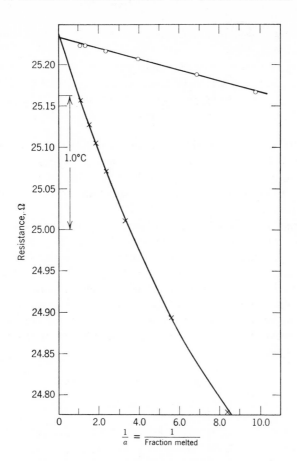

Figure X.11. Melting curve for *n*-pentane (20). (○), *n*-pentane, 99.80% pure; (×) same + 2.26 mole-% *iso*-octane.

synthetic *n*-pentane-*iso*-octane mixture are given in the resistance (temperature) versus time curve in Figure X.11.   For purity determination, these data have been converted to the fraction melted after each equilibration period by allowing for the heat necessary to raise the temperature of the solid and liquid and for the amount of heat leak from radiation and conduction.   The heat of fusion determined from this work was 2090 calories per mole, which gave a purity of 99.79 mole-% for the *n*-pentane.

To the sample of pure *n*-pentane, 2.40 mole-% of *iso*-octane was added. The melting curve so obtained showed considerable curvature, while that for the pure *n*-pentane was a straight line.   From the slope of the line, a purity of 97.58 mole-% was obtained compared to a theoretical value of 97.53 mole-%.

The purity of two samples of pentaborane was also determined with this instrument. Sample I was 99.99 mole-% pure, while Sample II analyzed as 99.91 mole-%. On mixing the two samples together to obtain a sample purity of 99.94 mole-%, the experimental calorimetric purity was 99.949 mole-%.

The purity of several highly reactive substances, such as titanium(IV) chloride, was obtained in a special freezing-point apparatus developed by Glasgow and Tenenbaum (28). The freezing point of titanium(IV) tetrachloride under saturation pressure with zero impurity was calculated to be $-24.10 \pm 0.01°C$.

The thermal analysis of a number of normal alkanes was studied by Mazee (23). In the case of a binary mixture in which the components are completely miscible both in the liquid and solid states, the curves in Figure X.12 were obtained. Curve (a) is the temperature-composition phase diagram, while (b) is the heating curve so obtained on a 50-mole-% n-C_1H_{24}–50-mole-% n-$C_{23}H_{48}$ synthetic mixture. The heating curve is simple and easy to interpret and leaves no room for uncertainties. The amount of "impurity," in this case the amount of the second component, can be calculated with sufficient accuracy.

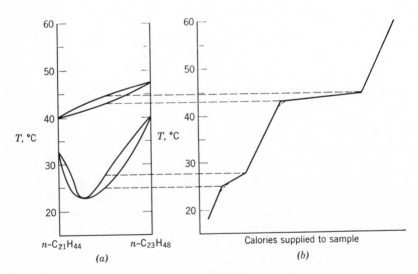

Figure X.12. (a) T-x diagram of mixtures of n-$C_{21}H_{44}$ and n-$C_{23}H_{48}$; (b) heating curve for 50 mole-% n-$C_{21}H_{44}$ and 50 mole-% n-$C_{23}H_{48}$ (23).

When certain organic salts, such as cyclohexylamine stearate, cyclohexyl-amine palmitate, and others, are melted, they undergo the probable double decomposition reaction

$$CS + DP \rightleftharpoons DS + CP$$

where CS and DP are the amine salts.   The system consisting of these four substances is a ternary system of the reciprocal or metathetical type.   Systems of this type have been investigated by thermal analysis by Skau et al. (36).

F. Purity Determinations by Differential Scanning Calorimetry

1. INTRODUCTION

The methods of thermal analysis (cryoscopy), when applicable, has long been acknowledged as the best method for analyzing pure substances because total impurity is measured directly (43).   This method depends only on the physicochemical behavior of the compound, and no reference standard is necessary.   For routine laboratory purity determinations, differential scan-ning calorimetry (DSC) (40–48) has been employed, thus eliminating the the complex equipment, large samples, and long analysis time.   It is esti-mated that over 75% of crystalline organic compounds can be analyzed by the DSC method if they are sufficiently pure (43).   Because total impurity is measured, a relatively large experimental error does not appreciably alter the purity value in the first decimal place.   Also, an experienced analyst can estimate the purity of an unweighed sample to within about 0.2 mole-% by visual inspection of the DSC curve produces in a three-minute run.

2. THEORY

The theoretical basis for the purity determinations is, of course, the van't Hoff equation from which the following melting depression–impurity content relationship can be derived (42, 44):

$$T_s = T_0 - \frac{RT_0^2 X}{\Delta H_f} \cdot \frac{1}{F} \tag{X.33}$$

where T_s is the instantaneous temperature of the sample (°K); T_0 is the melting point of the infinitely pure sample (solvent, °K); R is the gas con-stant (1.987 cal/mole, °K); ΔH_f is the heat of fusion of the sample (solvent, cal/mole); X is the mole fraction of impurity; and F is the fraction of the total sample melted at T_s.   Thus, a plot of T_s versus $1/F$ should give a straight line of slope $RT_0^2 X/\Delta H_f$, with an intercept of T_1.   The value for ΔH_f is obtainable from the integration of the current peak; T_s is measurable from the curve also, as is $1/F$ (the reciprocal of the partial area of the curve up to T_s divided by the total area).

3. Instrumental Considerations

The application of this equation to purity determination requires the following corrections for two instrument variables: (a) thermal resistance of the instrument and (b) undetected melting. Both corrections have been discussed in detail (42, 44, 47, 49). Barrall and Diller (42) correct for the thermal resistance, that is, the resistance between the heater/sensor platform and the bottom of the sample pan, as illustrated in Figure X.13. To a first approximation, it is necessary to melt a compound of very high purity (>99.99%) and record the melting curve. The leading edge slope is directly proportional to the thermal resistance. To correct any temperature in a recorded transition it is necessary to superimpose this slope on the curve and extrapolate to the isothermal base line. For maximum precision, the pure material (usually indium metal) must melt in or near the range for which it is being used as a correction. The thermal resistance increases as the total

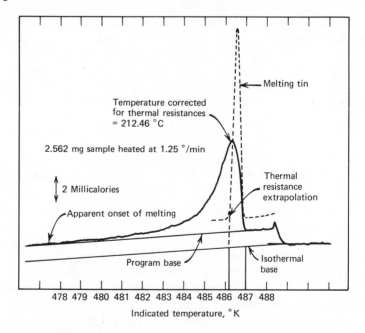

Figure X.13. Correction of a DSC curve for thermal resistance; 9,10-dichloroanthracene containing 1.10 mole-% anthracene (42).

temperature increases, as was demonstrated with high-purity indium, tin, and lead (42).

Other corrections in the DSC curve which have been discussed are those for base-line shift due to heat-capacity change (47, 50) and thermal lag (49).

The effect of these small corrections are illustrated on the DSC curve of cyclohexane as shown in Figure X.14a. The correct sample temperature, T_s, is the curve extrapolated to a true base line (44). A small-area measurement (ABCD) must be added to the partial area under consideration. A typical plot of T_s versus $1/F$ is illustrated in Figure X.14b. The raw data, as shown in the curve, do not define a straight line (44). This is because some peak area is missed before the instrument deviates measurably from the base line due to the noise levels and sensitivity involved. Trial-and-error additions of small increments to both the partial area and the total area are carried out until a straight line is obtained.

One of the most serious limitations of the method is the assumption that no solid solutions are formed. For solid solution systems, equation (X.33) can be modified to give (14, 44)

$$T_s = T_0 - \frac{RT_0^2 X}{\Delta H_f} \cdot \frac{1}{K/(1 - K) + F} \qquad (X.34)$$

where the distribution ratio of the impurity between the liquid and solid phases is $K = k/k'$, and is zero in the absence of solid solution formation. There is no criterion, however, which permits solid solutions to be detected in the DSC curve. Computer programs have been developed to simplify these calculations.

To obtain useable data from DSC measurements, Barrall and Diller (42) recommended the following guidelines:

(1) Sample size less than 3 mg.

(2) Heating rate $< 1.25°/min$.

(3) Encapsulation in a volatile sample sealer modified to maintain good thermal contact.

(4) Precise calibration of the temperature axis. The area considered for $1/F$ calculation must start at the first detectable melting and finish with a point at the endothermal minimum and contain at least six points.

(6) Heating rate and sample size must be adjusted so that the slope of the endotherm never exceeds the slope of the pure standard at half the peak height of the standard.

(7) The thermal lag must be measured with a standard which melts near the sample.

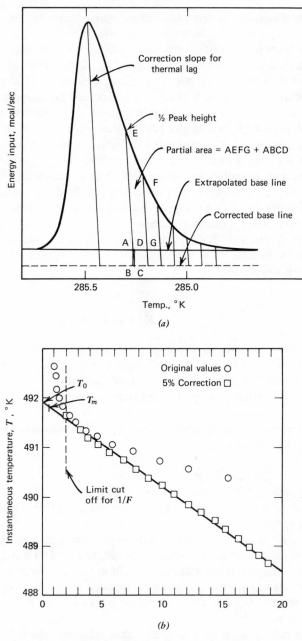

Figure 14. (*a*) Typical scan for purity determination of cyclohexane (44). (*b*) Plot of sample temperature versus $1/F$ for 5,7-dimethyl–1,3-adamantane diol (44).

424

TABLE X.3

Comparison of GC Assay and DSC Purity Values for Some Zone-Refined Hydrocarbons
(41)

Compound	J. T. Baker ULTREX Lot No.	GC assay (area-%)	DSC purity (mole-%)
Acenaphthene	UHC 322	99.99 (170)	99.99; 99.99
Anthracene	UHC 323	99.99 (220)	99.95; 99.96
Bibenzyl		98.24 (125)	99.97; 99.96
Biphenyl	UHC 324	99.99 (170)	99.99; 99.96
			99.99; 99.99
Durene	UHC 325	99.99 (75)	99.96; 99.98
Naphthalene	UHC 326	99.99 (140)	99.96; 99.93
	UHC 327	99.99 (140)	99.99; 99.97
Phenanthrene		99.99 (210)	99.75; 99.77
		99.99 (210)	99.73; 99.92
Pyrene	UHC 328	99.97 (250)	99.94; 99.94
	UHC 329	99.99 (250)	99.97; 99.98
trans-Stilbene		99.77 (200)	99.97; 99.94
p-Terphenyl	UHC 330	99.99 (240)	99.97; 99.96

These seven steps are all due to the inherent thermal lag and response time of the calorimeter and the requirements of a close approach to thermal equilibrium. The formation of an eutectic mixture melting near the principal component will decrease the accuracy of the method. However, at low concentrations sufficient accuracy is maintained to make the method relatively satisfactory. Materials which form cocrystals can be studied by DSC in some cases, but the van't Hoff method cannot be applied satisfactorily to these materials. A calibration curve constructed on the basis of endothermal minimum as a function of the minor component concentration is very sensitive to purity at concentrations of <5 mole% of impurity.

Joy et al. (41) compared the purity determination results obtained by DSC with those obtained by gas chromatography (GC) and titrimetric assay values. For a number of polycyclic hydrocarbons purified by multipass zone refining followed by simple sublimation, the results obtained by DSC and GC assay are given in Table X.3. As can be seen, in most cases, the agreement between the two methods is good considering the quite different underlying phenomena. It should be noted that a value of 99.99 area-% was assigned for the GC assay whenever no impurity peak was detected.

4. Applications

Differential scanning calorimetry as a purity determination technique has been applied to a large number of substances (41). Compounds studied include aliphatic hydrocarbons (44), amides, amines and carbamates (43, 51), benzene derivatives (43, 51, 52), halogenated compounds (41, 43, 44), malic acid (52), organophosphates (43), pesticidal chemicals (43), pharmaceuticals (51, 52), steroids (52), benzoic acid (41), polycyclic hydrocarbons (41), urea (41), cholesterol (41), liquid-crystal-forming materials (46, 53, 54), and numerous others. One such investigation (43) determined the purity and heat of fusion of 95 high-purity organic compounds.

5. Assessment

In view of its rapidity, use of milligram quantities of samples, and application to the purity region from 98.0 to 99.95 mole-%, DSC is a most valuable tool for characterization of organic compounds (41). For a thermally stable compound, a *low* purity value, based on a satisfactory run of the instrument, is clear evidence that the compound is not of high purity. In contrast, a *high*-purity value cannot be taken as conclusive evidence that the compound is indeed of high purity. Above 99.90 mole-% purity, the premelting behavior on which the DSC calculation is based, becomes progressively smaller and the purity value assigned becomes strongly dependent on the assumption made in the calculation. The *practical* upper limit for *absolute* DSC measurements may therefore be about 99.95 mole-% with the presently available instrument and technique. However, it is possible to detect differences in impurity content of as little as 0.005 mole-%. If replicates are run, the *relative* purity of two lots of a single compound can be assessed up to 99.98 mole-%.

References

1. Smit, W. M., *Z. Elektrochem.*, **66**, 779 (1962).
2. Sturtevant, J. M., "Calorimetry," in *Techniques of Organic Chemistry*, 2nd ed., Vol. I, A. Weissberger, ed., Interscience, New York, 1949, Part 1, p. 731.
3. Cines, M. R., "Solid-Liquid Equilibria of Hydrocarbons," in *Physical Chemistry of Hydrocarbons*, Vol. I., A. Farkas, ed., Academic, New York, 1950, Chap. 8.
4. Mathiew, M. P., *Acad. Roy. Belg. Classe Sci. Mem.*, **28**, No. 2 (1953).
5. Smit, W. M., *Thermal Analysis*, Elsevier Press, Amsterdam, 1959.
6. White, W. P., *J. Phys. Chem.*, **24**, 393 (1920).
7. Andrews, D. H., G. T. Kohmann, and J. Johnston, *J. Phys. Chem.*, **29**, 914 (1925).
8. Skau, E. L., *J. Am. Chem. Soc.*, **57**, 243 (1935).
9. Mair, B. J., A. R. Glasgow, and F. D. Rossini, *J. Res. Natl. Bur. Std. (U.S.)*, **26**, 591 (1941).
10. Glasgow, A. R., A. J. Streiff, and F. D. Rossini, *J. Res. Natl. Bur. Std. (U.S.)*, **35**, 355 (1945).
11. Malotaux, R. N. M. A., and J. Straub, *Rec. Trav. Chim.*, **52**, 275 (1933).
12. Thomas, S. B., and O. S. Parks, *J. Phys. Chem.*, **35**, 2091 (1931).
13. Rossini, F. D., *Chemical Thermodynamics*, Wiley, New York, 1950.

14. Mastrangelo, S. V. R., and R. W. Dornte, *J. Am. Chem. Soc.*, **77**, 6200 (1955).
15. Badley, J. H., *J. Phys. Chem.*, **63**, 1991 (1959).
16. Carleton, L. T., *Anal. Chem.*, **27**, 845 (1955).
17. McCullough, J. P., and G. Waddington, *Anal. Chim. Acta*, **17**, 80 (1957).
18. Glasgow, A. R., G. S. Ross, A. T. Horton, D. Enagonio, H. D. Dixon, C. P. Saylor, G. T. Furukawa, M. L. Reilly, and J. M. Henning, *Anal. Chim. Acta*, **17**, 54 (1957).
19. Stull, D. R., *Anal. Chim. Acta*, **17**, 133 (1957).
20. Clarke, J. T., H. L. Johnston, and W. De Sorbo, *Anal. Chem.*, **25**, 1156 (1953).
21. Aston, J. G., and H. L. Fink, *Anal. Chem.*, **19**, 218 (1947).
22. Pilcher, G., *Anal. Chim. Acta*, **17**, 144 (1957).
23. Mazee, W. M., *Anal. Chim. Acta*, **17**, 97 (1957).
24. Ruehrwein, R. A., and H. M. Huffman, *J. Am. Chem. Soc.*, **65**, 1620 (1943).
25. Smit, W. M., and G. Kateman, *Anal. Chim. Acta*, **17**, 161 (1957).
26. Smit, W. M., *Chem. Weekblad*, **36**, 750 (1939).
27. Smit, W. M., *Rec. Trav. Chim.*, **75**, 1309 (1956).
28. Glasgow, A. R., and M. Tenenbaum, *Anal. Chem.*, **28**, 1907 (1956).
29. Handley, R., *Anal. Chim. Acta*, **17**, 115 (1957).
30. Barnard-Smith, E. G., and P. T. White, *Anal. Chim. Acta*, **17**, 125 (1957).
31. Smit, W. M., *Anal. Chim. Acta*, **17**, 23 (1957).
32. Schwab, F. W., and E. Wichers, *Temperature—Its Measurement and Control in Science and Industry*, Reinhold, New York, 1941, p. 256.
33. Herington, E. F. G., *Anal. Chim. Acta*, **17**, 15 (1957).
34. Saylor, C. P., *Anal. Chim. Acta*, **17**, 36 (1957).
35. Kienitz, H., *Anal. Chim. Acta*, **17**, 43 (1957).
36. Skau, E. L., F. C. Magne, and R. R. Mod, *Anal. Chim. Acta*, **17**, 107 (1957).
37. Gunn, S. R., *Anal. Chem.*, **34**, 1292 (1962).
38. Glasgow, A. R., and G. S. Ross, in *Treatise on Analytical Chemistry*, I. M. Kolthoff and P. J. Elving, eds., Interscience, New York, 1968, Part I, Vol. 8, p. 4991.
39. Skau, E. L., and J. C. Arthur, in *Physical Methods of Chemistry*, A. Weissberger and B. W. Rossiter, eds., Wiley-Interscience, New York, 1971, Vol. 1, Part V, p. 105.
40. Barrall, E. M., and J. F. Johnson, in *Purification of Inorganic and Organic Materials*, M. Zief, ed., Marcel-Dekker, New York, 1969, p. 77.
41. Joy, E. F., J. D. Bonn and A. J. Barnard, *Thermochim. Acta*, **2**, 57 (1971).
42. Barrall, E. M., and R. D. Diller, *Thermochim. Acta*, **1**, 509 (1970).
43. Plato, C., and A. R. Glasgow, *Anal. Chem.*, **41**, 330 (1969).
44. Driscoll, G. L., I. N. Duling and F. Magnotta, in *Analytical Calorimetry*, R. S. Porter and J. M. Johnson, eds., Plenum, New York, 1968, Vol. I, p. 271.
45. Driscoll, G. L., I. N. Duling, and F. Magnotta, *Sun Oil Quart. Rev.*, No. 3, 24 (1969).
46. Barrall, E. M., and M. J. Vogel, *Thermochim. Acta*, **1**, 127 (1970).
47. Heuvel, H. M., and K. C. J. B. Lind, *Anal. Chem.*, **42**, 1044 (1970).
48. Sondack, D. L., *Anal. Chem.*, **44**, 888 (1972).
49. Gray, A. P., *Perkin-Elmer Instrument News*, **16**, No. 3, 9 (1966).
50. Brennan, W. P., B. Miller, and J. C. Whitwell, *Ind. Eng. Chem. Fundam.*, **8**, 314 (1969).
51. Reubke, R., and J. A. Mollica, *J. Pharm. Sci.*, **56**, 822 (1957).
52. De Angelis, N. J., and G. J. Papariello, *J. Pharm. Sci.*, **57**, 1868 (1968).
53. Barrall, E. M., J. F. Johnson, and R. S. Porter, *Mol. Cryst.*, **8**, 27 (1969).
54. Ennulat, R. D., *Mol. Cryst.*, **8**, 247 (1969).

CHAPTER XI

MISCELLANEOUS THERMAL ANALYSIS TECHNIQUES

A. Introduction

Although the principal thermal analysis techniques are thermogravimetry and differential thermal analysis, there are a number of other thermal techniques which are extremely useful for solving certain types of chemical systems. Many of these techniques are of recent development but possess the potential for fairly wide use in the future. Some of these techniques are used to supplement or complement the TG or DTA data that are available, and hence yield another insight into the investigation under study.

Almost all of the analytical techniques that produce temperature-dependent data may perhaps be classified as thermal techniques. This would include ultraviolet and visible spectroscopy, infrared spectroscopy, nuclear magnetic resonance, electron spin resonance, X-ray diffraction, electron diffraction, and many others. Obviously, space limitations prevents taking such a broad viewpoint of thermal techniques. Thus, only the more conventional thermal methods, such as dilatometry (thermomechanical analysis), thermo-luminescence, electrical conductivity, emanation thermal analysis, and so on, will be discussed here. The treatment for each technique is, unfortunately, rather brief in scope; an entire monograph could be written on dilatometry alone.

B. Dilatometry (Thermomechanical Analysis)

1. INTRODUCTION

The determination of the change in length or volume of a sample as a function of temperature constitutes the technique called *dilatometry*. This technique has long been used in ceramics and metallurgy and has recently been applied with great enthusiasm to polymeric materials. In the case of polymers, there are more capabilities of measurement than just length or volume changes. Measurements can be made of the softening point, glass transitions, heat distortion under load, tensile modulus, and others if the appropriate probe or instrumentation mode is applied. For this reason, the term dilatometry is not often used, but rather a broader, more definite name is applied: *thermomechanical analysis* (TMA). Other terms which are used for the dilatometry of solids are *thermodilatometric analysis* (TDA) and *thermal dimensional analysis* (TDA).   Both terms, dilatometry and thermo-mechanical analysis, will be used here.

Classical dilatometry is generally used to detect volume or length changes caused by phase transitions of various types.   The most common phase transition that is determined is the *solid*$_1$ → *solid*$_2$, although *solid* → *liquid* and *solid* → *gas* transitions can be determined.   The technique can also be used to detect shrinkage and sintering of a sample upon heating to elevated temperatures.   The linear coefficient of thermal expansion of a sample can be determined; other uses include the determination of the glass transition temperature, T_g, softening temperatures, distortion temperatures, and so on. Temperature ranges employed vary from −150 to 2200°C, and the sample can be heated or cooled during the measurement.

Gray (1) discussed the recording of the derivative of the TMA curve, or *derivative thermomechanical analysis* (DTMA).   This mode of recording is of value particularly in expansion measurements.   The variation in length of a sample as a function of temperature is commonly expressed by the equation

$$l = l_0(1 + \alpha T) \tag{XI.1}$$

where l is the sample length at T, l_0 is the sample length at 0°C, T is the temperature in °C, and α is the linear coefficient of thermal expansion.   In the TMA mode, l is recorded as a function of temperature, and in the absence of a phase transition and assuming α is constant, the curve will be a straight line with slope α.   If equation (XI.1) is differentiated with respect to time, the following expression is obtained:

$$\frac{dl}{dt} = l_0\alpha\left(\frac{dT}{dt}\right) \tag{XI.2}$$

where dT/dt is the heating rate and is constant.   Thus, in the DTMA mode of thermal expansion measurements, the pen displacement is directly proportional to the coefficient of thermal expansion.   By appropriate calibration, α can be read directly from the curve.   Also, a first-order transition would appear as a peak in the curve the area of which is Δl.

2. INSTRUMENTATION

The instrumentation involved in the dilatometry or TMA technique is quite simple.   Changes in volume (ΔV) or length (Δl) of a sample are detected by a mechanical, optical, or electrical transducer, and recorded as a function of temperature (or time).   A simple apparatus (2) is shown in Figure XI.1. Changes in length of the sample are detected via a pushrod by a linear variable differential transformer (LVDT).   The transformer armature is attached directly to the pushrod, and the displacement of it results in an

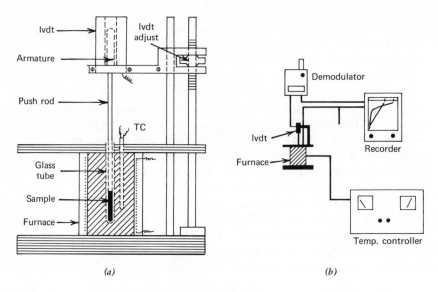

Figure XI.1. Apparatus for dilatometry measurements (2): (*a*) Dilatometer detail; (*b*) complete apparatus.

output voltage which is proportional to the linear displacement. A displacement of ± 0.001 in. resulted in an output voltage of ± 1.11 mV. Thus, very small changes in length of the sample can be detected.

A somewhat more complete apparatus was described by Barrall et al. (3, 4) in which dilatometric measurements were made parallel to DTA studies. This apparatus is shown in Figure XI.2. The sample length was detected by a movable core transformer, the armature of which was attached to the balance beam. In order to obtain a twofold mechanical amplification, the balance beam was lengthened on the transformer side by a factor of 2. The apparatus had a maximum usable sensitivity of 2.5×10^{-5} in. of sample expansion per inch of recorder deflection. According to Barrall et al., the ultimate sensitivity of the apparatus was limited by the vibration of the room and floor.

One of the several thermomechanical analyzers that are commercially available is shown in Figure XI.3. In the penetration and expansion modes, the sample is placed on the platform of a quartz sample tube. The appropriate quartz probe is connected to the armature of a LVDT transformer. Any change in the position of the armature results in a output voltage from the transformer which is then recorded. The probe assembly includes a weight tray, which permits a choice of loadings on the sample surface. All of the components on the assembly are supported by a plastic float rigidly

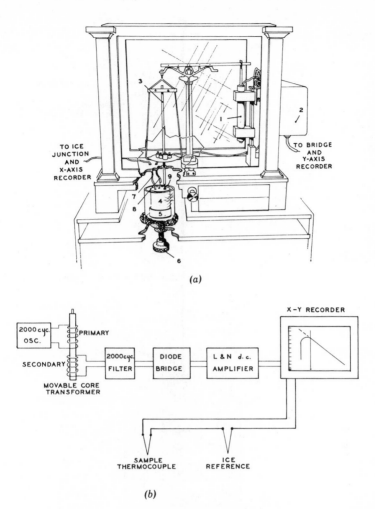

(a)

(b)

Figure XI.2. Thermal expansion apparatus of Barrall *et al.* (3, 4): (*a*) Balance and sample holder; 1, Movable core transformer in rack and pinion mount; 2, 100 turns per inch vernier; 3, quartz extension rod; 4, aluminum furnace block surrounded by liquid nitrogen Dewar flask; 5, quartz base plate; 6, sample position vernier; 7, temperature program thermocouple; 8, sample temperature thermocouple; 9, quartz foot on sample. (*b*) schematic diagram of apparatus.

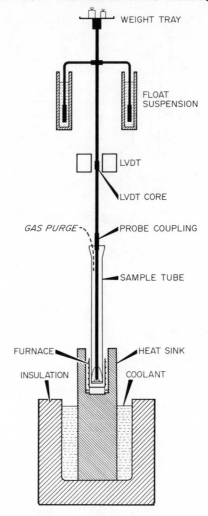

Figure XI.3. Thermomechanical analyzer.

fixed to the shaft and totally immersed in a high-density fluid. This method of support has the advantage that true loading on the sample is essentially independent of the probe position over the range where the float remains totally immersed. Sensitivity of the apparatus is 4×10^{-5} in./in. on a 10 mV recorder. Two furnaces are employed to cover the ranges from -150 to $325°C$ and 25 to $700°C$.

Sample probes that are available for use in TMA are illustrated in Figure XI.4. Probe (a) is used for measurements of thermal expansion, while (b)

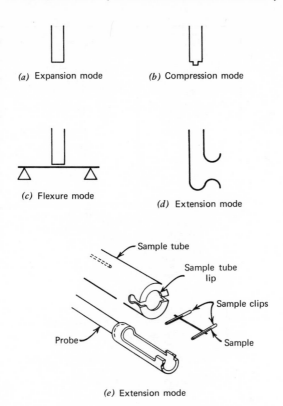

(a) Expansion mode   (b) Compression mode

(c) Flexure mode

(d) Extension mode

Sample tube

Sample tube
lip

Sample clips

Probe

Sample

(e) Extension mode

Figure XI.4. TMA sample probes.

is used to determine sample penetration, softening points, and so on. For Vicat softening or heat distortion under load, the probe in (c) is used. Special-purpose probes for textile fibers and films are shown in (d) and (e). In (e), the sample is placed between slots in the clips, which are then crimped closed. Very thin samples can be conveniently handled and studied under uniform tension.

Simultaneous dilatometric studies with other thermal techniques have been described. Examples include dilatometry measurements in derivatography (5) (see Figure VIII.17), emanation thermal analysis (6–8), DTA and EC (9, 10), and others (3, 4, 11).

3. APPLICATIONS

The use of dilatometry to detect phase transitions (2) of various types is shown by the curves in Figure XI.5. *Solid$_1$* → *solid$_2$* phase transitions were

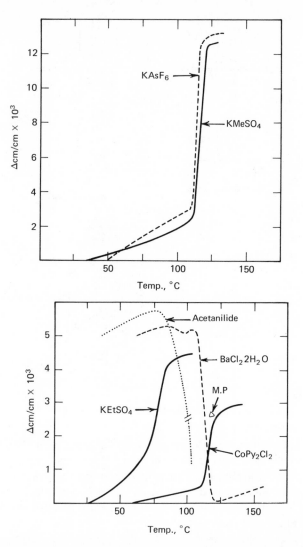

Figure XI.5. Typical dilatometric curves of various substances (2).

detected in $KAsF_6$, $KC_2H_5SO_4$, and $Co(py)_2Cl_2$, while a decomposition reaction, $BaCl_2 \cdot 2H_2O$ deaquation, and a *solid → liquid* transition, the fusion of acetanilide, are illustrated.

The dilatometry curve and the DTA and ETA curves of $Sr(OH)_2 \cdot 8H_2O$ are illustrated in Figure XI.6 (9). As might be expected, the loss of eight moles of water per mole of compound causes a substantial contraction of the

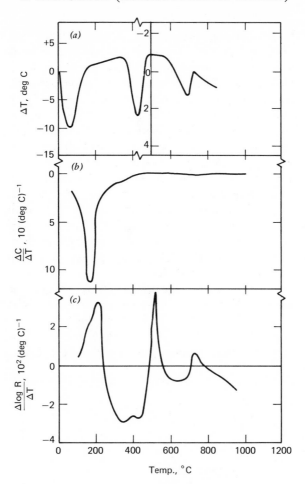

Figure XI.6. (a) Simultaneous DTA; (b) dilatometry; and (c) ETA curves of
$Sr(OH)_2 \cdot 6H_2O$ (9).

initial hydroxide; the density of $Sr(OH)_2 \cdot 8H_2O$ is 1.88 g cm^{-3}, while that of
$Sr(OH)_2$ is 3.62 g cm^{-3}. The rate of sample contraction slows down appreciably at ~220°C and it is possible that this can be attributed to the
formation and subsequent decomposition of $Sr(OH)_2 \cdot H_2O$. Other compounds recently investigated by this simultaneous technique include $Ca(OH)_2$,
$Ba(OH)_2$ (9), calcium, strontium, and barium carbonates (10), $CaC_2O_4 \cdot H_2O$
(5), $CaCO_2 \cdot MgCO_3$ (5), $CaSO_4 \cdot 2H_2O$ (5), kaolin (5), NH_4NO_3 (5),
$ZnO + Fe_2O_3$ (6), $FeC_2O_4 \cdot 2H_2O$ (8), basic iron carbonate (8), and many
others.

A comprehensive investigation of the thermal penetrations of various asphalt compositions has been reported by Schmidt and Barrall (4).   The glass softening point, T_{gsp}, was taken at the point where the curve departs from a straight line.   The thermal behavior of asphalts is similar to simple, linear low-molecular-weight viscoelastic polymers.

The most recent applications of dilatometry and TMA have been to polymeric materials.   A thorough investigation of glass transitions in polymers has been described by Miller (12).   A mechanical model was proposed for the glass transition in polymers by examining existing literature from this mechanical view, and correlating the observed volume, heat capacity, and mechanical data of inorganic glasses with those observed for organic polymers.

The linear expansion, heat capacity, and tensile strength of a general-purpose polystyrene are shown in Figure XI.7 (12).   The glass transition, observed at 75° by TMA, agreed well with the DTA method in which a temperature of 80°C was found.   The abrupt change in the TMA curve at 160° is due to the initiation of melting.

The extension mode of TMA is illustrated (13) by the curve for a 15-denier drawn nylon fiber in Figure XI.8.   To observe the melting at 246°C, a slight mass loading was applied.   The negative slope up to 0°C is the result of linear expansion, while shrinkage is observed at the glass transition near 20°C.   The T_g is lower than normal because of the effect of orientation.   A

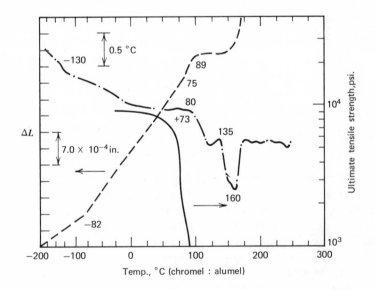

Figure XI.7.   Linear expansion and other curves of polystyrene (12).

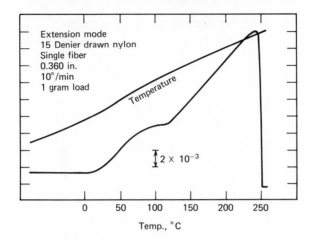

Figure XI.8. TMA extension mode for nylon fiber (13).

change in the rate of disorientation is observed between 60 and 120°C, probably due to the loss of water.

To illustrate the TMA and DTMA expansion modes, the curves in Figure XI.9 are given (14). The two glass transitions in this temperature range are more easily detectable in the DTMA curve than in the TMA curve.

Some of the more practical applications of TMA include the heat-seal characteristics of polyethylene film (15), quality control of brake linings (16), characterization of PVC plastisols (17) and PVC-nitrile rubber copolymer (18), linear expansion coefficients of epoxies (19), and numerous others.

C. Electrical Conductivity

The continuous measurement of the change of *electrical resistance* or *conductivity* of a sample as a function of temperature is a useful technique for the study of polymorphic phase transformation temperatures. This method may be used to supplement other thermal methods such as differential thermal analysis or calorimetry and may be more useful than the above methods in certain cases. In general, the instrumentation is more elementary than DTA, but the applicability is not as great as in the latter. A combined DTA–electrical conductivity apparatus has been described by Berg and Burmistrova (20).

The thermal coefficient of electrical resistance of a sample changes during a phase transformation because of a change in ionic mobility within the lattice (that is, fusion) or because of a change in the electronic energy levels (21). This change in resistance as a function of temperature may be easily measured by use of a dc or ac bridge circuit and appropriate recording equipment.

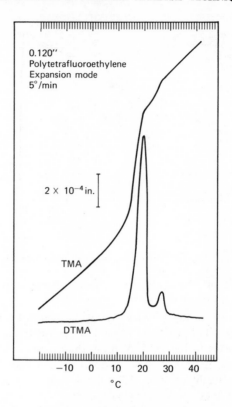

Figure XI.9. Illustration of TMA and DTMA curves of polytetrafluoroethylene in the expansion mode (14).

The bridge circuit and high-temperature sample holder described by Garn and Flaschen (21) are illustrated in Figure XI.10. The bridge circuit is rather standard in that bridge potentials from 0.0015 to 1.5 V can be obtained from a single cell. The output from the bridge is amplified by a dc microvolt amplifier and recorded on one channel of a Y_1-Y_2 strip-chart recorder. Sample temperature, as detected by a thermocouple, is recorded on the other channel of the recorder.

The sample is placed between two disk-shaped platinum electrodes. These electrodes are 9 and 12 mm in diameter, respectively, by 10 mil thick, and retain the sample by a weight-loaded contact for good electrical connections. Sample temperature was measured by a platinum–platinum–10% rhodium thermocouple welded to the upper disk.

There is a marked difference in the electrical resistance curves for pure crystalline potassium niobate and a powdered $KNb_{0.90}Ta_{0.10}$-O_3 mixture.

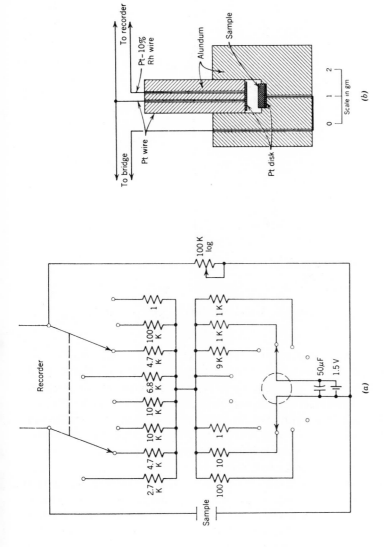

Figure XI.10. EC apparatus used by Garn and Flaschen (21): (*a*) Bridge circuit; (*b*) high-temperature cell.

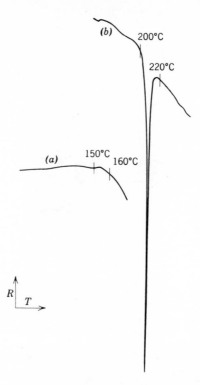

Figure XI.11. Phase transitions by electrical resistance change. (a) $KNb_{0.90}Ta_{0.10}O_3$ (ceramic); (b) $KNbO_3$ (single crystal) (21).

These differences are illustrated in Figure XI.11 (21). The pure $KNbO_3$ exhibits a sharp crystalline phase transition between 200 and 220°C while the mixture showed a gradual resistance change beginning at about 150°C.

The change in electrical conductivity of a sample, coupled with DTA, is a useful combination of techniques to study *solid → solid* reactions, as was shown by Berg and Burmistrova (20). There was a sharp increase in electrical conductivity at the onset of the solid-state reactions. Electrical conductivity was also used to study the fusion of pure substances and eutectic mixtures, and to detect polymorphic transitions of salts. The electrical conductivity of the fused salt can also be used to determine its purity (20).

An apparatus for simultaneous DTA–ATA measurements (where ATA = amperometric thermal analysis) has been described by David (22). This apparatus, as shown in Figure XI.12, consisted of a 20-gauge platinum wire electrode, H, coupled to a small platinum plated stainless-steel cup, F, which constituted the other electrode in the electrical conductivity system.

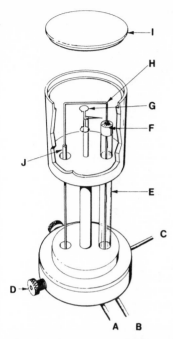

Figure XI.12. DTA-ATA sample holders described by David (22). A, electrometer input lead; B, negative voltage input lead; C, thermocouple leads; D, cooling water connectors; E, pyrex capillary voltage insulator; F, sample pan container; G, differential thermocouple; H, platinum electrode; I, sample holder cover; J, platinum electrode coupler.

The small platinized cup received the same type of sample as was utilized in ring-type thermocouples. The upper electrode, H, was movable so as to maintain continuous contact with the sample during first-order phase transformations in which a reduction in bulk volume frequently occurred. A suitable dc voltage from 1 to 1000 V was applied to the sample and the current flow through the system was measured by an electrometer connected in series with the voltage source. Loss of evolved water in the dehydration of $CuSO_4 \cdot 5H_2O$ was followed by this technique as well as the fusion behavior of potassium nitrate and certain polymeric materials.

Wendlandt (23) used electrical conductivity (EC) measurements to detect quadruple points in various metal salt hydrate systems. This was the first application of this technique to detect the presence of quadruple points which had previously been determined by DTA (24) techniques. Using DTA, the quadruple point is indicated usually as a shoulder peak on a larger endothermic peak. In the case of $CuSO_4 \cdot 5H_2O$, the four phases present at the quadruple point are $CuSO_4 \cdot 5H_2O$, $CuSO_4 \cdot 3H_2O$, H_2O (l), and H_2O (g).

The EC apparatus used by Wendlandt (23) is illustrated in Figure XI.13. It consisted of a recording micro-microammeter, an X-Y recorder, a power supply capable of furnishing 3–25 V dc, a sample holder and electrode probe, and a metal block furnace whose temperature rise was controlled by a simple

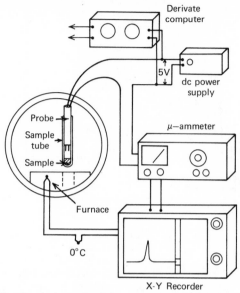

Figure XI.13. Electrical conductivity apparatus used by Wendlandt (23).

programmer. Powered samples of the metal salt hydrates were contained in Pyrex glass tubes, 5 mm in diameter by 50 mm in length.

The EC curve of $CuSO_4 \cdot 5H_2O$ is shown in Figure XI.14. This curve consisted of a single peak due to the evolution of a liquid water phase from the hydrate, which began at a temperature of 97°C. No other liquid phases were detected in the system in the temperature range investigated. Similar results were obtained for $BaCl_2 \cdot 2H_2O$ and $BaBr_2 \cdot 2H_2O$.

A rather elaborate thermal analysis system was described by Chiu (25, 26) in which parallel TG, DTG, DTA, and electrothermal analysis (ETA) measurements were recorded. The sample-handling system of this apparatus is shown in Figure XI.15. One electrode, M, in the sample holder, is a 0.003-in.-thick piece of platinum foil wrapped around the ceramic insulation, T, of the sample thermocouple, Z. The other electrode, L, is made from platinum foil in the form of a cylinder to fit inside the quartz tube, K. The sample is packed tightly between the two electrodes; a spacer, S, located at the bottom of the tube, is used to prevent accidental shorting of the electrodes. Current flowing through the system, under an applied dc potential of 1 to 2 V, is detected with an electrometer and recorded on the Y axis of an X-Y recorder. Sample currents from 10^{-10} to 10^{-5} A, in five decades, were recorded.

To study the EC of several metal oxide systems, Rudloff and Freeman (27)

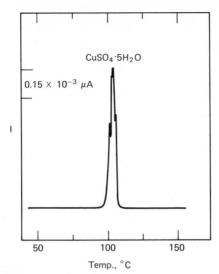

Figure XI.14. EC curve of $CuSO_4 \cdot 5H_2O$ (23).

used the apparatus illustrated in Figure XI.16. An electrode, E_1, inside of a gas-flow tube made of Vycor glass, is mounted on a glass disc fused to a sturdy glass capillary tube fixed at one side of the flow tube. The other electrode, E_2, is fixed to a similar disk–capillary tube combination. A spring provides adequate pressure of the electrodes to the single crystal or powder pellets for good electrical contact. The flow tube can be placed into a heated tube furnace. During operation, the entire system must be carefully shielded to prevent noise pickup from the surroundings.

An apparatus for simultaneous electrothermal analysis (ETA) and dilatometry has been described by Judd and Pope (9). It consisted of a thermal aluminous porcelain 525 tube mounted horizontally in a Kanthal wire-wound tube furnace capable of operation up to a temperature of 1250°C. The tube is reduced to a narrow neck at one end to which a vacuum line is connected. Fixed to the other end of the tube is a metal bracket fitted with two "O"-rings, giving a vacuum-tight connection. A spacer keeps the compacted sample in a fixed position near the center of the furnace. Electrical contact with the sample is made by means of two platinum disk electrodes pressed against the opposing faces.

Carroll and Mangravite (28) described an apparatus in which simultaneous EC and DSC measurements could be made on the same sample. They referred to their technique as simultaneous scanning calorimetry and conductivity (SSCC).

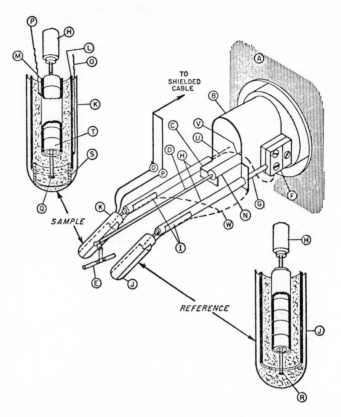

Figure XI.15. Apparatus used by Chiu (25) for parallel TG–DTG–DTA and ETA measurements. A, balance housing; B, balance beam sheath; C, beam stop; D. quartz beam; E, sample container; F, thermocouple block; G, sample measuring thermocouple; H, ceramic tubing; I, platinum jacket; J, reference quartz tube; K, sample quartz tube; L, outer platinum electrode; M, center platinum electrode; N, cold beam member; O, P, platinum lead wires; Q, sample thermocouple junction; R, reference thermocouple junction; S, spacer; T, ceramic insulation; U, V, sample thermocouple wires; W, platinum grounding wire.

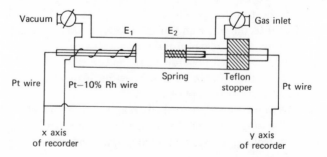

Figure XI.16. Electrical conductivity apparatus used by Rudloff and Freeman (27).

444

D. Emanation Thermal Analysis (ETA)

The technique of *emanation thermal analysis* (ETA) is based on the introduction of inert radioactive gases into a solid and the measurement of their subsequent liberation from the substance as it is heated (7, 29).   Release of the radioactive gas makes possible the monitoring of various types of changes taking place during the thermal cycle.   These include chemical reactions such as dehydration, thermal dissociation, and synthesis, and also polymorphic transformations, melting, conversion of metastable amorphous structures into crystalline compounds, and changes in the concentrations of lattice defects.   The ETA technique possesses several advantages over conventional TG and DTA.   Under dynamic conditions, it permits the study of structural changes of compounds even when these changes are not related to a thermal effect (for example, second-order phase transformations).   In other cases, when finely crystalline or amorphous phases are formed, ETA is more sensitive than X-ray methods.

The ETA apparatus (6), which also permits the recording of the DTA and dilatometric curves concurrently, is shown in Figures XI.17 and XI.18.   A labeled sample (generally 100 mg), a DTA reference material (Al_2O_3), and a sample for dilatometric measurements are placed in the heated chamber. Temperature measurement is by thermocouples embedded directly into the samples.   A heating rate of 8 to $10°C/min$ is normally used since this is an optimum rate for DTA as well as ETA measurements.   The radioactive gas released from the solid sample is carried by a carrier-gas stream which flows at a constant rate into the cells for gas radioactivity measurements (Figure XII.18).   The apparatus simultaneously registers the α-activity of radon and the β-activity of xenon introduced by ion bombardment.   An ETA curve is recorded together with the DTA and dilatometric curves using a multipoint recorder.

In the ETA method, the number of atoms emanated from a small crystallite can be written as the sum of two terms (30):

$$E = E_r + E_d$$

or

$$E = \left(\frac{r_0 S}{4m}\varphi\right) + \frac{(D_0)^{1/2}}{\lambda}\left(\frac{S}{m}\varphi\right)\exp\left(\frac{Q}{2RT}\right)$$

where E_r is the emanation released due to the recoiled emanation atoms, E_d is the diffusion part of the reduced emanation atoms, r_0 is the range of recoiling atoms, S is the specific surface, m is the crystalline mass, D_0 is the pre-exponential term in the expression

$$D = D_0 \exp\left(\frac{-Q}{RT}\right)$$

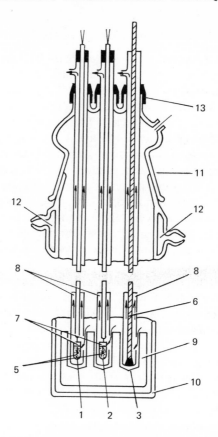

Figure XI.17. Reaction chamber for concurrent ETA, DTA, and dilatometric measurements (after Balek) (6). 1, Activated sample; 2, DTA standard; 3, dilatometer sample; 5, composite thermocouple; 6, quartz dilatometer rod; 7, quartz vessels; 8, supporting pipes; 9, metal block; 10, quartz outer vessel; 11, ground glass joint; 12, coolant tube.

D is the diffusion constant, Q is the activation energy of diffusion of the emanation in the solid, R is the gas constant, φ is the density, λ is the decay constant of emanation, and T is the absolute temperature.

The temperature dependence of the rate of emanation release from Fe_2O_3 is shown in Figure XI.19a. Alternatively, a semilogarithmic curve of the form of $E_d = f(1/T)$ may be constructed, as shown in Figure XI.19b, the quantity E_d being evaluated from E (emanation power at the relevant temperature) and E_r (value of E measured at room temperature). In Figure XI.19b, two slopes may be seen in the curve. A low-temperature section with a low value of $2 \log E_d / 2T$ and a high-temperature section with a larger

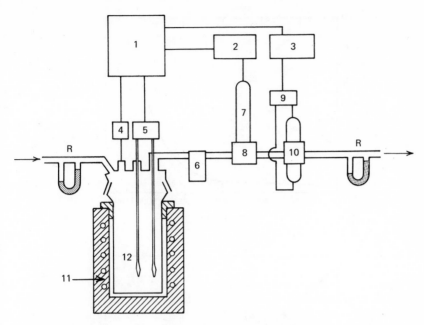

Figure XI.18. ETA apparatus (6). 1, Electronic potentiometer; 2, alpha count integrator; 3, beta integrator; 4, dilatometer pickup; 5, thermocouple; 6, gas dryer; 7, photomultiplier; 8, scintillation chamber; 9, cathode repeater; 10, β emission measurement chamber; 11, electric heater; 12, quartz reaction vessel; R, rheometer.

value of $2 \log E_d/2T$. The discontinuity on the curve lies at 693°C, that is, 0.53 of the absolute melting point of Fe_2O_3. With other crystalline, powdered inorganic substances, the discontinuity on the curve corresponds to 0.5 to 0.6 of their absolute melting point. This temperature is related to the beginning of sufficiently intensive motion of atoms or ions in the crystal lattice to cause an effective diffusion rate in the solid.

Based on the classical emanation method, it is therefore possible to determine this temperature range, which is of great practical importance since above this approximate temperature it is possible that solid-state reactions can occur by diffusion mechanisms. The slope of individual sections of the curve can be used to determine the activation energy of the emanation process in a solid for a specified temperature range. The E of randon diffusion in Fe_2O_3 was evaluated as $Q = 15 \pm 3$ kcal/mole in the range 600–700°C and $Q = 40 \pm 5$ kcal/mole in the range 850–1100°C.

The ETA and DTA curves (7) of $Th(C_2O_4)_2 \cdot 6H_2O$ are illustrated in Figure XI.20. The ETA curve reveals three processes taking place: (*1*) stepwise

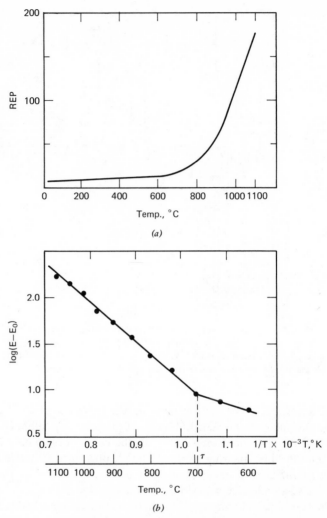

Figure XI.19. Emanation curve of Fe_2O_3: (a) in $E = f(T)$ coordinates; and (b) in log $(E - E_r) = f(1/T)$ coordinates, E_r being the value of E at room temperature.

dehydration in the temperature range 40–220°C (the DTA curve shows only two endothermic peaks in this range for the transitions of $6H_2O \rightarrow 2H_2O$ and $2H_2O \rightarrow 1H_2O$); (2) the flat peak in the 300–400°C range corresponds to decomposition of $Th(C_2O_4)_2 \cdot H_2O$; (3) the last peak, at 400–500°C, is due to the conversion of amorphous ThO_2 to crystalline. The curve of the rate of release of xenon which had been incorporated into the sample by ionic

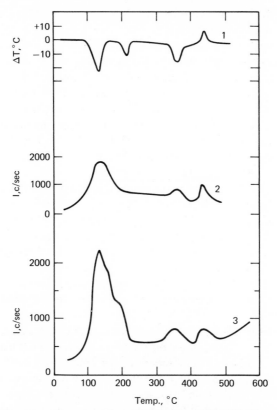

Figure XI.20. The thermal decomposition of $Th(C_2O_4)_2 \cdot 6H_2O$. (1) DTA curve; (2) temperature dependence of xenon release; (3) ETA curve (7).

bombardment fully confirms the reaction sequence indicated by the ETA curve.

E. Thermoluminescence

The technique of *thermoluminescence* involves the measurement of the emitted light energy of a sample as a function of temperature as it is heated to elevated temperatures at a slow, constant rate, below that of incandescence. The curve so obtained, called a glow curve, consists of a series of peaks or maxima which are caused by the emission of light energy at various elevated temperatures. The intensity and temperatures at which the peaks appear may be used to characterize or identify the sample or may be used for other applications. These applications include the study of geological activity, dating of ancient pottery, lunar materials, evaluation of catalysts, solid-state radiation dosimetry, and others. The technique has been found to be extremely useful in geochemical investigations of limestone and dolomite deposits as well as for lava flows.

The early applications of thermoluminescence to the analysis and identification of rocks have been made by Deribere (31), Garlick (32), Kohler and Leitmeier (33), Royer (34), Saurin (35), and Northrup and Lee (36); the more recent applications include those by Saunders (37), Parks (38), Daniels et al. (39), Lewis (40), and Bose et al. (41). The applications of thermoluminescence to problems of geological thermometry and age determination have been discussed by Ingerson (42) and Zeller (43). Excellent reviews on the general subject of thermoluminescence include those by Daniels et al. (39), Bose et al. (41), and others (44–48, 57).

The property of thermoluminescence is exhibited by crystalline substances, for example, the alkali halides, which have been exposed to radioactivity of various types or to X-rays. The radiation causes the dislocation of electrons in the crystal, which may result in an F-center; the release of these various F-centers or electron traps to lower-energy states is brought about by increasing the crystal temperature and is accompanied by the emission of light energy. The emission of light energy is dependent upon changes in crystalline structure, concentration of impurities in the crystal, and past physical treatment.

A simple model of the thermoluminescence (TL) process is shown in Figure XI.21 (57). In (a), ionizing radiation creates an electron and a hole in the crystal. The electron wanders in the conduction band until it falls into an electron trap, forming an F-center, and the hole moves in the valence band until it falls into a hole trap, forming a V-center. Upon heating, the electron (hole) is excited into the conduction (valence) band, where it wanders until it recombines with the hole (electron), giving off light. This simple model, however, does not take into account the dominant influence of impurities on the TL process (57). In the commonly used phosphor, LiF, for example, impurities are necessary to give highly sensitive TL.

Figure XI.21b shows schematically the TL process involving a divalent impurity. Again, electrons and holes are formed upon irradiation with ionizing radiation. The hole can now be trapped by the impurity cation. When the electron is released upon heating, it recombines with the hole at the impurity ion, exciting a characteristic luminescence. The impurity ions can also form complexes such as Z-centers that can trap electrons. As the crystal is heated, the traps are emptied—first the shallow traps at low energy and then deeper traps at higher energies, producing a number of TL peaks.

Other models and theories have been discussed by a number of investigators (39, 59, 60).

The kinetics of the isothermal decay of the thermoluminescence of single

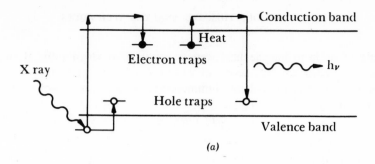

(a)

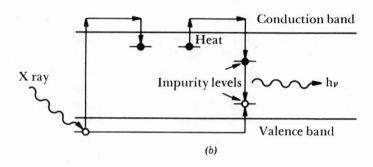

(b)

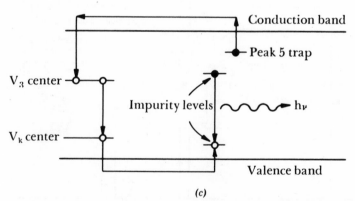

(c)

Figure XI.21. Schematic band models for thermoluminescence (57): (*a*) Simple model; (*b*) model with impurity present; (*c*) model involving electron and hole traps and impurity ion recombination centers.

crystals of lithium fluoride has been determined by Boyd (49). Using a model analogous to the theory of absolute reaction rates for chemical reactions, the intensity of thermoluminescence, I, as a function of time, was shown to be

$$I = \frac{\bar{E}_1 k_3 [K/(K+1)]}{[k_3[(K/(K+1))]t + 1/n_3^0]^2} \tag{XI.3}$$

where $\bar{E}_1$ is the average energy involved in the transition, k_3 is the rate constant, and $K = n_2/n_1$, where n_1, n_2, and n_3 are the numbers of molecules in the initial, activated, and final states, respectively.

Various instruments have been described for the measurement of the thermoluminescence of solid samples. Instrumental arrangements have been described by Urbach (50), Randall and Wilkins (51), Boyd (49), Daniels et al. (39), Saunders (37), Parks (38), and Lewis (40). Basically, the apparatus consists of the following components: (a) a heated sample block, (b) a sample-temperature programmer or power supply, (c) a photomultiplier tube and power supply, and (d) a recorder, either of the two-pen strip-chart or the

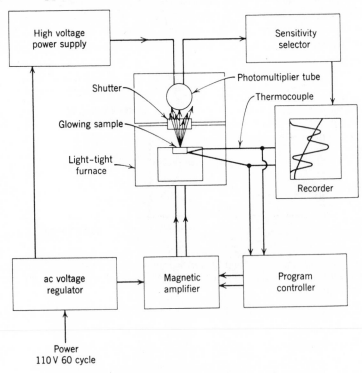

Figure XI.22. Schematic diagram of a TL apparatus (40).

X-Y function type. Such an apparatus, as described by Lewis (40), is illustrated in Figure XI.22. The glow curves are normally obtained at a heating rate of 1°C/min. As the sample temperature is increased, the sample emits light which is detected by the photomultiplier tube whose output current varies linearly with the intensity of the emitted light. This output current is converted into a voltage signal and recorded on one channel of the strip-chart recorder. The other channel of the recorder is connected to a thermocouple embedded in the sample chamber and hence records the sample temperature. Thus, both light intensity emitted and temperature of the sample are recorded as a function of time. The temperature range of the apparatus is from ambient to 500°C. Glow curves can be obtained on 50 mg of finely powdered sample.

A specialized TL apparatus used for radiation dosimetry is shown schematically in Figure XI.23. The apparatus is divided into two separate units—the TL detector (as shown in the figure) and an automatic integrating picoammeter. The latter is used to present the data from the detector in a more convenient form for the operator. The temperature range of the detector unit is from 50 to 400°C; the temperature increases linearly at a preadjusted rate to the maximum temperature set point, where it will remain until a timer shuts off the circuit. Glow curves are obtained at rather rapid heating rates, for example, a 240°C maximum temperature in 50 sec.

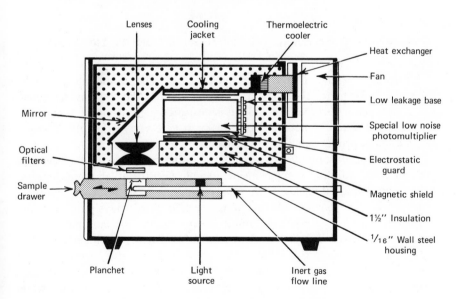

Figure XI.23. Harshaw Model 200 TL analyzer.

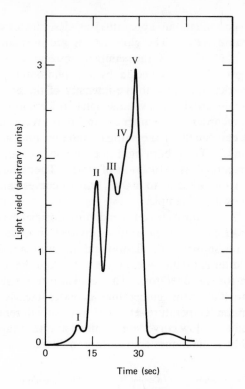

Figure XI.24. Typical glow curve. I, 120°; II, 140°; III, 200°; IV, 240°; V, 340°C.

A typical glow curve is shown in Figure XI.24.   The compound is a doped lithium fluoride material, TLD-100, manufactured by the Harshaw Chemical Co.   The intensity of the emitted light is measured in arbitrary units (but can be converted into microlamberts or lumens, if necessary), while the X axis is in time units, proportional to the temperature as indicated.

The types of substance which are thermoluminescent, either in their natural state or after radiation bombardment, include (39) the alkali metal halides, calcite, dolomite, fluorite, aluminum oxide, magnesium oxide, gypsum, quartz, glass, feldspars, feldspathoids, certain dried clays, and ceramic materials.   Of over 3000 rock samples examined for thermoluminescence, some 75% showed visible light emission (39).   Nearly all limestones and acid igneous rocks are naturally thermoluminescent, due mainly to the presence of trace elements of uranium, thorium, and so on.   Calcium and magnesium carbonates show light yellow to orange light emission, while potassium and sodium feldspars exhibit white to blue-violet light emission.

The glow curves obtained are characteristic of a specific sample or substance and yield information concerning specific impurities present or indicate that the sample has had certain heat treatments or physical histories. However, glow curves are not suitable for the analysis of chemical compounds but are useful for identification and control purposes only.   This is illustrated by the glow curves for several dolomite and calcite samples in Figure XI.25. The curves definitely show differences based upon the composition of the sample but could hardly be used for the analysis of, say, the magnesium or calcium contents.

Saunders (37) has shown that the intensities of the peaks in the glow curves increased with increase in depth in a Niagara limestone deposit. The glow curves were useful for studying the various strata of the deposit. Parks (38) reported a similar study relating glow curves of the samples with the identification and characterization of a formation, the identification of the top and bottom of limestone formations, and the characterization of erosion or nondeposition of zones.

Garino-Canina and Cohen (52) used glow curves to characterize germanium oxide–aluminum oxide mixtures. It was found that after excitation by ultraviolet radiation, the peak positions, between 50 and 70°C, and the peak intensities of the curves varied with a change in alumina content introduced into the germanium oxide glass.   The amount of alumina varied from 0 to 5%.

A rather interesting application of this technique is in the evaluation of the efficiency of surface catalysis (39, 53).   The glow curves for three commercial alumina catalysts are given in Figure XI.26; the relationship between glow curve peak area and catalyst activity are given in Figure XI.26b.   The total glow curve is composed of two peaks; the areas under peak number 2 were related to catalytic activity.   As can be seen from Figure XI.26b, an excellent correlation exists between the peak areas and the activity of the catalyst.   It should be noted that many catalysts do not give any thermoluminescence, and in other cases there is no apparent correlation.   A number of catalysts that have been examined do exhibit such a correlation and hence suggest thermoluminescence as a useful tool for catalyst evaluation.

The glow curves of samples of lunar material appear at 350°C for lunar fines and 400°C for crystalline rocks (61).   The TL is probably the result of an equilibrium between gains from radiation by cosmic rays bombarding the moon's surface combined with the possible presence of radioactivity in the lunar soil and losses from the thermal environment of the moon.   Thermoluminescent results indicate that the temperature variation penetrates about 10.5 cm beneath the lunar surface.

Another application of a geological type is that of dating lava flows (62).

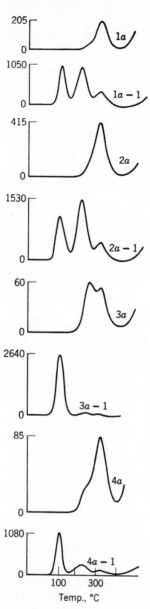

Figure XI.25. Natural and γ-ray activated glow curves of some calcite–dolomite samples. The -1 indicates γ-ray irradiated (40).

Sample	Dolomite, %	Calcite, %	Carbonate, %
1a	89.7	10.3	100.0
2a	92.2	7.9	92.5
3a	0.0	100.0	100.0
4a	48.6	51.4	100.0

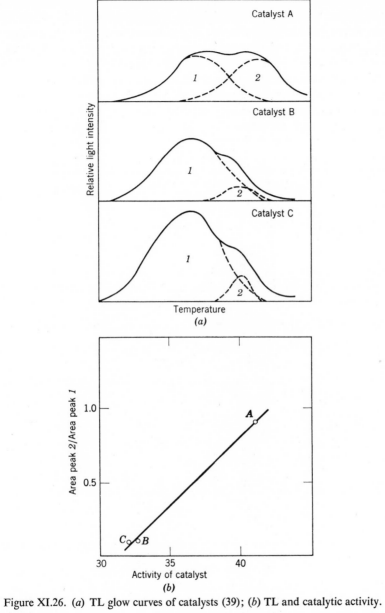

Figure XI.26. (a) TL glow curves of catalysts (39); (b) TL and catalytic activity.

At the time of a volcanic eruption, hot lava removes any past thermo-luminescence, and after the lava cools the natural radioactivity present in the environment irradiates the lava over a period of years.   The thermolumines-cence intensity can then be recorded to determine the age of the lava flow, a calculation which is based on a number of factors.

Thermoluminescence measurements have been used to date pottery sherds. After a piece of pottery has been shaped, it is fired, a process which removes all the previously accumulated thermoluminescence in the quartz crystals that are mixed with the clay of the pottery (57).   Over a period of years the

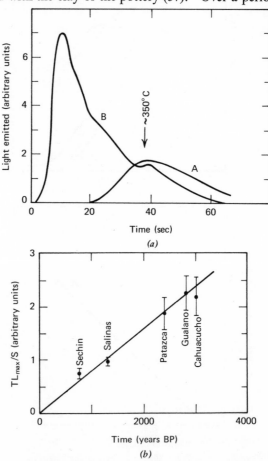

Figure XI.27. (*a*) Glow curves from a pottery sherd, A, compared to the thermo-luminescence induced by a standard *X*-irradiation, B, of a typical sherd; (*b*) ratio of natural thermoluminescence to sensitivity (TL_{max}/S) plotted versus age of the pottery (63).

quartz crystals in the pottery store luminescence from radiation caused by the very small amount of natural radioactivity present in the clay, as well as the radioactivity in the soil.   The date of firing, and thus the age of the pottery, can be determined by the intensity of the thermoluminescence (63, 64).

The glow curves of a pottery sherd are shown in Figure XI.27 (63); the ratio of natural thermoluminescence to sensitivity (TL_{max}/S) plotted versus the age of the pottery is also shown.   The ratio of natural thermolumines-cence to sensitivity is a fairly accurate indicator of archeologic age, especially when averaging is used to obtain a better mean for sherds from the same time level.

Perhaps one of the most important applications of thermoluminescence has been to the measurement of ionizing radiation (57, 60).   Since the thermoluminescence intensity is proportional to the intensity of radiation striking the solid, crystals or powders can act as radiation dosimeters.   The advantages of this method of radiation measurement are the small size and the large dose range of such solids.   A small crystal of a thermoluminescent material is embedded in a small badge which is worn by the personnel.   At periodic intervals, the crystal is placed in a detector (thermoluminescence apparatus) and heated.   From the integrated intensity of the light emission, the amount of ionizing radiation can be determined.

Thermoluminescence has been used to study radiation damage (54) and also the radioactivity of certain minerals (33, 38, 55, 56).

F. Thermomagnetic Analysis

The measurement of the magnetic susceptibility of a compound as a function of temperature is frequently useful for the detection of changes of oxidation state, reduction and oxidation reactions, ferromagnetic and anti-ferromagnetic behavior, and so on.   Without temperature-dependent studies the prediction of the number of unpaired electrons, the oxidation state, and the stereochemistry of magnetically concentrated systems must be suspect (65).

Numerous instruments permitting variable temperature control of the sample chamber have been described.   One of the more recent instruments employs the variable-temperature accessories for a nmr or epr spectrometer (65) to obtain measurements by either the Gouy or the Faraday method. Mulay and Keys (66) described a helical-spring microbalance for Faraday-type magnetic susceptibility as well as for adsorption and TG measurements of the dehydration of $CuSO_4 \cdot 5H_2O$, as shown by the data in Table XI.1, illustrate the application of the balance for TG and thermomagnetic analysis (TMA).   Four moles of water were evolved from 25 to 100°C under a system pressure of $\approx 10^{-6}$ Torr.   The change in susceptibility is due not only

TABLE XI.1
TG and TMA data for $CuSO_4 \cdot 5H_2O$ (66)

Temp., °C	Expected formula	χ_g ($\times 10^{-6}$ c.g.s. units)	χ_m ($\times 10^{-6}$ c.g.s. units)	χ_m ($\times 10^{-6}$ c.g.s. units)
25	$CuSO_4 \cdot 5H_2O$	5.860	1129.6	
100	$CuSO_4 \cdot H_2O$	6.632	1178.4	−49[a]
$p = 10^{-6}$ torr				

[a] Corresponds to 4 moles of water if $\chi_m(_2O) = -12.9 \times 10^{-6}$ c.g.s. units.

to the loss of water molecules but also to a change in paramagnetism of the sample. The latter factor is easily accounted for by calculating theoretically the susceptibility for $CuSO_4 \cdot 5H_2O$ at 100°C, on the basis of the Curie-Weiss law. Another application of the apparatus was to study the adsorption of oxygen on γ-alumina and the changes in magnetic susceptibility during adsorption. Measurements indicate the formation of a dimeric O_4 species with a very small magnetic susceptibility.

Simmons and Wendlandt (67) have described a Faraday-type instrument in which TG and TMA measurements can be obtained from room temperature to 500°C. The apparatus is illustrated in Figure XI.28. A two-channel recorder was employed in which one channel recorded the temperature and the other the TG curve. Superimposed on the TG curve were the deflections of the sample caused by the introduction of the magnetic field about the sample at periodic intervals. The TG-TMA curve of $[Co(NH_3)_6]Cl_3$ is shown in Figure XI.29. A more useful parameter of this system is the mole-% of cobalt(III) reduced, as a function of temperature, as is shown in Figure XI.30. This parameter is calculated from the following equation (3):

$$\text{Mole-\% reduced} = 100 \frac{[(\Delta m - \Delta m_c)/(\Delta m_s - \Delta m_c)]m_s X_s(2.84)^2(T - \theta)]}{\mu_p^2}$$

Since

$$100 \frac{m_s X_s(2.84)^2}{\mu_p^2(\Delta m_s - \Delta m_c)} = \text{constant} = K$$

the former reduces to

$$\text{Mole-\% reduced} = K(\Delta m - \Delta m_c)(T + \theta)$$

where Δm is the apparent mass change, and the subscripts s and c refer to sample and empty container, respectively.

The reactions of sodium metal with Fe_3C, $Fe_{20}C_9$, and $Fe_{2.82}Mn_{0.18}C$ were investigated at elevated temperatures using TMA (68). It was possible by

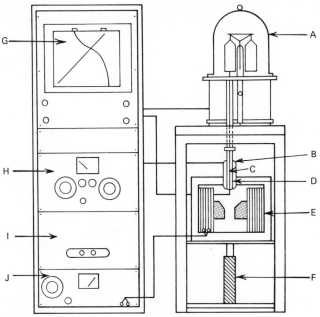

Figure XI.28. TG–TMA apparatus of Simmons and Wendlandt (67). A, Ainsworth semimicro recording balance; B, tube furnace; C, sample container; D, furnace thermocouple; E, electromagnet; F, hydraulic piston; G, two-pen recorder; H, furnace temperature programmer; I, voltage regulator; J, magnet power supply.

this technique to follow the disappearance of iron carbide when in contact with the liquid sodium metal, as well as the formation of iron metal or other ferromagnetic species as reaction products.

The TMA curves (68) of Fe_3C with excess sodium or lithium metal are illustrated in Figure XI.31. Decomposition of Fe_3C alone is rapid above 600°C, while the reaction with sodium metal indicates that no direct reaction such as

$$2Fe_3C + 2Na \rightarrow Na_2C_2 + 6Fe$$

can take place in the temperature range at which the dissociation reaction is important. Both the Curie point and the extent of interaction with the magnetic field at room temperature were found to be unaffected by the heating.

G. Torsional Braid Analysis

The technique of *torsional braid analysis* (TBA) was introduced by Gillham (69, 70) for the investigation of the mechanical properties of polymeric

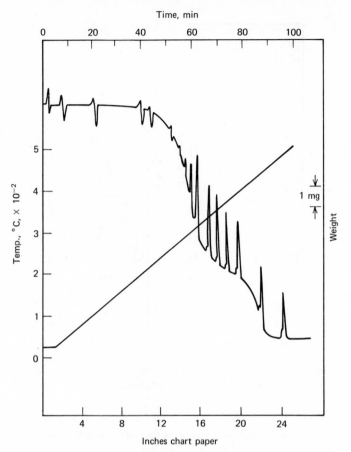

Figure XI.29. TG–TMA curve of [Co(NH$_3$)$_6$]Cl$_3$ (67).

substances. It permits thermomechanical "fingerprints" of polymer transitions in the temperature range -190 to $500°C$ in controlled atmospheres. The sample is prepared by impregnating a glass braid or thread substrate with a solution of the material to be tested, followed by evaporation of the solvent. During the heating of the sample impregnated braid, it is subjected to free torsional oscillations. From these oscillations, the relative rigidity parameter $1/p^2$, where p is the period of oscillation, is used as a measure of the shear modulus. The mechanical damping index, $1/n$, is used as a measure of the logarithmic decrement, where n is the number of oscillations between two arbitrary but fixed boundary conditions in a series of waves. Changes in the relative rigidity and damping index are interpreted as far as possible in

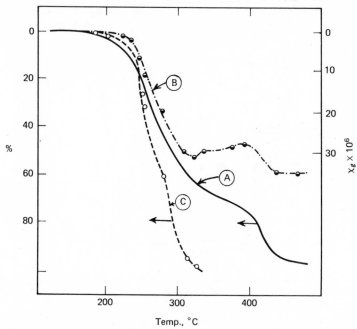

Figure XI.30. TG, mass susceptibility and molar percent reduction of $[Co(NH_3)_6]Cl_3$. A, TG curve; B, mass susceptibility curve; C, mole-% cobalt(III) reduced curve (67).

terms of changes in the polymer. Major and secondary transitions, such as melting or glass transitions, are readily revealed, as are the effects of many other chemical and degradative reactions. The technique has been the subject of a review by Gillham (71).

The apparatus used in TBA is shown in Figure XI.32. The frequency (less than 1 cps) and decay of the freely oscillating pendulum provide information on the modulus and mechanical damping of the polymer under examination. An electrical analog of the decaying pendulum oscillation is obtained by attenuating light with a circular transmission disk, which features a linear relationship between light transmission and displacement angle.

The TBA curves of cellulose triacetate (69) are illustrated in Figure XI.33. The composite specimen of glass braid and cellulose triacetate was prepared from a 7% solution of the highly acetylated polymer in methylene dichloride. Prior to analysis, the solvent was removed by heating to 150°C and then cooling to 25°C. The curves show the glass transition at about 190°C where there is a sudden decrease in rigidity and a maximum in damping. The

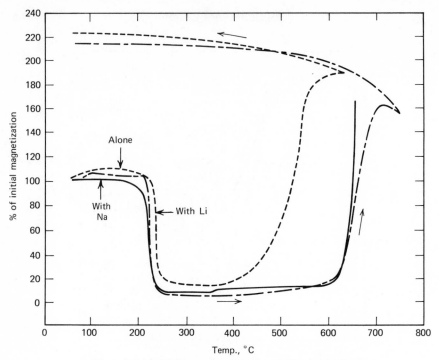

Figure XI.31. TMA curves of Fe₃C with Na or Li (68).

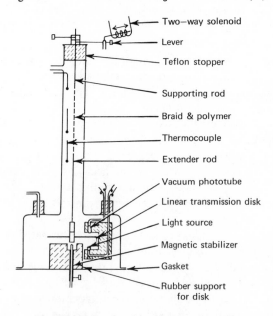

Fig. XI.32. Torsional braid apparatus (71).

464

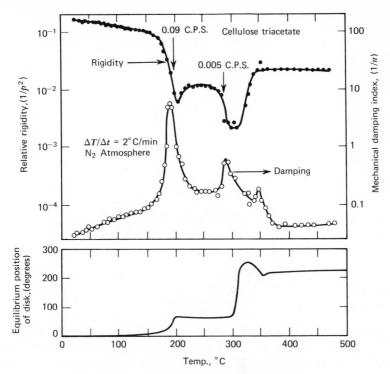

Figure XI.33. TBA curves of cellulose triacetate (69, 71).

decrease in rigidity above 200°C is attributed to crystallization and/or cross-linking. At the melting transition, 290°C, there is a rapid decrease in rigidity and a maximum in damping. Further increase in rigidity and decrease in damping are attributed to cross-linking and/or chain-stiffening processes.

Other compounds that have been investigated (71) by this technique include gelatin, a glass resin, polybenzimidazole, and others.

H. Oxyluminescence

Closely related to thermoluminescence is the technique of *oxyluminescence*. This term was adopted by Ashby (72) to describe the emission of light when certain polymers are heated in air or oxygen to 150–200°C. Oxygen must be present for light emission to take place; the intensity of the light is proportional to the concentration of the oxygen in contact with the polymer surface. Experimentally, the same type of apparatus is used as was described

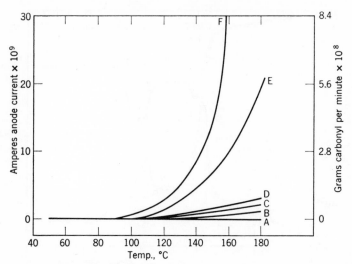

Figure XI.34. Effect of oxygen concentration in the intensity of oxyluminescence of polypropylene. A, Argon; B, 2.6% oxygen in nitrogen; C, 6.4% oxygen in nitrogen; D, 11.0% oxygen in nitrogen; E, air; and F, oxygen (72).

for thermoluminescence studies. The amount of light emitted is quite weak and varies from 10^{-10} to 10^{-8} lumens on the polymer samples studied. With nylon, the strongest emitter thus far examined, the emitted light at 200°C was just bright enough to be seen by the human eye. Preliminary studies on polypropylene showed that half of the emitted light was in the 4200–5150-A wavelength range and the other half in the 3000–4200-A wavelength range.

The intensity of the oxyluminescence is affected by the concentration of the oxygen in the gas in contact with the sample surface, as is shown with polypropylene in Figure XI.34. The temperature of the polymer was increased from ambient to 180°C at a heating rate of 6°C/min.

Since the presence of antioxidants decreases the intensity of the emitted light, the technique is useful for the screening of compounds for antioxidant action. Any substance which decrease the initial light emission is a potential antioxidant.

References

1. Gray, A. P., *Perkin-Elmer Instr. News*, **20**, No. 1, 10 (1969).
2. Wendlandt, W. W., *Anal. Chem. Acta*, **33**, 98 (1965).
3. Barrall, E. M., R. S. Porter, and J. F. Johnson, *Anal. Chem.*, **36**, 2316 (1964).
4. Schmidt, R. J., and Barrall, E. M., *J. Inst. Petro.*, **51**, 162 (1965).
5. Paulik, F., J. Paulik, and L. Erdey, *Mikrochim. Acta*, **1966**, 894.
6. Balek, V., *Glasnik Hemijskog Drustva*, **34**, 345 (1969).

7. Balek, V., *J. Materials Sci.*, **4**, 919 (1969).
8. Balek, V., *J. Materials Sci.*, **5**, 166 (1970).
9. Judd, M. D., and M. I. Pope, *J. Appl. Chem.*, **20**, 380 (1970).
10. Judd, M. D., and M. I. Pope, *J. Appl. Chem.*, **20**, 384 (1970).
11. Berger, C., L. Eyraud, M. Richard, and R. Riviere, *Bull. Soc. Chim. France*, **1966**, 628.
12. Miller, G. W., in *Thermal Analysis*, R. F. Schwenker and P. D. Garn, eds., Academic, New York, 1969, Vol. I, p. 435.
13. *Perkin-Elmer Instrument News*, **20**, No. 4, 6 (1970).
14. *Perkin-Elmer Instrument News*, **20**, No. I, 10 (1970).
15. *DuPont Application Brief*, No. 17, E.I. DuPont de Nemours & Co. Inc., Instrument Products Div., Wilmington, Dela.
16. Ref. 15, No. 22.
17. Ref. 15, No. 16.
18. Ref. 15, No. 14.
19. Ref. 15, No. 5.
20. Berg, L. G., and N. P. Burmistrova, *Russ. J. Inorg. Chem.*, **5**, 326 (1960).
21. Garn, P. D., and S. S. Flaschen, *Anal. Chem.*, **29**, 268 (1957).
22. David, D. J., *Thermochim. Acta*, **1**, 277 (1970).
23. Wendlandt, W. W., *Thermochim. Acta*, **1**, 11 (1970).
24. Borchardt, H. J., and F. Daniels, *J. Phys. Chem.*, **61**, 917 (1957).
25. Chiu, J., *Anal. Chem.*, **39**, 861 (1967).
26. Chiu, J., *J. Polymer Sci.*, **C8**, 27 (1965).
27. Rudloff, W. K., and E. S. Freeman, *J. Phys. Chem.*, **74**, 3317 (1970).
28. Carroll, R. W., and R. V. Mangravite, in *Thermal Analysis*, R. F. Schwenker and P. D. Garn, eds., Academic, New York, 1969, Vol. I, p. 189.
29. Balek, V., *Anal. Chem.*, **42**, 16A (1970).
30. Balek, V., and K. B. Zaborenko, *Radiokhimiia* **10**, 450 (1968).
31. Deribere, M., *Argile*, **188**, 5 (1938); *Rev. Sci.*, **76**, 383 (1938).
32. Garlick, G. F., *Luminescent Materials*, Oxford University Press, London, 1949.
33. Kohler, A., and H. Leitmeir, *Z. Krist.*, **87**, 87 (1934).
34. Royer, L., *Compt. Rend.*, **204**, 602, 991 (1937).
35. Saurin, E., *Compt. Rend. Soc. Geol. France*, **1939**, 209.
36. Northrup, M. A., and O. I. Lee, *J. Opt. Soc. Am.*, **30**, 206 (1940).
37. Saunders, D. F., *Bull. Am. Assoc. Petro. Geol.*, **37**, 114 (1953).
38. Parks, J. M., *Bull. Am. Assoc. Petrol. Geologists*, **37**, 125 (1953).
39. Daniels, F., C. A. Boyd, and D. F. Saunders, *Science*, **117**, 343 (1953).
40. Lewis, D. R., *J. Phys. Chem.*, **60**, 698 (1956).
41. Bose, S. N., J. Sharma, and B. C. Dutta, *Trans. Bose Res. Inst.*, *Calcutta*, **20**, 117 (1955).
42. Ingerson, E., *Econ. Geol.*, *50th. Anniv. Volume*, **1956**, 341.
43. Zeller, E. J., *Congr. Geol. Intern. Compt. Rend. 19e Algiers*, **12**, 365 (1952).
44. Mott, N. F., and R. W. Gurney, *Electronic Processes in Ionic Crystals*, Oxford University Press, London, 1940.
45. Fonda, G. R., and F. Seitz, *Cornell Symposium of the American Physical Society*, Wiley, New York, 1948.
46. Kroger, F. A., *Some Aspects of the Luminescence of Solids*, Elsevier, Amsterdam, 1948.
47. Leverenz, H. W., *An Introduction to the Luminescence of Solids*, Wiley, New York, 1950.
48. Pringsheim, P., *Fluorescence and Phosphorescence*, Interscience, New York, 1949.

49. Boyd, C. A., *J. Chem. Phys.*, **17**, 1221 (1949).
50. Urbach, F., *Sitzber. Akad. Wiss. Wein. Math. Naturw. Kl.*, *Abt. IIa*, **139**, 363 (1930).
51. Randall, J. T., and M. H. F. Wilkins, *Proc. Roy. Soc.* (London), **A184**, 347 (1945).
52. Garino-Canina, V., and S. Cohen, *J. Am. Ceram. Soc.*, **43**, 415 (1960).
53. Boyd, C. A., and J. Hirschfelder, U. S. Patent No. 2, 573, 245, October 30, 1951.
54. Morehead, F. F., and F. Daniels, *J. Phys. Chem.*, **56**, 546 (1960).
55. Ellsworth, H. V., *Canad. Dept. Mines Geol. Survey, Econ. Geol. Ser. No.* **11**, 55 (1932).
56. Alt, M., and H. Steinmetz, *Z. Angew. Mineral.*, **2**, 153 (1940).
57. DeWard, L. A., and T. G. Stoebe, *Am. Scientist*, **60**, 303 (1972).
58. Mayhugh, M. R., *J. Appl. Phys.*, **41**, 4776 (1970).
59. Christy, R. W., N. M. Johnson, and R. R. Wilbarg, *J. Appl. Chem.*, **38**, 2099 (1967).
60. Cameron, J. R., N. Suntharalingam, and G. N. Kenney, *Thermoluminescent Dosimetry*, University of Wisconsin Press, Madison, 1968.
61. Dalrymaple, G. B., and R. R. Doell, *Science*, **167**, 715 (1970).
62. McDougall, D. J., Ed., *Thermoluminescence of Geological Materials*, Academic, New York, 1968.
63. Mazess, R. B., and D. W. Zimmerman, *Science*, **152**, 347 (1966).
64. Zimmerman, D. W., *Archeometry*, **13**, 29 (1971).
65. Hyde, K. E., *J. Chem. Educ.*, **49**, 69 (1972).
66. Mulay, L. N., and L. K. Keys, *Anal. Chem.*, **36**, 2383 (1964).
67. Simmons, E. L., and W. W. Wendlandt, *Anal. Chim. Acta*, **35**, 461 (1966).
68. Charles, R. G., L. N. Yannopoulas, and P. G. Haverlack, *J. Inorg. Nucl. Chem.*, **32**, 447 (1970).
69. Gillham, J. K., *Appl. Polymer Sym.*, **2**, 45 (1966).
70. Lewis, A. F., and J. K. Gillham, *J. Appl. Polymer Sci.*, **6**, 422 (1962).
71. Gillham, J. K., in *Proceedings of the Second Toronto Symposium on Thermal Analysis*, H. G. McAdie, ed., Chemical Institute of Canada, Toronto, 1967, p. 79.
72. Ashby, G. E., *J. Polymer Sci.*, **50**, 99 (1961)

CHAPTER XII

THE APPLICATION OF DIGITAL AND ANALOG COMPUTERS TO THERMAL ANALYSIS

A. Introduction

Although there are numerous applications of digital and analog computers to most chemical laboratory techniques (53), there have been relatively few applications to thermal analysis. All the applications that have been described are of a *passive* type in which there is no significant computer control of the experiment. There have been no applications of the *active* type in which the computer is involved to some extent in the control of the instrument.

This chapter attempts to summarize the important applications of computers to thermal analysis techniques. Because of the difficulty in searching the literature for this type of information, no attempt has been made to write a comprehensive treatment of the subject. Rather, it is hoped that this chapter will include the more important investigations. For convenience, the discussion is divided into applications in thermogravimetry (TG), differential thermal analysis (DTA), differential scanning calorimetry (DSC), and miscellaneous techniques.

B. Thermogravimetry (TG)

One of the first applications of a digital computer to calculations of thermogravimetric data was that by Soulen (1). Since the amount of computation required to obtain kinetics constants from TG is large, a computer program was developed for the calculation of temperature, mass, and rate of reaction from the dc voltage generated by the thermobalance. A Remington Rand Univac computer was used, employing a Math-Matic compiling system, in which a 23-sentence English language program was used to compute 60 values each of temperature, mass, cumulative mass-loss, and rate of reaction, and to store these for subsequent computation of the kinetics constants. Instead of a data-logging system, numerical values were manually taken from the strip-chart recordings at one-minute intervals and used as input into the computer. It was stated that an English language program was rather inefficient for this type of program and that a more efficient program could no doubt be developed using machine language.

469

Almost all of the other applications of computers to thermogravimetry involved calculations pertaining to reaction kinetics. Schempf et al. (2) developed a program, POLY 2, for the determination of the preexponential factor of the Arrhenius equation and the activation energy. This program, designed to accept sample mass (w) and sample temperature (T) values as a function of time (t) was written for first-order reactions only, although with slight modifications it could be used for any order of reactions. It made use of a least-squares-of-polynomial fit of the time-sample mass values to the equation.

$$w = \sum_{i=0}^{n} C_{(i+1)} t^{(i)} \tag{XII.1}$$

where n is the desired order polynomial, C the coefficient of the polynomial, and t the time. From the w-t curve thereby generated, an additional Fortran subroutine, FREEB, calculated the reaction rate constant for any point on the TG curve using the equation

$$k = \frac{-\sum_{i=0}^{n-1} [i+1] C_{(i+2)} t^{(i)}}{\sum_{i=0}^{n} C_{(i+1)} t^{(i)}} = \frac{dw/dt}{w} \tag{XII.2}$$

where n is the desired order polynomial. A least-squares analysis of the values of $\log k$ versus $1000/T$ was obtained for the following first-order polynomial:

$$\log k = \log A - Ea/2{,}303 \, R \, (1000/T) \tag{XII.3}$$

The complete program is illustrated by the flow diagram in Figure XII.1. The accuracy of the computer fit of the TG curve was 0.2 mg, while the limit of accuracy for reading a weight value was 0.1 mg.

Two programs for the algorithmization of kinetic-data computations from TG curves were developed by Sestak et al. (3). They make use of the basic equation

$$\frac{d\alpha}{dt} = \exp\left(\frac{-E}{RT}\right)(1-\alpha)^n \tag{XII.4}$$

where α is the degree of decomposition, and n the order of reaction. The kinetics parameters, E, Z, and n, were evaluated by use of two differential methods. The first method utilized a least-squares polynomial fit of the TG curve with a jth-order polynomial:

$$\alpha = A_0 + A_1 x + \cdots + A_j x^j \tag{XII.5}$$

where j is about 13 and the A's are constants obtained from the least-squares fit of the experimental data. The second method attempted to use the simplest

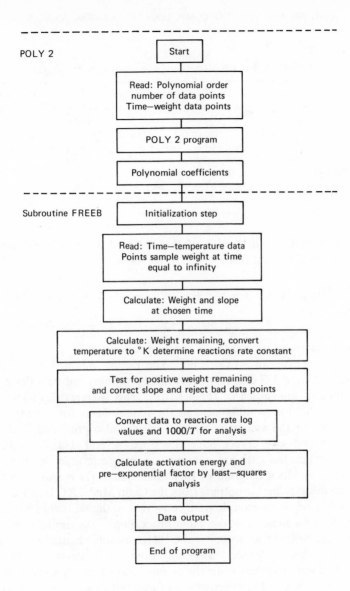

POLY 2

Start

Read: Polynomial order
number of data points
Time—weight data points

POLY 2 program

Polynomial coefficients

Subroutine FREEB

Initialization step

Read: Time—temperature data
Points sample weight at time
equal to infinity

Calculate: Weight and slope
at chosen time

Calculate: Weight remaining, convert
temperature to °K determine reactions rate constant

Test for positive weight remaining
and correct slope and reject bad data points

Convert data to reaction rate log
values and $1000/T$ for analysis

Calculate activation energy and
pre—exponential factor by least—squares
analysis

Data output

End of program

Figure XII.1. Flow diagram for calculation (2).

TABLE XII.1
Comparison Between Manual and Computer Calculations (3)

Experimental data used	Manual results			Computer program	
	Derivative (4)	Integral (5)	Freq. factor (6)	1	2
$CaC_2O_4 \rightarrow CaCO_3 + CO(g)$ (from Ref. 4) $E = 74$ kcal, $n = 0.7$	$E = 67 \pm 15$	74.1 ± 3.5	74	72.1	58.67
	$n = 0.6$	1	1	1	0.591

means of obtaining a derivative of the observed TG curve by numerical derivation using the first three terms of the series.

$$\left(\frac{d\alpha}{dt}\right)_i = \{\tfrac{1}{2}(w_{j+1} + w_{j-1}) - \cdots + \tfrac{1}{60}(w_{j+3} - 4w_{j+2} + 5w_{j-1} - 5w_{j-1}$$

$$+ 4w_{j-2} - w_{j-3}) - \cdots + \cdots\}/Qw_{max} \quad (XII.6)$$

where w is the mass-loss and Q is a constant time interval of scanning. This program was written in ALGOL 60. The results obtained by computing data obtained from TG curves, with various programs and with those calculated manually, are shown in Table XII.1. The discrepancies which occur were said to be due to differences in the requirements for the input data. A flow chart for the second program used was also presented in detail.

Gallagher and co-workers have described several TG data-collection systems in which the data are obtained on magnetic tape or on punched paper tape. A block diagram of their first system (7) is shown in Figure XII.2. In this system, the outputs from the Cahn Model RG balance and the Chromel-Alumel thermocouple were converted to digital form and punched on paper tape for subsequent computer processing. The timing cycle for the counter was normally set to count the thermocouple channel for 1 sec and the mass channel for 99 sec. Switching time was relatively instantaneous and the data were punched while the counter was operating so that the dead time was negligible. The effective use of averaging each reading over these times leads to a reduction of noise, which is important for the computation of the time derivative.

The digital data were transferred from punched paper tape to cards and the EMF versus temperature tables for the compensated thermocouple were fitted by a least-squares technique to the equation

$$°C = 22.2877 + 25.7003 \ (mV)^2 + 0.0017 \ (mV)^3 \quad (XII.7)$$

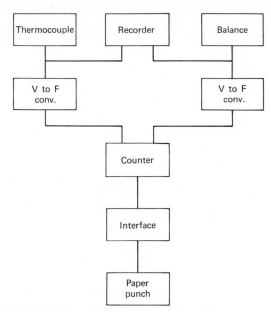

Figure XII.2. Block diagram of digital TG system of Gallagher and Schrey (7).

which was satisfactory to $\pm 1°C$ over the temperature region 200 to 1000°C. A program was developed to compute the average temperature for each pair of consecutive temperature readings and associate this temperature with the average mass readings in the interval between the thermocouple readings. A General Electric Model 600 computer then tabulated and plotted both the percent mass-loss and the rate of mass-loss (mg/min) as a function of temperature. The rate of mass-loss was obtained from the difference in mass of consecutive readings (100-sec intervals) and corrected to give milligrams per minute. No further refinement or smoothing of the differential data was necessary.

For isothermal measurements, using a Cahn Model RG thermobalance, the data acquisition system shown in Figure XII.3 was employed (11, 12). The system accepted up to four analog input signals, of which two were used for mass and temperature, respectively. The voltages were converted to frequency using a voltage-to-frequency converter, and four channels were simultaneously counted on four scalers for a predetermined time interval. The magnetic tape interface served as the control center. In the automatic mode, the data were scanned repeatedly at a preset time interval and placed on the magnetic tape along with channel identification numbers. A fifth channel could be created to insert a six-digit number for labeling or control

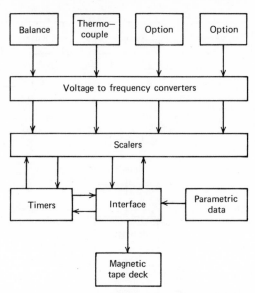

Figure XII.3. Data acquisition system of Gallagher and Johnson (11, 12).

purposes.   Data processing consisted of transferring the data from tape onto the disk storage of a Honeywell Model 635 Computer in appropriate arrays corresponding to each channel.   Computer-generated plots of each array as a function of time were then made with subsequent data processing as previously described (12).

Gallagher and co-workers (8–10, 13) also described a modification of the Perkin-Elmer thermobalance to obtain the data in digital form.   In this instrument, the platinum furnace heater winding serves also as the temperature sensor.   It forms one side of a bridge circuit, while the other side is driven by the output voltage from the programming potentiometer.   This same voltage is used to supply the temperature portion of the digital equipment and is directly related to temperature by use of magnetic (Curie point) TG calibration standards.

The two input voltages were converted to frequency and counted for a predetermined time in the sequence shown in Figure XII.4.   The temperature signal was counted for 0.1 sec and then automatically switched to the mass signal for 10 sec.   The output data were constantly punched on paper tape for input into the computer.

The first stage in the computer processing of the punched-tape data was to transfer the data to cards and to use these cards for the three steps in processing (9).   The first step was to obtain a graphical output of the mass as a function of time, as shown in Figure XII.5.   The second step of data handling

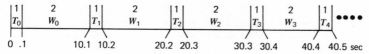

Figure XII.4. Timing sequence employed by digital thermobalance (8).

consisted of utilizing the initial and final mass for each interval to determine values of α, the fraction reacted. The computer, having calculated the values of α for each point, then plotted these to conform to the 18 equations given in Table XII.2. Appropriate equations were determined by visual inspection of the computer output plots for their linearity. One such set of curves for the plot of $-\ln(1 - \alpha)$ versus time is shown in Figure XII.6. The choice of equation was then based on the exact degree of fit determined by the standard deviation arising from the calculation of k in the third stage

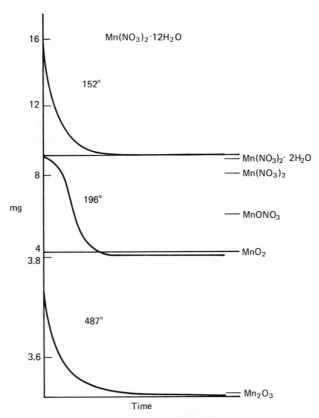

Figure XII.5. Mass versus time curves as reported by Gallagher and Johnson (9).

TABLE XII.2
Kinetic Equations used in the Computer Analysis (9)

1. Power law	α^n; $n = \frac{1}{4}, \frac{1}{3}, \frac{1}{2}, 1$ and 2
2. Contracting geometry	$1 - (1 - \alpha)^{1/n}$; $n = 2$ and 3
3. 2D diffusion controlled	$(1 - \alpha)\ln(1 - \alpha) + \alpha$
4. Erofeev	$[-\ln(1 - \alpha)]^{1/n}$; $n = 1, \frac{3}{2}, 2, 3$ and 4
5. 3D diffusion controlled	$(1 - \frac{3}{2}\alpha) - (1 - \alpha)^{2/3}$
6. Jander	$[1 - (1 - \alpha)^{1/3}]^2$
7. Prout-Thompkins	$\ln\left(\dfrac{\alpha}{1 - \alpha}\right)$
8. Second order	$\dfrac{1}{1 - \alpha} - 1$
9. Exponential	$\ln \alpha$

of processing. This third stage consisted of the selection of the most likely kinetic equation or equations and the plotting of the best values of $\log k$ versus the reciprocal of the absolute temperature, and a least-squares fit to determine the best straight line. The resulting activation energies, E, and the preexponential terms were printed out along with the plot.

Vachuska and Vobori (20) developed a program called VACHVO II (21) for use on the GIER computer, in which the first-order and also the second-order derivations of the time dependencies of both sample mass and temperature are computed numerically with respect to time from experimental values

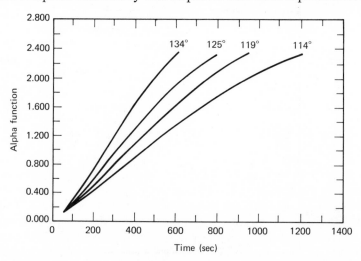

Figure XII.6. Plots of $-\ln(1 - \alpha)$ versus time for the dehydrations of aqueous manganese(II) nitrate (9).

of these quantities. A newer version of this program is VYRVACHVON (22), in which a certain polynomic function is laid through the experimental points and its course is determined by a least-squares method. The computer then calculates the "corrected" input data from a given expressed function and, using these data, numerically differentiates. Both programs were written in the GIER–ALGOL language.

Although the programs or techniques were not discussed in detail, a digital computer was used to analyze the kinetics data obtained from TG by a number of investigators (14–19). One of these studies (19) used a Hewlett-Packard Model 9100A programmable calculator.

Hughes and Hart (23) have developed an analog simulation program, BASE, which was used for the prediction of a TG curve. The calculation involved the plotting of the TG curve from the equation.

$$f(\alpha) = \frac{A}{a} \int_{T_1}^{T_2} \exp\left(\frac{-E}{RT}\right) dt \qquad \text{(XII.8)}$$

where $f(\alpha)$ represents some description of the rate law for the fractional decomposition (α) of the solid; A is the preexponential factor; a is the heating rate (dT/dt), and E is the activation energy. In order to write a patch diagram for the program, they set

$$y = \exp\left(\frac{-E}{RT}\right) = e^z \qquad \text{(XII.9)}$$

where $z = -E/RT$ and $dz = (E/RT^2) dT$. Then

$$dy = e^z dz \qquad \text{(XII.10)}$$

$$dy = e^z \frac{E}{RT^2} dT = 2.4 \, e^z \left(\frac{E}{RT^2}\right) dT \qquad \text{(XII.11)}$$

or

$$\dot{y} = 2.4 \exp\left(\frac{-E}{RT}\right)\left(\frac{E}{RT^2}\right) \qquad \text{(XII.12)}$$

Since the integral of $\dot{y}$ is y and because $2.4 \exp(-E/RT)E/RT^2$ is identical with y, it is assumed that $E/RT^2 y$, where $y = \exp(-E/RT)$ and generate $f(y)$ (that is, $2.4y \, E/RT^2$), the integral of this function will be y. This process is represented by the patch diagram in Figure XII.7. The computations were applied to the dehydration of $CaC_2O_4 \cdot H_2O$, a system which has been well studied by a number of investigators. Using the data given by Freeman and Carroll (4), the calculated and experimental curves are given in Figure XII.8. It is interesting to note that the curve calculated by integrated

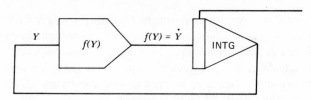

Figure XII.7. Patch diagram of Hughes and Hart (23).

methods using Akahira's tables and the experimental parameters gave a curve which coincided nicely with the computed one.

The in-house differences between results for duplicate samples tested by different laboratories under supposedly similar conditions led to interest in the effects of the rate and mode of heating, and also of fluctuations in temperature after heating, on the mass-loss curves.   Because these problems did not lend themselves to direct solution with the experimental equipment on hand, Gayle and Egge (24) applied analog computation to study the importance of these variables.   The calculations were performed on an analog computer where the heating-rate curves were programmed as the corresponding differential equations and the temperature integral of these used as input

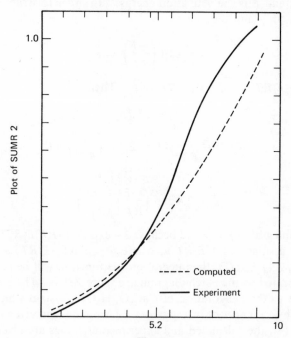

Figure XII.8. Calculated and experimental TG curves for $CaC_2O_4 \cdot H_2O$ (23).

into the Arrhenius equation. Integration of the latter provided the corresponding mass-loss curves. The treatment provided an estimate of the influence of constant thermal errors and of fluctuations about the programmed temperature level. It is noteworthy that symmetrical fluctuations did not result in a cancellation of errors when the rate behavior was exponential rather than a linear function of temperature. The analog computer provided a graphic, reasonably accurate picture of the magnitude of such effects.

C. Differential Thermal Analysis (DTA) and Differential Scanning Calorimetry (DSC)

Nearly all of the computer applications to DTA have been concerned with the calculations of reaction kinetics where they find the ideal means of simulating the DTA curve of a chemical reaction of known kinetics. One of the first of these applications was that by Reed et al. (25) in which the quantitative determination of kinetics by the methods of Borchardt and Daniels (26) and Kissinger (27) were evaluated and compared. The DTA curve was generated numerically by use of equations such as

$$-\theta_r' \frac{d\psi}{d\theta_r} = \zeta \psi^n e^{\epsilon - /\theta} \qquad \text{(XII.13)}$$

where ϵ is the activation energy and $\theta_r' \, d\psi/d\theta_r$ the reaction order, ψ is N/N_0 (number of moles), $\zeta = \alpha A (N_0/V)^{n-1}$, and θ_r' is the dimensionless heating rate; in finite-difference form,

$$\psi(\theta_r) - \psi(\theta_r + h) = \frac{h\zeta}{\theta_r'} \left[\frac{\psi(\theta_r + h) + \psi \theta_r}{2} \right]^n$$

$$\times \exp\left(\frac{-2\epsilon}{\theta}(\theta_r + h) + \theta(\theta_r) \right) \qquad \text{(XII.14)}$$

where h represents the mesh spacing, $\Delta\theta_r$. A typical computer-generated curve, in which the effect of activation energy ϵ on the DTA curve is plotted, is shown in Figure XII.9. Both the location and the shape of the curves is affected, but the dependence is inverse to that observed for the changes of the frequency factor.

The fraction of sample decomposed (α) from DTA curves was calculated by an alogorithm made in ALGOL 60 language for a NCR Elliot Model 4130 computer by Skvara and Satava (28). This alogorithm calculated α and $\log g(\alpha)$ and plotted the latter as a function of temperature. Comparison of the computed DTA data with the experimental values for several dissociation reactions indicated a good agreement and applicability of the method.

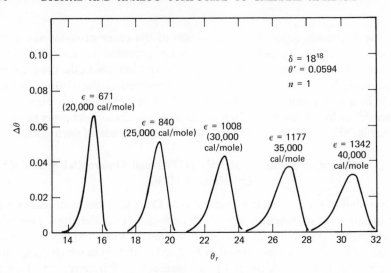

Figure XII.9. Effect of activation energy on DTA curve (25).

The use of a systems analog to improve the performance of a DTA apparatus and also to study the thermal effects in the DTA curve was investigated by Wilburn et al. (29, 30). A finite-difference procedure was used to relate the thermal gradients within the samples and to generate or absorb heat according to a known equation. The influence of such physical properties on the shape and peak temperature of a typical DTA curve was calculated on an ICT Model 1909 computer.

The application of computer calculations to DTA studies of the crystallization kinetics of polymers was described by Gornick (54). Calculations were made of the temperature of a polymeric sample during the cooling process using a Burroughs Model 5500 computer. Morie et al. (55) used an IBM Model 1130 computer to prepare standard vapor pressure plots of $\ln P$ versus $1/T$, the vapor pressure data being obtained from DTA or DSC curves. The heat of vaporization was calculated by the Haggenmacher method as modified by Fishtine.

David et al. (31) used a digital temperature readout device in conjunction with an analog recorder for transition temperature measurements. Temperature resolution was about 0.05°C at a heating rate of 10°C/min.

Amstutz (32) described the Mettler data-acquisition system which is capable of handling eight-digit numbers of any format type or voltage level. A schematic diagram of the system is shown in Figure XII.10. Expansion capabilities include digital and analog multiplexers, keyboards and switch banks for manual entry of data, timers, and programmers. Applications to

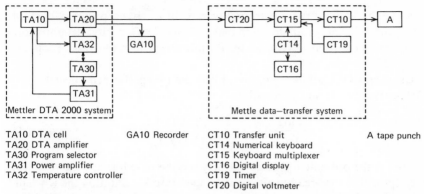

TA10 DTA cell   GA10 Recorder   CT10 Transfer unit   A tape punch
TA20 DTA amplifier   CT14 Numerical keyboard
TA30 Program selector   CT15 Keyboard multiplexer
TA31 Power amplifier   CT16 Digital display
TA32 Temperature controller   CT19 Timer
                                           CT20 Digital voltmeter

Figure XII.10. Mettler data-transfer system connected to a Mettler DTA 2000 system (32).

DTA include off-line recording of raw data on punched paper tape or magnetic tape, and on-line processing, ranging from simple peak area calculations by means of programmable desk-top calculators to the more complex numerical determinations of heat or reaction, kinetics, and purity analysis.

One of the major applications of computers to differential scanning calorimetry (DSC) is in the determination of the purity of organic and inorganic compounds. The precision and accuracy of purity determinations by this technique have been reviewed by Joy et al. (33). One of the first programs for purity determinations using DSC data was that developed by Driscoll et al (34). Required input data are the sample mass and molecular weight, instrument constants, a reference temperature at a point where the curve is still on the base line, and ordinate and abscissa measurements on the curve. One measurement should be at the melting curve peak, but the intervals need not be of uniform size. A maximum of 99 pairs of readings can be accommodated by the program.

The program divides the curve into 99 equal temperature intervals and integrates to obtain the ΔH_f. Temperature correction and base-line area correction are determined for each interval, and the partial area is calculated. The program then applies successive $\frac{1}{2}\%$ area corrections on each partial area and the total area and calculates the $1/F$ values. A least-squares regression analysis is used in each corrected line until a minimum standard deviation of the points about the calculated line is reached. The "best values" are then used in the subsequent calculations. Output from the computer includes the ΔH_f, T_0, and T_m, the mole-% impurity, the $1/F$ limits used, the percent correction applied, and the cryoscopic constant. A corrected mole percent impurity assuming solid-solution behavior is also calculated. The "linearization" of the T_s versus $1/F$ curve has been discussed (35). Joy et al. (33)

rewrote the above program from Fortran into a basic program operable on a time-shared computer terminal. Other DSC purity determination programs have been developed by Barrall and Diller (36) and Scott and Gray (27).

Using an IBM Quiktran program, Ellerstein (38) performed calculations of DSC-curve data from the equation.

$$\frac{\Delta \log l}{\Delta \log A_r} = \frac{E}{2.303R} \frac{[\Delta(l/\gamma)]}{[\Delta \log A_r]} + n \qquad \text{(XII.15)}$$

where l is the ordinate displacement between the base line and curve and A_r corresponds to the area remaining at temperature T. Plotting the left-hand side versus the bracketed expression gives a curve whose slope is $E/2.303R$, and the intercept will be equal to n, the order of reaction. Results from the Quiktran program are then fed into an IBM Quiktran common library program (FITLIN) which gives a "best" line fit of the calculated points.

Gray (39) developed a program which accepts the DSC sample and base-line data, matches the "isothermal," performs cumulative and total area integrations in units of cal/g, corrects the temperature for thermal lag, and tabulates and plots ordinate values in specific-heat units as well as cumulative area in enthalpy units. The analog data from the DSC instrument are digitized and transferred to paper tape with the use of the Perkin-Elmer ADS VI Analytical Data System for Thermal Analysis. The data are digitized every two seconds or every 0.133°. A computer plotter then plots the DSC curve and also the cumulative peak area in specific enthalpy units, cal/g.

Crossley et al. (40) used a computer reduction technique for the DSC isothermal curve which was developed to replace the use of a planimeter. The data reduction was divided into two phases: (a) mechanism-independent solutions for the reactant fraction, α, and various functions of α (where α is the reactant fraction remaining at time t), and (b) solutions for mechanism-dependent rate constants. For the first phase, the DATAR program was developed, which consisted of the following: Ordinal points referred to a "coarse data," and evenly spaced in time over the time span of the DSC curve, are read directly into the computer. Up to 1000 points may be read, but 40–50 are usually sufficient for acceptable accuracy. The resultant fraction remaining at time t is calculated by the equation

$$\alpha(t) = \frac{\displaystyle\int_{t_{max}}^{t} (\text{ORD})\, dt}{\displaystyle\int_{t_{max}}^{0} (\text{ORD})\, dt} \qquad \text{(XII.16)}$$

A Simpson's Rule procedure modified to handle odd numbers of intervals is used to calculate the integrals. The program calculates and prints the time in sec and in min, α, ln α, $1/\alpha^2$, $(1 - \alpha)\alpha$, and log 100 $[(1 - \alpha)/\alpha]$. For the second stage, the PARACT program is used to determine the true rate constants, k_1 and k_2.

Other programs which can be used to calculate reaction kinetics from DSC data were formulated by Kauffman and Beech (41) and Rogers and Smith (42). Heuvel and Lind (43) used a computer to correct DSC data for effects due to thermal lag and heat capacity changes, while Sondack (44) developed a simple equation for linearization of data in DSC purity determinations.

D. Miscellaneous Thermal Techniques

Analysis of isoperibol calorimetric data requires lengthy graphical procedures and tedious calculations to obtain corrected resistance changes for the reaction and calibration experiments. The reaction experiment graphically resolves into two linear portions, the initial rating period (IRP) and the final rating period (FRP), connected by a curve for which no analytical equation is known. A program was developed for these calculations by Gayer and Bartel (45).

Friedman and co-workers (46–48) have developed a digital converter for recording evolved gas analysis (EGA)–mass spectrometric (MS) data on punched paper tape. The data collection system is shown in Figure XII.10. It is based on a very stable programmed power supply that steps the gate to preselected discrete ionic mass peak locations. In practice, analog gate voltages are determined for about 20 peak centers from m/e 1 to about 203, as monitored by a digital voltmeter. These are then analyzed by a least-squares shared-time program using a polynomial equation that includes five constants. If all of the calculated points are within values equivalent to 10 nsec of the observed time delay, the fit is accepted and a printout is obtained for all analog gate voltages as a function of mass number, by interpolation and extrapolation. The punched paper tapes generated during a run are read by an optical reader and stored in a small computer. A large computer then sorts the data by mass number and plots the data on a graphic plotter. The plotted data are corrected for background and instrument sensitivity, editorial corrections are made, and the data are normalized to 1 mg initial sample mass.

Gibson (49) described a TG-MS system which contained a PDP 8/L computer interfaced to the mass spectrometer for on-line control of the mass spectral data. A schematic diagram of the system is shown in Figure XII.11. The output analog signal from the mass spectrometer is integrated and converted to digital form by a 12-bit converter and transferred to magnetic tape.

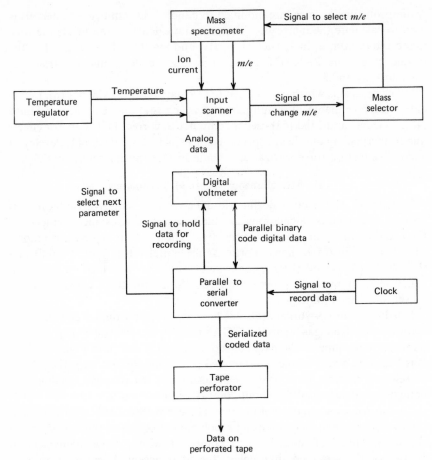

Figure XII.10. Data collection system for EGA–MS after Friedman et al. (47).

Two separate modes are possible: (*1*) data logging and (*2*) spectrum control; they differ only in the manner in which they acquire data.   During operation of the mass spectrometer, the plotter gives a real-time gas release curve, which is simply a plot of the ion current from the electron multiplier at each recorded spectrum scanned.   Other output routines include.

(*1*) Reconstruction of a gas release pattern which normalizes the largest gas release peak or region to 100 and indexes the mass spectra during the run.

(*2*) Printed or plotted spectra from any mass spectrum collected.

(*3*) Plots of individual mass (for example, $m/e = 18$) peaks as a function of temperature.

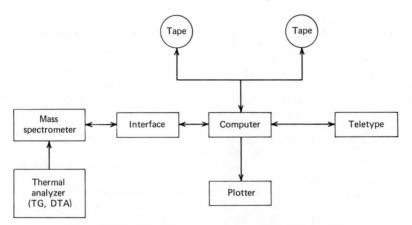

Figure XII.11. TG–MS computer system after Gibson (49).

(4) Spectrum plots or printouts with positively identified mass scales and numeric ion intensities.

(5) Spectrum plots or printouts with the background spectra subtracted to eliminate the effect of contaminants or backgrounds.

Kinetics calculations on poly(methylmethacrylate) using mass-spectrometric thermal analysis (MTA) were described by Sakamoto et al. (56). Using the activation energy calculated from the experimental data, the computer plots the logarithm of the reduced rate, $dc/d\theta$, versus the logarithm of the reduced time, θ. A comparison is then made between this curve and a calculated theoretical curve for various reaction mechanism, and the curve which fits best is that of the first-order reaction. Pfeil (51) discussed the application of digital computers to the statistical analysis of the TG measurements of edema. Computer graphics can also provide a useful and much-needed service in the thermal analysis of biological systems.

It should be noted that computer data-acquisition systems for thermal analysis instruments are available from the Perkin-Elmer Corporation (37), the DuPont Company, and the Mettler Instrument Company (32). A data-acquisition system for a thermal analyzer has also been described by Yui et al. (52).

References

1. Soulen, J. R., *Anal. Chem.*, **34**, 136 (1962).
2. Schempf, J. M., F. E. Freeburg, D. J. Roger, and F. M. Angeloni, *Anal. Chem.*, **38**, 520 (1966).
3. Sestak, J., A. Brown, V. Rihak, and G. Berggren, *Thermal Analysis*, R. F. Schwenker and P. D. Garn, eds., Academic, New York, Vol. 2, 1969, p. 1035, 1085.
4. Freeman, E. S., and B. Carroll, *J. Phys. Chem.*, **62**, 394 (1958).
5. Coates, A. W., and J. P. Redfern, *Nature*, **201**, 88 (1964).

6. Satava, V., and J. Sestak, *Silikaty* **8**, 134 (1964).
7. Gallagher, P. K., and F. Schrey, *Thermal Analysis*, R. F. Schwenker and P. D. Garn, eds., Academic, New York, Vol. 2, 1969, p. 929.
8. Gallagher, P. K., and F. Schrey, *Thermochim. Acta*, **1**, 465 (1970).
9. Gallagher, P. K., and D. W. Johnson, *Thermochim. Acta*, **2**, 413 (1971).
10. Gallagher, P. K., and D. W. Johnson, *Thermochim. Acta*, **3**, 303 (1972).
11. Gallagher, P. K., and D. W. Johnson, *Thermochim. Acta*, **5**, 455 (1973).
12. Johnson, D. W., and P. K. Gallagher, *J. Phys. Chem.*, **75**, 1179 (1971).
13. Gallagher, P. K., F Schrey, and B. Prescott, *Thermochim. Acta*, **2**, 405 (1971).
14. Ozawa, T., *J. Thermal Anal.*, **2**, 301 (1970).
15. Zsako, J., *J. Phys. Chem.*, **72**, 2406 (1968).
16. Zsako, J., C. Varhelyi, and M. Agosescu, *Studia Babes-Bolyai Univ.*, **2**, 33 (1970).
17. Zsako, J., Ref. 16, p. 113.
18. Sharp, J. H., and S. A. Wentworth, *Anal. Chem.*, **41**, 2060 (1969).
19. Reich, L., W. Gregory, and S. S. Stivala, *Thermochim. Acta*, **4**, 493 (1972).
20. Vachuska, J., and M. Voboril, *Thermochim. Acta*, **2**, 379 (1971).
21. Vachuska, J., and N. Rykalova, *GIER Computer Library*, *NRI, Czech. Acad. Sci.*, No. 1064.
22. Rykalova, N., and J. Vachuska, Ref. 21, No. 1399.
23. Hughes, M. A., and R. Hart, ICTA III, Davos, Switzerland, August 23–28, 1971, paper No. I-21.
24. Gayle, J. B., and C. T. Egger, *Anal. Chem.*, **44**, 421 (1972).
25. Reed, R. L., L. Weber, and B. S. Gottfried, *I & EC Fundamentals*, **4**, 38 (1965).
26. Borchardt, H. J., and F. Daniels, *J. Am. Chem. Soc.*, **79**, 41 (1957).
27. Kissinger, H. E. *Anal. Chem.*, **29**, 1702 (1957).
28. Skvara, F., and V. Satava, *J. Thermal Anal.*, **2**, 325 (1970).
29. Wilburn, F. W., J. R. Hesford, and J. R. Flower, *Anal. Chem.*, **40**, 777 (1968).
30. Melling, R., F. W. Wilburn, and R. M. McIntosh, *Anal. Chem.*, **41**, 1275 (1969).
31. David, D. J., D. A. Ninke, and B. Duncan, *Am. Lab.*, January, 31 (1971).
32. Amstutz, D., ICTA III, Davos, Switzerland, August 23–28, 1971, paper I-39.
33. Joy, E. F., J. D. Bonn, and A. J. Barnard, *Thermochim. Acta*, **2**, 57 (1971).
34. Driscoll, G. L., I. N. Duling, and F. Magnotta, in *Analytical Calorimetry*, R. S. Porter and J. F. Johnson, eds., Vol. 1, Plenum, New York, 1968, p. 271.
35. Driscoll, G. L., I. N. Duling, and F. Magnotta, *Sun Oil Quart.*, **3**, 24 (1969).
36. Barrall, E. M., and R. D. Diller, *Thermochim. Acta*, **1**, 509 (1970).
37. Scott, L. R., and A. P. Gray, *Perkin-Elmer Instrument News*, **19**, No. 3, p. 1 (1969).
38. Ellerstein, S. M., in *Analytical Calorimetry*, R. S. Porter and J. F. Johnson, eds., Vol. 1, Plenum, New York, 1968, p. 279.
39. Gray, A. P., *Perkin-Elmer Instrument News*, **20**, 8 (1969).
40. Crossley, R. W., E. A. Dorko, and R. L. Diggs, in *Analytical Calorimetry*, R. S. Porter and J. F. Johnson, eds., Vol. 2, Plenum, New York, 1970, p. 429.
41. Kauffman, G. B., and G. Beech, *Thermochim. Acta*, **1**, 99 (1970).
42. Rogers, R. N., and L. C. Smith, *Thermochim. Acta*, **1**, 1 (1970).
43. Heuvel, H. M., and K. C. J. B. Lind, *Anal. Chem.*, **42**, 1044 (1970).
44. Sondack, D. L. *Anal. Chem.*, **44**, 888 (1972).
45. Gayer, K. H., and J. Bartel, *Thermochim. Acta*, **3**, 337 1972).
46. Freidman, H. L., *J. Macromol. Sci.*, **A1**, 57 (1967).
47. Friedman, H. L., G. A. Griffith, and H. W. Goldstein in *Thermal Analysis*, R. F. Schwenker and P. D. Garn, eds., Vol. 1, Academic, New York, 1969, p. 405.

48. Friedman, H. L., *Thermochim. Acta*, **1,** 199 (1970).
49. Gibson, E. K , *Thermochim. Acta*, **5,** 243 (1973).
50. Ingraham, T. R., in *Proceedings of the Second Toronto Symposium on Thermal Analysis*, H. G. McAdie, ed., Chemical Institute of Canada, Toronto, 1967, p. 21.
51. Pfeil, R. W., in *Proceedings of the Third Toronto Symposium on Thermal Analysis*, H. G. McAdie, ed., Chemical Institute of Canada, Toronto, 1969, p. 187.
52. Yui, H., R. Kato, H. Okamoto, and A. Maezono, *7th Japanese Calorimetry Conference*, November 25–26, 1971, Nagoya, Japan.
53. Perone, S. P., *Anal. Chem.*, **43,** 1288 (1971).
54. Gornick, F., *J. Polymer Sci.*, Part C, **25,** 131 (1968).
55. Morie, G. P., T. A. Powers, and C. A. Glover, *Thermochim. Acta*, **3,** 259 (1972).
56. Sakamoto, R., T. Ozawa, and M. Kanazaski, *Thermochim. Acta*, **3,** 291 (1972).

CHAPTER XIII

THERMAL ANALYSIS NOMENCLATURE

A. Introduction

In 1967, McAdie (1) reported the recommendations of the committee on standardization of the International Confederation of Thermal Analysis for reporting DTA or TG data. To accompany each DTA or TG curve, the following information should be reported:

1. Identification of all substances (sample, reference, diluent) by a definitive name, an empirical formula, or equivalent compositional data.

2. A statement of the source of all substances, details of their histories, pretreatments, and chemical purities, so far as these are known.

3. Measurement of the average rate of linear temperature change over the temperature range involving the phenomena of interest.

4. Identification of the sample atmosphere by pressure, composition, and purity; whether the atmosphere is static, self-generated, or dynamic through or over the sample. Where applicable, the ambient atmospheric pressure and humidity should be specified. If the pressure is other than atmospheric, full details of the method of control should be given.

5. A statement of the dimensions, geometry, and materials of the sample holder, and the method of loading the sample where applicable.

6. Identification of the abscissa scale in terms of time or temperature at a specified location. Time or temperature should be plotted to increase from left to right.

7. A statement of the methods used to identify intermediates or final products.

8. Faithful reproduction of all original records.

9. Wherever possible, each thermal effect should be identified and supplementary supporting evidence stated.

In the reporting of TG data, the following additional details are also necessary:

10. Identification of the thermobalance, including the location of the temperature-measuring thermocouple.

11. A statement of the sample weight and weight scale for the ordinate. Weight-loss should be plotted as a downward trend, and deviations from this practice should be clearly marked. Additional scales (such as fractional

decomposition or molecular composition) may be used for the ordinate where desired.

12. If derivative thermogravimetry is employed, the method of obtaining the derivative should be indicated and the units of the ordinate specified.

When reporting DTA curves, these specific details should also be presented:

10. Sample weight and dilution of the sample.

11. Identification of the apparatus, including the geometry and materials of the thermocouples and the locations of the differential and temperature-measuring thermocouples.

12. The ordinate scale should indicate deflection per °C at a specified temperature. Preferred plotting will indicate upward deflection as a positive temperature differential, and downward deflection as a negative temperature differential, with respect to the reference. Deviations from this practice should be clearly marked.

In 1969, Mackenzie (2), Chairman of the ICTA Nomenclature Committee, published the first definitive nomenclature report. These recommendations should be adhered to in all English-language publications in thermal analysis. The recommendations are as follows.

B. General Recommendations

(a) *Thermal analysis* and not "thermography" should be the acceptable name in English, since the latter has at least two other meanings in this language, the major one being medical. The adjective should then be thermoanalytical (cf. physical chemistry and physicochemical); the term "thermoanalysis" is not supported (on the same logical basis).

(b) *Differential* should be the adjectival form of difference; *derivative* should be used for the first derivative (mathematical) of any curve.

(c) The term "analysis" should be avoided as far as possible, since the methods considered do not comprise analysis as generally understood chemically; terms such as *differential thermal analysis* are too widely accepted however, to be changed.

(d) The term *curve* is preferred to "thermogram" for the following reasons:

1. "Thermogram" is used for the results obtained by the medical technique of thermography.
2. If applied to certain curves (such as thermogravimetric curves), "thermogram" would not be consistent with the dictionary definition.
3. For clarity there would have to be frequent use of terms such as "differential thermogram," "thermogravimetric thermogram," and so on, which are not only cumbersome but also confusing.

(*e*) In multiple techniques, *simultaneous* should be used for the application of two or more techniques to the same sample at the same time; *combined* would then indicate the use of separate samples for each technique.

(*f*) *Thermal decomposition* and similar terms are being further considered by the committee.

C. Terminology

Acceptable names and abbreviations, together with names which were for various reasons rejected, are listed in Table XIII.1. The committee is in accord with the suggestion, made during discussion of the report, that the limited number of abbreviations considered permissible should be adopted internationally, irrespective of language.

The committee did not wish to pronounce on nomenclature in borderline techniques (such as thermometric titrimetry or calorimetry) which are, to its knowledge, being considered by other bodies. Consideration of techniques not yet extensively employed has been deferred.

D. Definitions and Conventions

1. General

Thermal analysis. A general term covering a group of related techniques whereby the dependence of the parameters of any physical property of a substance on temperature is measured.

2. Methods Associated with Weight Change

a. Static

Isobaric weight-change determination. A technique of obtaining a record of the equilibrium weight of a substance as a function of temperature (T) at a constant partial pressure of the volatile product or products.

The record is the isobaric weight-change curve; it is normal to plot weight on the ordinate with weight decreasing downwards, and T on the abscissa increasing from left to right.

Isothermal weight-change determination. A technique of obtaining a record of the dependence of the weight of a substance on time (t) at constant temperature.

The record is isothermal weight-change curve; it is normal to plot weight on the ordinate with weight decreasing downwards, and t on the abscissa increasing from left to right.

TABLE XIII.1
Recommended Terminology

Acceptable name	Acceptable abbrev.[a]	Rejected name(s)
A. General		
Thermal analysis		Thermography
		Thermoanalysis
B. Methods associated with weight change		
1. Static		
Isobaric weight-change determination		
Isothermal weight-change determination		Isothermal thermogravimetric analysis
2. Dynamic		
Thermogravimetry	TG	Thermogravimetric analysis
		Dynamic thermogravimetric analysis
Derivative thermogravimetry	DTG	Differential thermogravimetry
		Differential thermogravimetric analysis
		Derivative thermogravimetric analysis
C. Method associated with energy change		
Heating curves[b]		Thermal analysis
Heating-rate curves[b]		Derivative thermal analysis
Inverse heating-rate curves[b]		
Differential thermal analysis	DTA	Dynamic differential calorimetry
Derivative differential thermal analysis		
Differential scanning calorimetry	DSC	
D. Methods associated with evolved volatiles		
Evolved gas detection	EGD	Effluent gas detection
Evolved gas analysis[c]	EGA	Effluent gas analysis
		Thermovaporimetric analysis
E. Methods associated with dimensional change		
Dilatometry		
Derivative dilatometry		
Differential dilatometry		
F. Multiple techniques		
Simultaneous TG and DTA, etc.		DATA (Differential and thermogravimetric analysis)
		Derivatography
		Derivatographic analysis

[a] Abbreviations should be in capital letters without full stops, and should be kept to the minimum to avoid confusion.

[b] When determinations are performed during the cooling cycle these become cooling curves, cooling-rate curves, and inverse cooling-rate curves, respectively.

[c] The method of analysis should be clearly stated and abbreviations such as MTA (mass-spectrometric thermal analysis) and MDTA (mass spectrometry and differential thermal analysis) avoided.

b. Dynamic

Thermogravimetry (TG). A technique whereby the weight of a substance, in an environment heated or cooled at a controlled rate, is recorded as a function of time or temperature.

The record is the thermogravimetric or TG curve; the weight should be plotted on the ordinate with weight decreasing downwards, and t or T on the abscissa increasing from left to right.

Derivative thermogravimetry (DTG). A technique yielding the first derivative of the thermogravimetric curve with respect to either time or temperature.

The curve is the derivative thermogravimetric or DTG curve; the derivative should be plotted on the ordinate with weight-losses downward, and t or T on the abscissa increasing from left to right.

E. Methods Associated with Energy Changes

Heating curves. These are records of the temperature of a substance against time, in an environment heated at a controlled rate.

T should be plotted on the ordinate increasing upwards, and t on the abscissa increasing from left to right.

Heating-rate curves. These are records of the first derivative of the heating curve with respect to time (that is, dT/dt) plotted against time or temperature.

The function of dT/dt should be plotted on the ordinate, and t or T on the abscissa increasing from left to right.

Inverse heating-rate curves. These are records of the first derivative of the heating curve with respect to temperature (that is, dt/dT) plotted against either time or temperature.

The function of dt/dT should be plotted on the ordinate, and t or T on the abscissa increasing from left to right.

Differential thermal analysis (DTA). A technique of recording the difference in temperature between a substance and a reference material against either time or temperature as the two specimens are subjected to identical temperature regimes in an environment heated or cooled at a controlled rate.

The record is the differential thermal or DTA curve; the temperature difference (ΔT) should be plotted on the ordinate with endothermic reactions downwards, and t or T on the abscissa increasing from left to right.

Derivative differential thermal analysis. A technique yielding the first derivative of the differential thermal curve with respect to either time or temperature.

The record is the derivative differential thermal or derivative DTA curve; the derivative should be plotted on the ordinate, and t or T on the abscissa increasing from left to right.

Differential scanning calorimetry (DSC). A technique of recording the energy necessary to establish zero temperature difference between a substance and a reference material against either time or temperature as the two specimens are subjected to identical temperature regimes in an environment heated or cooled at a controlled rate.

The record is the DSC curve; it represents the amount of heat applied per unit time as ordinate against either t or T as abscissa.

F. Methods Associated with Evolved Volatiles

Evolved gas detection (EGD). This term covers any technique of detecting whether or not a volatile product is formed during thermal analysis.

Evolved gas analysis (EGA). A technique of determining the nature and amount of volatile product or products formed during thermal analysis.

G. Methods Associated with Dimensional Change

Dilatometry. A technique whereby changes in dimension(s) of a substance are measured as a function of temperature. The record is the dilatometric curve.

Derivative dilatometry: differential dilatometry. These terms carry the connotations given in I(*b*) above.

H. Multiple Techniques

This term covers simultaneous DTA and TG and other techniques, and definitions follow from the above.

Mackenzie et al. (3) protested the use of the term "thermohygrometric analysis" (THA) as used by Still and Cluley (4). They suggested that this technique was simply a branch of evolved-gas analysis, for which the abbreviation "EGA" has been internationally accepted. The ICTA Nomenclature Committee regards the coining of abbreviations for what are variants of an accepted term as completely indefensible, and would urge all scientists to give serious consideration to the implications of introducing and publishing new abbreviations or complicated terminology.

In a second report of the Nomenclature Committee by Mackenzie et al. (5), the following more definitive recommendations were published:

I. DTA

The *sample* is the actual material investigated, whether diluted or undiluted.

The *reference material* is a known substance, usually inactive thermally over the temperature range of interest.

The *specimens* are the sample and reference material.

The *sample holder* is the container or support for the sample.

The *reference holder* is the container or support for the reference material.

The specimen-holder assembly is the complete assembly in which the specimens are housed.   Where the heating or cooling source is incorporated in one unit with the containers or supports for the sample and reference material, this would be regarded as part of the specimen-holder assembly.

A *block* is a type of specimen-holder assembly in which a relatively large mass of material is in intimate contact with the specimen holders.

The *differential thermocouple* or ΔT *thermocouple* is the thermocouple system used to measure temperature difference.   Should another thermo-sensing device be used, its name should replace "thermocouple."

J. TG

A *thermobalance* is an apparatus for weighing a sample continuously while it is being heated or cooled.

The *sample* is the actual material investigated, whether diluted or undiluted. Samples used in TG are normally not diluted, but in simultaneous TG and DTA diluted samples might well be used.

The *sample holder* is the container or support for the sample.

K. DTA and TG

The *temperature thermocouple* or *T thermocouple* is the thermocouple system used to measure temperature; its position with respect to the sample should always be stated.   Should another thermosensing device be used, its name should replace "thermocouple."

The *heating rate* is the rate of temperature increase, which is customarily quoted in degrees per minute (on the Celsius or Kelvin scales).   Correspondingly, the cooling rate is the rate of temperature decrease.   The heating or cooling rate is said to be *constant* when the temperature-time curve is linear.

In simultaneous DTA–TG, definitions follow from those given for DTA and TG separately.

The following definitions are to be used in conjunction with those previously reported (1, 2) for DTA and TG data.

L. DTA

In DTA it must be remembered that although the ordinate is conventionally labeled ΔT, the output from the ΔT thermocouple will in most instances vary with temperature, and the measurement recorded is normally the e.m.f. output, E, [the conversion factor, b, in the equation $\Delta T = bE$ is not constant since $b = f(T)$].   A similar situation occurs with other sensor systems.

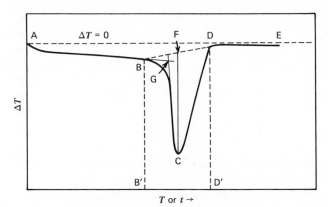

Figure XIII.1. Formalized DTA curve (5).

All definitions refer to a single peak such as that shown in Figure XIII.1; multiple-peak systems, showing shoulders or more than one maximum or minimum, can be considered to result from superposition of single peaks.

The *base line* (Figure XIII.1, AB and DE) corresponds to the portions of the DTA curve for which ΔT is approximately zero.

A *peak* (Figure XIII.1, BCD) is that portion of the DTA curve which departs from, and subsequently returns to, the base line.

An *endothermic peak* or *endotherm* is a peak where the temperature of the sample falls below that of the reference material (that is, ΔT is negative).

An *exothermic peak* or *exotherm* is a peak where the temperature of the sample rises above that of the reference material (that is, ΔT is positive).

Peak width (Figure XIII.1, B'D') is the time or temperature interval between the points of departure from the return to the base line.

Peak height (Figure XIII.1, CF) is the distance, vertical to the time or temperature axis, between the interpolated base line and the peak tip Figure XIII.1, C.

Peak area (Figure XIII.1, BCDB) is the area enclosed between the peak and the interpolated base line.

The *extrapolated onset* (Figure XIII.1, G) is the point of intersection of the tangent drawn at the point of greatest slope on the leading edge of the peak (Figure XIII.1, BC) with the extrapolated base line (Figure XIII.1, BG).

M. TG

All definitions refer to a single-stage process such as that shown in Figure XIII.2; multistage processes can be considered as resulting from a series of single-stage processes.

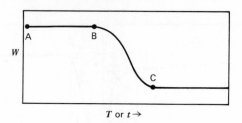

Figure XIII.2. Formalized TG curve (5).

A *plateau* (Figure XIII.2, AB) is that part of the TG curve where the weight is essentially constant.

The *initial temperature*, T_i (Figure XIII.2, B) is that temperature (on the Celsius or Kelvin scales) at which the cumulative weight change reaches a magnitude that the thermobalance can detect.

The *final temperature*, T_f (Figure XIII.2, C) is that temperature (on the Celsius or Kelvin scales) at which the cumulative weight change reaches a maximum.

The *reaction interval* is the temperature difference between T_f and T_i as defined above.

Recently, McAdie (6) published the recommendations of ICTA concerning evolved-gas detection (EGD) or evolved-gas analysis (EGA) curves. These recommendations are as follows:

1. Identification of all substances (sample, reference, diluent) by a definitive name, an empirical formula, or equivalent compositional data.

2. A statement of the source of all substances, and the details of their histories, pretreatments, and chemical purities, so far as these are known.

3. A clear statement of the temperature environment of the sample during reaction.

4. Measurement of the average rate of linear temperature change over the temperature range involving the phenomena of interest. Nonlinear temperature programming should be described in detail.

5. A statement of the dimensions, geometry, and materials of the sample holder, and the method of loading the sample where applicable.

6. Identification of the abscissa scale in terms of time or temperature at a specified location. Time or temperature should be plotted to increase from left to right.

7. Identification of the ordinate scale in specific terms where possible. In general, increasing concentration of evolved gas should be plotted upwards. For gas density detectors, increasing gas density should also be plotted upwards. Deviations from these practices should be clearly marked.

8. A statement of the methods used to identify intermediate or final products.

9. Faithful reproduction of all original records.

10. Identification of the sample atmosphere by pressure, composition, and purity, and by whether the atmosphere is self-generated or dynamic through or over the sample. The flow rate, total volume, construction, and temperature of the system between the sample and detector should be given, together with an estimate of the time delay within this system.

11. Identification of the apparatus used by type and commercial name, together with details of the location of the temperature-measuring thermocouple and the interface between the systems for sample heating and detecting or measuring evolved gases.

12. In the case of EGA, when exact units are not used, the relationship between signal magnitude and concentration of species measured should be stated. For example, the dependence of the flame ionization signal on the number of carbon atoms and their bonding, as well as on concentration, should be given.

References

1. McAdie, H. G., *Anal. Chem.*, **39**, 543 (1967).
2. Mackenzie, R. C., *Talanta*, **16**, 1227 (1969).
3. Mackenzie, R. C., C. J. Keattch, A. A. Hodgson, and J. P. Redfern, *Chem. Ind.*, **1970,** 272.
4. Still, J. E., and H. J. Cluley, *Chem. Ind.*, **1969,** 1777.
5. Mackenzie, R. C., C. J. Keattch, D. Dollimore, J. A. Forrester, A. A. Hodgson, and J. P. Redfern, *Thermochim. Acta*, **5**, 71 (1972).
6. McAdie, H. G., *Anal. Chem.*, **44**, 640 (1972).

INDEX